Progress in Stellar Spectral Line Formation Theory

NATO ASI Series

Advanced Science Institutes Series

A series presenting the results of activities sponsored by the NATO Science Committee, which aims at the dissemination of advanced scientific and technological knowledge, with a view to strengthening links between scientific communities.

The series is published by an international board of publishers in conjunction with the NATO Scientific Affairs Division

A	Life Sciences	Plenum Publishing Corporation
B	Physics	London and New York
C	Mathematical and Physical Sciences	D. Reidel Publishing Company Dordrecht, Boston and Lancaster
D	Behavioural and Social Sciences	Martinus Nijhoff Publishers
E	Engineering and Materials Sciences	The Hague, Boston and Lancaster
F	Computer and Systems Sciences	Springer-Verlag
G	Ecological Sciences	Berlin, Heidelberg, New York and Tokyo

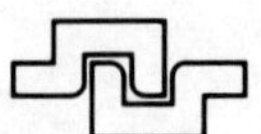

Series C: Mathematical and Physical Sciences Vol. 152

Progress in Stellar Spectral Line Formation Theory

edited by

John E. Beckman

Department of Physics, Queen Mary College, London, U.K. and
Instituto de Astrofisica de Canarias, La Laguna, Tenerife, Spain

and

Lucio Crivellari

Osservatorio Astronomico di Trieste, Trieste, Italy

D. Reidel Publishing Company

Dordrecht / Boston / Lancaster

Published in cooperation with NATO Scientific Affairs Division

Proceedings of the NATO Advanced Research Workshop on
Progress in Stellar Spectral Line Formation Theory,
Grignano-Miramare, Trieste, Italy
September 4-7, 1984

Library of Congress Cataloging in Publication Data

NATO Advanced Research Workshop on Progress in Stellar Spectral Line Formation Theory (1984: Trieste, Italy)
Progress in stellar spectral line formation theory.

(NATO ASI series. Series C, Mathematical and physical sciences; vol. 152)
"Proceedings of the NATO Advanced Research Workshop on Progress in Stellar Spectral Line Formation Theory, Grignano-Miramare, Trieste, Italy, September 4–7, 1984"-T.p. verso.
"Published in cooperation with NATO Scientific Affairs Division."
Includes index.
1. Stars-Spectra-Congresses. 2. Astronomical spectroscopy-Congresses. 3. Spectral line formation-Congresses. 4. Radiative transfer-Congresses. I. Beckman, J. E. II. Crivellari, Lucio, 1950– III. North Atlantic Treaty Organization, IV. Title. V. Series: NATO ASI series. Series C, Mathematical and physical sciences; vol. 152.
QB871.N38 1984 523.8'7 85–2425
ISBN 90–277–2007–X

Published by D. Reidel Publishing Company
P.O. Box 17, 3300 AA Dordrecht, Holland

Sold and distributed in the U.S.A. and Canada
by Kluwer Academic Publishers,
190 Old Derby Street, Hingham, MA 02043, U.S.A.

In all other countries, sold and distributed
by Kluwer Academic Publishers Group,
P.O. Box 322, 3300 AH Dordrecht, Holland

D. Reidel Publishing Company is a member of the Kluwer Academic Publishers Group

Printed in The Netherlands.

TABLE OF CONTENTS

LIST OF PARTICIPANTS

ABBOTT, D.C., JILA, University of Colorado, USA
ANDERSON, L., Ritter Observatory, University of Toledo, USA
ARRIBAS, S., Instituto de Astrofisica de Canarias, Spain
AYDIN, C., Astronomi Institüsü, Turkey
BARYLKO, M., Inst. of Theoretical Physics and Astrophysics, Poland
BECKMAN, J.E., Queen Mary College, University of London, UK
BRUCA, L., International School for Advanced Studies, Miramare, Italy
CARLSSON, M., Institute of Theoretical Astrophysics, Oslo, Norway
CASTELLI, F., Osservatorio Astronomico di Trieste, Italy
CATALA, C., DESPA, Observatoire de Paris-Meudon, France
CATALANO, S., Istituto di Astronomia, Univ. di Catania, Italy
CRIVELLARI, L., Osservatorio Astronomico di Trieste, Italy
DENIZMAN, L., International School for Advanced Studies, Miramare, Italy
DRAKE, S.A., JILA, University of Colorado, USA
FAUROBERT, M., Observatoire de Nice, France
FARAGGIANA, R., Osservatorio Astronomico di Trieste, Italy
FREIRE FERRERO, R., Observatoire de Strasbourg, France
FRISCH, H., Observatoire de Nice, France
GIAMPAPA, M., NOAO -NSO, Tucson, USA
GOUTTEBROZE, P., LPSP, Verrieres-Le-Buisson, France
HACK, M., Osservatorio Astronomico di Trieste, Italy
HAMANN, W.-R., Institute für Theoretische Physik und Sternwarte der Universität, Kiel, FRG
HEMPE, K., Hamburg Sternwarte, FRG
HUSFELD, D., Institut für Astron. und Astrophysik der Universität, München, FRG
JAEGLE, P., Laboratoire de Spectroscopie Atomique at Ionique, Orsay, France
JAMELOT, G., Laboratoire de Spectroscopie Atomique at Ionique, Orsay, France
KALKOFEN, W., Center for Astrophysics, Cambridge MA, USA
KRSLJANIN, V., Astronomical Observatory Belgrade, Yugoslavia
KUNASZ, P.B., University of Colorado, USA
LINSKY, J.L., JILA, University of Colorado, USA
LITES, B., HAO/NCAR, Boulder CO, USA
MALAGNINI, M.-L., Osservatorio Astronomico di Trieste, Italy

MATHYS, G., Institut für Astronomie ETH-Zentrum, Switzerland
MOLARO, P., International School for Advanced Studies, Miramare, Italy
MOROSSI, C., Osservatorio Astronomico di Trieste, Italy
NORDLUND, A., Astronomical Observatory, Copenhagen, Denmark
OXENIUS, J., Univ. Libre de Bruxelles, Belgium
PRADERIE, F., DESPA, Observatoire de Paris-Meudon, France
RAMELLA, M., Osservatorio Astronomico di Trieste, Italy
RYBICKI, G. Center for Astrophysics, Cambridge MA, USA
SCHARMER, G., Stockholm Observatory, Sweden
SCHMIDT, M., Institut für Theoretische Astrophysik der Universität, Heidelberg, FRG
SELVELLI, P.L., Osservatorio Astronomico di Trieste, Italy
SEVERINO, G., Osservatorio Astronomico di Capodimonte, Italy
SIKORSKI, J., Inst. of Theoretical Phys. and Astrophys., Univ. of Gdansk, Poland
SIMONNEAU, E., Institute d'Astrophysique, Paris, France
TALAVERA, A., ESA IUE Obs., VILSPA, Spain
VINCE, I., Astronomical Observatory Belgrade, Yugoslavia
VLADILO, G., Osservatorio Astronomico di Trieste, Italy
WEHRSE, R., Institut für Theoretische Astrophysik der Universität, Heidelberg, FRG

Scientific Committee.

J.E. Beckman (Director)
L. Crivellari
M. Hack (President)
F. Praderie
E. Simonneau

Local Organizing Committee.

L. Bruca, L. Crivellari, P. Molaro, G. Vladilo.

1 Jamelot
2 Wherse
3 Hamann
4 Selvelli
5 Vince
6 Faraggiana
7 Giampapa
8 Jaeglé
9 Vladilo
10 Husfeld
11 Kalkofen
12 Hempe
13 Drake
14 Carlsson
15 Krisljanin
16 Arribas
17 Crivellari
18 Canziani
19 Scharmer
20 Simonneau
21 Linsky
22 Schmidt
23 Talavera
24 Beckman
25 Oxenius
26 Marmolino
27 Freire-Ferrero
28 Catalano
29 Rybicki
30 Faurobert
31 Kunasz
32 Frisch
33 Gouttebroze
34 Mathys
35 Nordlund
36 Molaro
37 Anderson
38 Bruca
39 Lites
40 Barytko
41 Catala
42 Sikorski
43 Castelli

FOREWORD

Spectral line formation theory is at the heart of astrophysical diagnostic. Our knowledge of abundances, in both stellar and interstellar contexts, comes almost entirely from line analysis, as does a major fraction of our ability to model stellar atmospheres. As new facets of the universe become observable so the techniques of high resolution spectroscopy are brought to bear, with great reward. Improved instruments, such as echelle spectrographs, employing detectors of high quantum efficiency, have revolutioned our ability to observe high quality line profiles, although until now this ability has been confined to the brightest stars. Fabry-Perot interferometers and their modern derivatives are bringing new ranges of resolving power to studies of atomic and ionic interstellar lines, and of course radio techniques imply exceedingly high resolution for the cool interstellar medium of molecules and radicals.

Telescopes in space are extending the spectral range of these types of observations. Already the Copernicus and IUE high resolution spectrographs have given us a tantalizing glimmer of what it will be like to obtain ultraviolet spectra with resolution and signal to noise ratio approaching those obtainable on the ground. Fairly soon Space Telescope will be producing high resolution spectroscopic data of unparalleled quality and distance range. As often happens in astrophysics the challenge is now coming from the observers to the theorists to provide interpretational tools which are adequate to the state of the data. This situation, now prevalent in stellar atmospheres, is being repeated in studies of interstellar and intergalactic media, and in an especially exciting if frustrating way in studies of the spectra of active galaxies and QSO's.

It seemed to the organizers of the workshop on Stellar Spectral Line Formation Theory that 1984 was an appropriate juncture to review progress in the field, while beginning to prepare the community of stellar spectroscopists for the new epoch of observations. It was our intention to bring together many interpretationalists, to lace the cocktail with a number of experienced observers, and to add for good measure a dash of laboratory plasma physics. The outcome is the present volume which does reflect that balance.

It was inevitable that a major body of work should be concerned with the distribution of photons under a line profile for the standard case of a static atmosphere far from LTE. The application of the concept and techniques of partial redistribution to line profiles formed partially in the chromosphere and partially in the photosphere, especially to resonance lines such as Ly α, H and K of Ca II and h and k of Mg II was in a sense a major theoretical triumph of the 1960's and 1970's. At times during the later part of the '70's one had the impression that the practicioners of line transfer theory were so relieved to be able to reproduce profiles which gave a general impression of matching to observations, that the impulse to improve the theory had gone. It was as if the remaining difficulties constituted a mopping-up operation which could be left to a fews, while others went off in search of new problems. In particular it appeared that even the relatively straightforward question of the formation of a line with PRD, taking into account the kinetic aspects of both transition levels, remained unsolved except via rather gross approximation, and it was hard to see why it had been neglected.

This impression was to some degree deceptive. Significant progress in the underlying atomic physics of the intrinsic line profile, especially kinematics, along semiclassical lines, was being made by several workers, especially in Europe. At the sometime a great deal of effort was going in two relevant directions, and the results surfaced in the present workshop. One set of workers were examining problems of line formation in expanding atmospheres, while another was looking into ways of using modern computational techniques to add dimensionality. It now appears that a barrier to realism, in the sense of realistic multilevel atoms with many simultaneous transitions, had been crossed and that solutions with many sub-levels per transition, and many angles per radiative beam can also be contemplated, thanks to the high speed and parallel processing of matrix manipulations with some of the newer computers. Another barrier, that of introducing velocity fields for both microturbulence and streaming in atmospheres, is also in the process of being overcome.

There is still some conflict of philosophy between those who embrace the power of the computer in this way, and those who seek approximate solutions which serve to limit the need for codes of heroic proportions. Those who look to elegant approximations contend that they can maintain a firm grasp on the physics in this way, which allows the conceptualizing of distinctive terms in equations as separate processes within the overall system of transfer.

The idea of a deep dichotomy is in fact an exaggeration since everyone has to approximate, and everyone writes major codes, and a full gamut of intermediate methods for the non-dynamic case is presented here. In addition, the need to use multi-level, multi-dimensional models in expanding atmospheres is well brought out in several of the Workshop presentations.

In spite of impressive progress, observers can still justifiably complain that the problems studied by theorists are those which please the latter, and not problems of line formation in real stars. The presenter of the first invited paper, who was requested to give an overview of the whole field, adopted a reconciliatory approach in his hystorical review of attempts to match chromospheric profiles by increasingly sophisticated PRD approaches. His conclusion is that stellar atmospheric models play a critical role, such that without a study of extended atmospheres and of expansion effects attempts to invert line profiles to obtain model information cannot expected to yeld success. To this list of model-dependent properties one should add the effects of horizontal as well as vertical inhomogeneities. Nevertheless the availability of powerful spectroscopic diagnostic tools is a necessary if insufficient condition for progress.

In another observationally oriented article it is stated bluntly that the major problems in interpreting stellar spectral lines are to explain superionization and stellar winds, with the clear implication that much of the mathematical and computational effort represented at the Workshop is tangential to the key problems occupying the observationalist. It may be contentious to single out two problems in this way, but there is no doubt that observers are obtaining data from a wide range of stars, especially hot stars, which are dominated by dynamic effects, and depend for their interpretation on an understanding of how energy can be deposited in the outer atmosphere of a star by non-radiative processes, as well as non-LTE radiative processes. There is still a major gap between what observers are noticing, and what theorists are able to explain. However the Workshop showed that this is less a matter of lack of sympathy, or even a chasm of ignorance, and more a matter of both parties doing their best to converge on common problems within pratical limits set by their respective capabilities. Once again, in science, one tackles those problems which one is able to solve, rather than those which are necessarily the most important.

One attractive feature of the Workshop was that it illustrated an effective balance of effort in the field between workers in Europe and the United States. Traditionally the U.S. side, stimulated perhaps by a greater flux of data, has led the way in many aspects of line formation theory. It is now clear that Europeans are making significant advances, with notable foci of activity in Scandinavia, in West Germany, in Czechoslovakia, and in France, to match the major centres in Cambridge, Massachussetts, and Boulder Colorado. The atmosphere of personal give and take among all the participants was warm and productive, and the decision to attemp to form a common pool of computer codes was a characteristic outcome.

As organizers we were delighted with the level and quality of support we received from many sources. Financially we were in receipt of a grant under the NATO Advanced Research Workshop programme, and additional funding from the Regione Autonoma Friuli Venezia-Giulia, as well as from International School for Advanced Studies, Miramare, Trieste. We are very happy to thanks Professor Hack, under whose helpful auspices the meeting was held, and all the personnel of Osservatorio Astronomico di Trieste. We were most fortunate in the membership of the local organizing committee: Loredana Bruca, Paolo Molaro, and Giovanni Vladilo, whose impressive dedication before, during and after the Workshop could serve as a model for anyone wishing to organize such events. We believe that the enthusiasm shown by all those attending the Workshop is reflected in the collective quality of the papers, and that the cause of improving our understanding of line formation was well served. Any shortcoming in the present work may be fairly laid at the door of the editors.

Progress implies change. The present volume gives a snapshot of a subject which is changing, and which is due to receive great impetus within the next years. It will be illuminating to convene a similar Workshop after Space Telescope has been operational for a period, to see whether the observational-theoretical divide has widened or narrowed. It is our hope and intention to do so.

John Beckman.
Lucio Crivellari.

Colmenar Viejo, December 1984.

OBSERVED AND COMPUTED STELLAR LINE PROFILES: THE ROLES PLAYED BY PARTIAL REDISTRIBUTION, GEOMETRICAL EXTENT AND EXPANSION

Jeffrey L. Linsky*
Joint Institute for Laboratory Astrophysics, National Bureau of Standards and University of Colorado, Boulder, CO 80309

ABSTRACT. I review partial redistribution (PRD) radiative transfer with emphasis on the complex interaction of observations and theoretical predictions of spectral line shapes. Although the physics of PRD was understood and formal solutions to the radiative transfer equation including PRD were published as early as 1962, more than a decade elapsed before the clear inconsistency between observed and theoretical profiles calculated with the restrictive assumption of complete redistribution stimulated the development and use of PRD semiempirical spectral analysis for the solar Lα, Ca II, and Mg II resonance lines. I summarize the work that has led to "realistic" plane parallel static chromospheric models for the Sun and other late-type stars with emphasis on the determination of the temperature minimum at the base of the chromosphere and the physical basis for resonance line limb darkening, the brightness of the line wings, and the width of the emission features. I then discuss the various roles played by atmospheric extension and expansion (winds) in determining resonance line profile shapes, and summarize the existing PRD calculations for late-type stars.

1. INTRODUCTION

Progress in spectral line formation theory has been stimulated by observations of stellar line profiles, developments in atomic physics theory, and advances in computing capabilities. However, there are often long delays between the recognition of how to solve a major problem in principle and realistic computations, and the perspectives of a few individuals have often channeled the development of the field for long periods. Thus an attempt to present the development and application of ideas in this field in a straightforward way is difficult, and there have been few attempts to present developments in this field historically. The literature consists of journal articles,

---

*Staff Member, Quantum Physics Division, National Bureau of Standards.

J. E. Beckman and L. Crivellari (eds.), Progress in Stellar Spectral Line Formation Theory, 1–26.

often collected in series, a few review papers (e.g. Avrett and Loeser 1983), and very concise presentations in textbooks (e.g. Jefferies 1968; Athay 1972; Mihalas 1978). The development of the concept of partial frequency redistribution, or partial redistribution (PRD) as it is frequently called, and its application to the analysis of solar and stellar spectral line profiles is fraught with all of these problems; therefore, to understand this topic today it is essential to recognize its complex development.

The historical development of PRD can be divided conveniently into two phases with a climactic event in the middle. Prior to 1972 the concept of frequency redistribution during a resonant scattering event was already understood as noncoherent in the line core and nearly coherent in the line wings. The different types of frequency redistribution functions had already been written down and procedures for their numerical evaluation were partially developed. In addition, Hummer (1969) had obtained the first real solution of the radiative transfer equations for the R_{I-A} and R_{II-A} redistribution functions (see below) for isothermal slabs and semi-infinite media, and he had demonstrated some of the basic physical consequences of PRD. However, there were no applications as yet of this new theory to the analysis of observed spectral line profiles.

The climactic event in the development of this field occurred during IAU Colloquium No. 19 on Stellar Chromospheres held at NASA Goddard Space Flight Center on February 21-24, 1972. At this Colloquium, Jefferies and Morrison (1973) reviewed calculations of line profiles formed in plane-parallel atmospheres characterized by a sharp rise in temperature with height (i.e., chromospheres), showing that for strong resonance lines the chromospheric rise in temperature can produce emission in the line cores. Avrett (1973) then described computed Ca II H and K line profiles for a solar model and a model scaled to simulate a lower gravity star. Both sets of calculations assumed complete frequency redistribution (CRD) over the whole line profile, which implies a frequency-independent source function. Essentially all previous calculations for nonisothermal model atmospheres (e.g. Jefferies and Thomas 1959, 1960; Avrett 1965; Athay and Skumanich 1968a,b; Linsky 1968; Linsky and Avrett 1970) had employed the CRD assumption. Avrett (1973) showed that it is possible to choose a solar chromospheric model that leads to computed Ca II H and K profiles that are close to the observed profiles at the center of the solar disk. He called attention, however, to the inner wings of the line profile, where the scattering process should be nearly coherent in the observer's frame and the computed line profile properly taking coherence into account should be significantly darker than in the CRD case (cf. Linsky and Avrett 1970). He proposed that a good test of the importance of coherent scattering in the line wings would be to match solar plage profiles where the drop in intensity from the K_2 peak to the K_1 minimum is particularly sharp. Subsequently, however, Shine and Linsky (1974) had no difficulty fitting solar plage profiles assuming CRD. Avrett also noted that the computed K line profile for a solar model scaled to the gravity of a giant star showed a decrease in intensity from K_2 to K_1 that is much shallower than is generally observed.

He suggested that the cause for this discrepancy is likely the failure to include coherent scattering in the line wings in the CRD calculations. Subsequent work has shown that his suggestion was correct.

In his summary talk, Wilson (1973) did not actually summarize the Workshop discussions but instead give some sage advice and criticism of the radiative transfer theoreticans in the audience. He stated that the existing calculations of spectral lines formed in chromospheres were inadequate because they have very different shapes than the observed profiles -- they fail to match the sharp decrease in intensity from K_2 peak to K_1 minimum. Therefore, something important has been left out of the theory. He argued that the physics missing from the theory "... may be the velocity distribution of the radiating elements."

Immediately thereafter, Thomas (1973) criticized observers who criticize theoreticians, but Thomas did point out the need for a better physical basis for both nonLTE line formation and mechanical heating theory. He also said that "non-LTE theory is not conceptually that complex."

I have described this confrontation to point out that at that time all the essential elements for major progress were available but not yet put to use -- sharp disagreements between theory and observation, a recognition by some of the importance of coherent scattering, and sufficient computing resources to solve the equations. What was lacking then was a human factor -- a few individuals who would proceed to develop PRD theory and apply it to situations that could resolve the problem of the inner wings of the Ca II and other resonance lines. This in fact came quickly and the Goddard Workshop might have been instrumental in breaking the logjam.

2. PARTIAL REDISTRIBUTION CALCULATIONS PRIOR TO 1972

Hummer (1962) summarized some of the important earlier papers that first introduced and applied the concepts of noncoherent scattering (frequency redistribution) due to radiative and collisional broadening of atomic levels and Doppler motions of the excited atom. In particular, Henyey (1941), Holstein (1947), Unno (1952), Sobolev (1955), and Warwick (1955) discussed frequency redistribution when both natural broadening (due to the finite lifetime of the upper state) and Doppler motions of the radiating atom are included.

Hummer (1962) wrote redistribution functions for direct resonance scattering processes of the type $i \to j \to i$ (for initial state i and excited state j), and all authors have generally followed his terminology, except that in subsequent work primed quantities generally refer to quantities before scattering and unprimed quantities refer to quantities after scattering. In what follows I adopt this revised convention. His terminology is summarized in Table 1. He defined the term redistribution function $R(\nu',\vec{n}'\nu,\vec{n})$ as the probability (as seen in the observer's frame) that a photon with initial frequency ν' and direction $\vec{n}'$ will be absorbed, leading to the re-emission in the line of a $(\nu,\vec{n})$ photon. He then defined four redistribution functions (see

Table 1. Hummer's (1962) notation with indicies reversed

Quantity	Atom's frame	Observer's frame
Incoming frequency	ξ'	$\nu' = \xi' + (v_o/c)\vec{n}'\cdot\vec{v}$ $x' = (\nu'-\nu_o)/\Delta$
Outgoing frequency	ξ	$\nu = \xi + (v_o/c)\vec{n}\cdot\vec{v}$ $x = (\nu-\nu_o)/\Delta$
Absorption probability for photon $(\xi',\vec{n}')$	$f(\xi')\, d\xi' \frac{d\Omega'}{4\pi}$	
Probability of frequency redistribution	$p(\xi',\xi)d\xi$	
Probability of angle redistribution	$g(\vec{n}',\vec{n})d\Omega$	
Probability that an absorption at $(\xi',\vec{n}')$ leads to a re-emission at $(\xi',\vec{n})$	$f(\xi')p(\xi',\xi)g(\vec{n}',\vec{n}) \cdot \frac{d\Omega'}{4\pi} d\Omega d\xi' d\xi$	
(The redistribution function)		$R(\nu',\vec{n}';\nu,\vec{n})$

Table 2) for the cases of zero line width in the atom's frame (R_I), radiative damping of the upper state only (R_{II}), collisional and radiative damping of the upper state only (R_{III}), and radiative damping of both states (R_{IV}). Since it is difficult to solve the transfer equation including the full angle-dependent redistribution function, he wrote expressions for angle-averaged redistribution functions for the cases of the exact dipole phase function [e.g., $R_{II-B}(x',x)$] and the approximation of an isotropic scattering phase function [e.g. $R_{II-A}(x',x)$]. He then discussed some symmetry properties of the redistribution functions and showed that $R_{I-A}(x',x)$ and $R_{I-B}(x',x)$ are very similar but not identical. In most subsequent radiative transfer calculations, isotropic angle-averaged redistribution functions have been used.

Table 2. Different types of redistribution functions

Type of scattering	Atom's frame	Observer's frame
Coherent in atom's frame	$p(\xi',\xi) = \delta(\xi'-\xi)$	$R(\nu',\vec{n}',\nu,\vec{n})$
Zero line width in atom's frame	$f(\xi')d\xi' = \delta(\xi'-\nu_o)d\xi'$ (isotropic) (dipole)	$R_I(\nu',\vec{n}';\nu,\vec{n})$ $R_{I-A}(x',x)$ $R_{I-B}(x',x)$
Radiation damping (upper state only)	$f(\xi')d\xi' = \frac{\delta}{\pi}\frac{d\xi'}{(\xi'-\nu_o)^2+\delta^2}$	$R_{II}(\nu',\vec{n}';\nu,\vec{n})$ $R_{II-A}(x',x)$ $R_{II-B}(x',x)$
Collisions and natural broadening (upper state only)	$p(\xi',\xi)d\xi = f(\xi)d\xi'$ $= \frac{\delta}{\pi}\frac{d\xi}{(\xi-\nu_o)^2+\delta^2}$	$R_{III}(\nu',\vec{n}';\nu,\vec{n})$ $R_{III-A}(x',x)$ $R_{III-B}(x',x)$
Resonance scattering (both upper and lower states broadened)	$f(\xi')p(\xi',\xi) =$ $\frac{\delta_i\delta_j}{\pi^2} \cdot \frac{1}{[(\xi'-\xi)^2+\delta_i^2]}$ $\cdot \frac{1}{[(\xi-\nu_o)^2+\delta_j^2]}$	$R_{IV}(\nu',\vec{n}';\nu,\vec{n})$ $R_{IV-A}(x',x)$ $R_{IV-B}(x',x)$

Before proceeding further, I should mention that Heinzel (1981) argued that Hummer's (1962) R_{IV} function is based on an incorrect expression for scattering in the atomic frame. In its place, Heinzel (1981) derived a new redistribution function R_V for resonance scattering of subordinate lines assuming that both levels are radiatively broadened, and Heinzel and Hubený (1982) generalized this redistribution function to include collisional broadening of both the upper and lower states. The R_V function includes all of the other redistribution functions as special cases. Computations of the angle averaged $R_{V-A}(x',x)$ are given by Heinzel and Hubený (1983).

Now that we have defined the terminology that will be followed, it is important to clarify why CRD was used for most calculations until recently, despite the existence of a usable theory for PRD. The added complexity of PRD calculations and insufficient computational resources is only part of the explanation. A more important reason is that in one sense the CRD approximation had been shown to be adequate.

In the first of a series of papers entitled "Source function in a nonequilibrium atmosphere," Thomas (1957) asked the question whether the ratio of the line emission profile (j_ν) to the line absorption profile (ϕ_ν) depends on frequency for the case of coherent scattering in the atom's rest frame and Doppler redistribution as seen by the observer. He showed that whereas ϕ_ν decreases four orders of magnitude from line center to three Doppler widths away, the j_ν/ϕ_ν ratio remains within a factor of 4 of unity over this whole range. Therefore, in the Doppler core $j_\nu/\phi_\nu = 1$ is a decent approximation and the source function is frequency-independent (as first shown by Kenty 1932) with simple form

$$S(\nu) = (1-\varepsilon) \int_0^\infty \phi(\nu')J(\nu')d\nu' + \varepsilon B \quad . \qquad (1)$$

It is important to recognize that Thomas was primarily concerned about the flux in the core of the line and whether this flux can be used to infer the properties of a stellar atmosphere, in particular the existence and location of a chromosphere. From this point of view the CRD assumption is often justified as most of the flux in a stellar profile for a dwarf star like the Sun is often located in the Doppler core. Unfortunately, this justification of the use of CRD for a limited range of purposes was often misused as justification for the CRD assumption when analyzing the whole line profile including the inner line wings, where the probability of Doppler motions producing redistribution to the line core is extremely small!

In the sixth paper in this series, Jefferies and White (1960) reexamined the validity of the CRD assumption for the case of radiative and collisional damping in the atomic frame and Doppler redistribution. They concluded (see Fig. 1) that photons within about three Doppler half-widths of line center are completely redistributed in frequency, but that Doppler motions only slightly change the initial frequency of photons further from line center and rarely reshuffle these photons to the line core. In other words, scattering in the wings is nearly coherent over a very wide range of damping parameters. They therefore proposed that to a good approximation the redistribution function may be written as

$$R(x',x) = a(x')\delta(x'-x) + [1 - a(x')]\phi(x')\phi(x) \quad , \qquad (2)$$

an approximation often called "partially coherent scattering," where ϕ is the normalized absorption line profile and $a(x')$ is a function that is essentially zero for $x' < 3$ and nearly unity for $x' \geq 3$. Subsequently, Kneer (1975) pointed out that this approximation is neither normalized nor symmetric, and he proposed a revised form of equation (2) that meets both requirements. Then Basri (1980) argued that neither this equation nor Kneer's revision take into account Doppler diffusion in the line wings, which can be important in very thick, low density atmospheres, and that accurate radiative transfer solutions require use of the exact $R(x',x)$. Ayres (1984) has proposed a revision of equation (2) that includes the effects of Doppler diffusion.

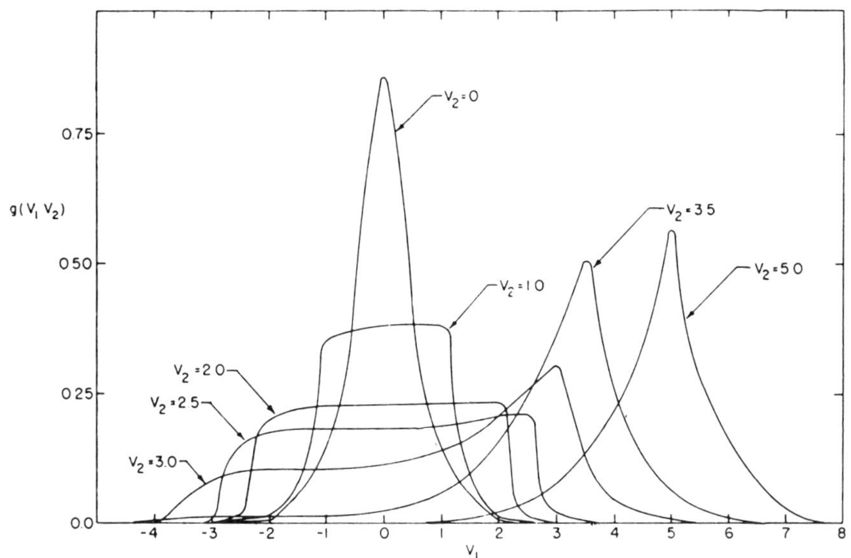

Fig. 1. The redistribution function for pure coherent scattering in the atom's rest frame $g(v_1,v_2) \equiv R_{I-A}(x',x)$ in the notation of this paper (from Jefferies and White 1960). Note that photons (in the observer's frame) at $v_2 < 3$ are non-coherently scattered and at $v_2 > 3$ are nearly coherently scattered about their initial frequency. Basri (1980) refers to the finite width of the redistribution function in the line wings ($v_2 >$ 3) as Doppler drifting of "coherently" scattered photons.

Jefferies and White also called attention to the depression of the line profiles in the inner wing that occurs as the scattering in the wing is taken to be more nearly coherent (a_{max} approaches 1.0). This is shown for an isothermal atmosphere in Figure 2, where for a_{max} = 1.0 the computed profile merges with the profile for coherent scattering in the observer's frame at x ($\equiv v_2$ in their notation) of about 4.6. They concluded that the emission features on either side of line center for the Ca II lines could in principle be produced in an isothermal atmosphere with coherent scattering in the line wings, as previously noted by Miyamoto (1953, 1954), or in a chromosphere if the whole line is formed in CRD. They were unable to choose among these two radically different explanations, although their calculations (Fig. 1) forcefully argue for the latter. Thus a proper understanding and treatment of spectral line formation in the line wings is essential for the determination of the temperature structure in a stellar atmosphere.

The first detailed solutions of the radiative transfer equation including PRD to go beyond the schematic partially coherent scattering calculations of Jefferies and White (1960) are to be found in a paper

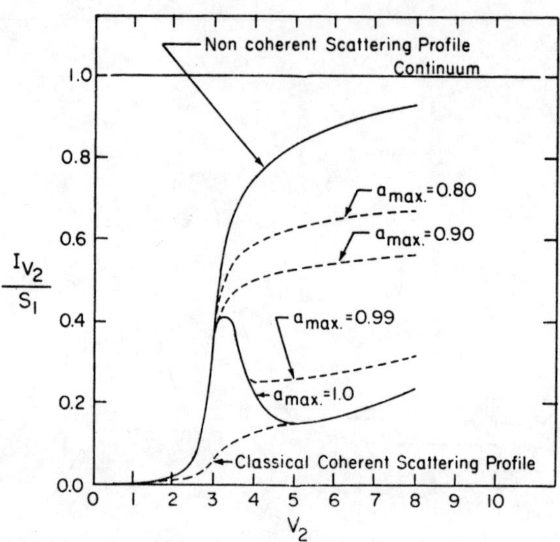

Fig. 2. Comparison of emergent line profiles for an isothermal atmosphere when the scattering is completely coherent in the atom's rest frame ($a_{max} = 1.0$), noncoherent ($a_{max} = 0$, upper solid line), and intermediate cases (from Jefferies and White 1960).

by Hummer (1969). These solutions assuming constant plasma parameters are for both plane parallel and semi-infinite geometries. The exact angle-dependent source function for plane-parallel geometry is

$$S(\nu,\vec{n}) = \frac{(1-\varepsilon)4\pi}{\phi(\nu)} \int_0^\infty \int_{4\pi} R(\nu',\vec{n}';\nu,\vec{n}) I(\nu',\vec{n}') d\nu' d\Omega' + \varepsilon B(\nu,T) \quad , \quad (3)$$

where $B(\nu,T)$ is the Planck function and ε the ratio of the collisional to radiative de-excitation rate for a two-level atom. In view of the complexity of retaining the full angular dependence of the redistribution functions, most calculations approximate $S(\nu,\vec{n})$ by dropping the angle dependence to compute the isotropic source function

$$S(\nu) = \frac{(1-\varepsilon)}{\phi(\nu)} \int_0^\infty R(\nu',\nu) J(\nu') d\nu' + \varepsilon B \quad ,$$

where

$$J(\nu') = \frac{1}{4\pi} \int I(\nu',\vec{n}') d\Omega' \quad . \quad (4)$$

This should be a good approximation to $S(\nu,\vec{n})$, except when the atmosphere is highly extended and the radiation field $I(\nu',\vec{n}')$ highly peaked in direction. Equation (4) reduces to the frequency-independent case (eq. (1)) when one assumes that the incoming and outgoing photon frequencies are uncorrelated, i.e.

$$R(\nu',\nu) = \phi(\nu')\phi(\nu) \quad , \tag{5}$$

which is a statement of complete frequency redistribution (CRD).

Hummer computed $S(\nu)$ for slabs with different line center thicknesses and ε values for $R_{I-A}(\nu',\nu)$ and $R_{I-B}(\nu',\nu)$. The latter differ only by a few percent, except near $x = 4$ for large optical depths T. He also computed $S(\nu)$ for $R_{II-A}(\nu,\nu')$ redistribution for several values of ε, A, and T, as well as for a semi-infinite geometry ($T = \infty$). He showed that $S(\nu)$ approaches arbitrarily close to the coherent scattering solution for sufficiently large values of ν, which means that the emergent intensity in the line wings can be substantially lower than the CRD solutions.

3. PARTIAL REDISTRIBUTION CALCULATIONS AFTER 1972

3.1. Resonance Line Transfer with Partial Redistribution Series (1973-1975)

About six months after the Colloquium on Stellar Chromospheres at Goddard, Vernazza (1972) presented calculations that showed that the flux and shape of the solar Lα line could be matched by a parametrized PRD calculation in which the core is computed with the CRD assumption and the scattering in the wings was taken to be 93% coherent and 7% CRD. These percentages were determined by computing profiles for a range of coherent scattering percentages without considering the various broadening rates. This result and the quantum mechanical calculation of Omont, Smith, and Cooper (1972) led Milkey and Mihalas (1973a) to propose a PRD computational scheme in which they formally divided the upper state of a two-level atom into frequency substates and then solved the statistical equilibrium equations for these substates including appropriate collisional couplings. These statistical equilibrium equations were then coupled into the transfer equation using a complete linearization scheme and solved in the *observer's frame*. Following Omont *et al*. (1972) they wrote

$$\begin{aligned} R(\nu',\nu) &= \frac{\Gamma_R + \Gamma_I}{\Gamma_R + \Gamma_I + \Gamma_E} R_{II}(\nu',\nu) + \frac{\Gamma_E}{\Gamma_R + \Gamma_I + \Gamma_E} \phi(\nu')\phi(\nu) \\ &= \gamma R_{II}(\nu',\nu) + (1-\gamma)\phi(\nu')\phi(\nu) \quad , \end{aligned} \tag{6}$$

where Γ_R, Γ_I, and Γ_E are the broadening widths due to radiative decays, inelastic collisions (i.e. collisional de-excitation), and elastic collisions (i.e. Stark and van der Waals collisions). The quantity γ measures the probability that incident photons are coherently scattered in the atom's frame, and thus the probability for coherent scattering in the far line wings in the observer's frame. They then solved the coupled statistical equilibrium-radiative transfer equations using the Vernazza, Avrett, and Loeser (VAL) (1973) solar model and the fixed electron densities of the model.

Milkey and Mihalas (1973a) did not attempt a detailed match to the solar Lα profile by choosing an optimum turbulent velocity and temperature distribution. They did, however, arrive at several important results. (1) A fully self-consistent PRD calculation using depth-dependent coherence fractions computed from the broadening rates is feasible. (2) The computed line wings lie between the CRD and pure coherent scattering limits and are similar to the 93% coherency calculation of Vernazza (1972). (3) A proper inclusion of PRD effects in the wings of Lα is needed to compute the electron density properly. This is because hydrogen is the dominant electron donor above 8000 K in the solar chromosphere, the ionization of hydrogen is controlled by photoionization from the second level, and the population of the second level depends on the Lα radiation field, which at some locations in the chromosphere is controlled by the loss of photons in the line wings. In a subsequent paper Milkey and Mihalas (1973b) were able to match the shape of the observed Lα profile by including a depth-dependent line profile function.

Despite the large improvement of the Milkey and Mihalas (1973a,b) Lα PRD calculations over the previous CRD calculations, Roussel-Dupré (1983) called attention to two major problems in their method -- the Omont et al. (1972) redistribution function used by Milkey and Mihalas is applicable only in the impact regime, which is not valid for the Lα wings, and it ignores the ℓ-degeneracy of the hydrogen energy levels. Both restrictions were relaxed by Yelnik et al. (1981). Using their formulae for Γ_E, Roussel-Dupré (1983) computed somewhat different wing intensities although similar wing shapes to those computed using the previous Omont et al. (1972) formula for Γ_E. Thus the Milkey-Mihalas formulation of PRD was a major step forward, but it should not be viewed as the final answer to a complex radiative transfer problem.

Milkey and Mihalas (1974) next applied their PRD formulation to the resonance lines of Mg II, treating each transition (the h and k lines) as equivalent to a two-level atom. An important aspect of these calculations is that they demonstrated the frequency dependence of $S(\nu)$ and its thermalization properties as a function of frequency. In the wings $S(\nu)$ behaves like the pure coherent scattering case with a monotonic decrease toward the surface since the chromosphere is optically thin in τ_ν (see Fig. 3, top). In the line core $S(\nu)$ is independent of ν and lies close to but slightly above the CRD case because thermalization involves redistribution to the optically thin wings where photons are lost from the atmosphere. Since redistribution to the line wings is made artificially large by the CRD assumption, there is less redistribution to the wings and a shorter thermalization length in the line core for PRD than CRD, so that $S_{PRD} > S_{CRD}$ in the line core. A second important point is that the h and k line profiles differ systematically in the sense that the h line wing lies above the k line wing and the h_1 minimum feature is brighter than k_2. Both effects naturally follow from the opacity in the h line being half that in the k line. The h line wing is brighter because at a given Δλ optical depth unity occurs deeper in the photosphere where B(T) is larger. The h_1 feature is brighter because the smaller h line opacity means that h_1 lies closer to line center than k_1, where the redistribution is less

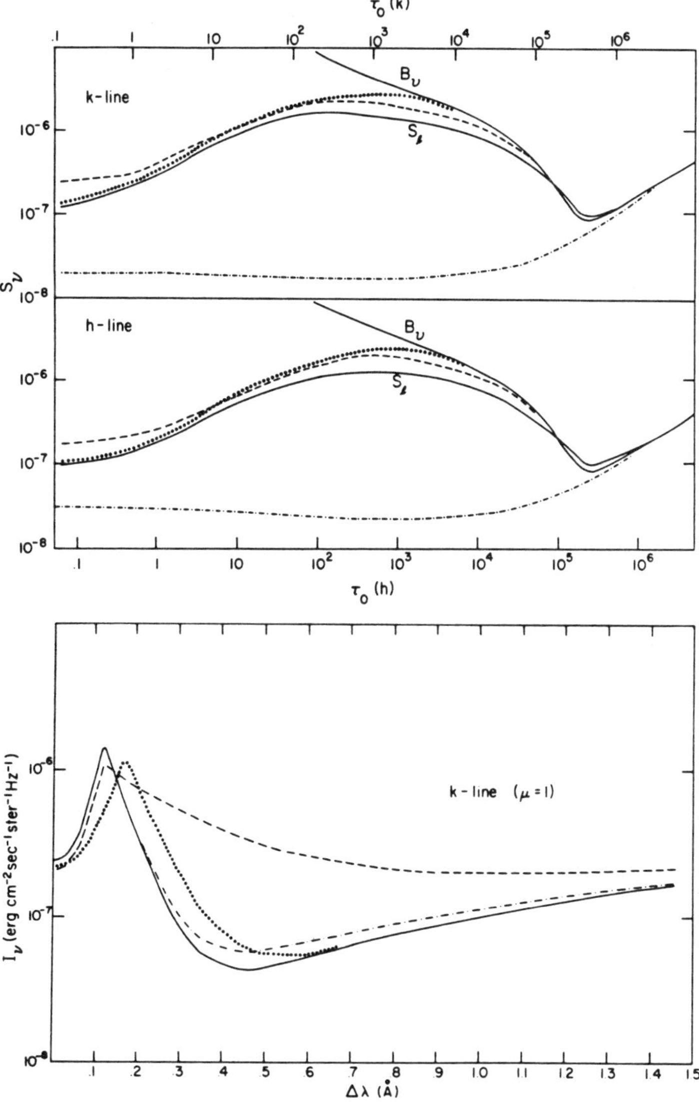

Fig. 3. Top -- Source functions (S_ν) for the Mg II h and k lines compared with B_ν: S_ℓ, complete redistribution; dots, line center ($\Delta\lambda = 0$); dashes, k_2 ($\Delta\lambda$ = 0.12 Å), h_2 ($\Delta\lambda$ = 0.109 Å); dot-dashes, k_1 ($\Delta\lambda$ = 0.462 Å), h_1 ($\Delta\lambda$ = 0.365 Å) (from Milkey and Mihalas 1974). Bottom -- Emergent Mg II k-line profiles for: dashes, complete redistribution; solid line, partial redistribution; dots, partial redistribution with enhanced v_t; dot-dashes, partial redistribution with enhanced Γ_{vdW} (from Milkey and Mihalas 1974).

coherent. A final important point made in the paper is that coherent scattering depresses the intensities $I(h_1)$ and $I(k_1)$ (see Fig. 3, bottom), so that one cannot derive the temperature minimum located between the photosphere and chromosphere by setting the Planck function equal to the intensity at h_1 or k_1, i.e., $B_\nu(T_{min}) > I(h_1), I(k_1)$.

Milkey, Ayres, and Shine (1975) then considered how stellar gravity might alter the Mg II line shapes. They computed PRD profiles of the Mg II lines for a solar model (log g = 4.44) and for a solar $T(\tau_{5000})$ distribution but log g = 2.0 to simulate an early G bright giant star. The effect of lowering the gravity is to decrease pressures in the photosphere. Since the major continuous opacity source (H^-) is proportional to P^2, while the Mg II optical depth scale is proportional to P, $\tau_{Mg\ II}$ (τ_{5000}) increases systematically with decreasing gravity. Since the temperature minimum occurs at larger $\tau_{Mg\ II}$ in a giant, the wavelengths of $(\Delta\lambda_{h_1}, \Delta\lambda_{k_1})$ of the h_1 and k_1 features increase. Ayres, Linsky, and Shine (1975) and Ayres (1979) have used this argument to explain in part the Wilson-Bappu effect. In addition, the Mg II line wings darken with decreasing gravity since lower photospheric densities result in more nearly coherent scattering.

The next extension of the theory by Milkey, Shine, and Mihalas (1975a) was to atoms in which two allowed transitions have common upper states. The application they had in mind was the Ca II ion in which the 4^2P upper states radiatively decay into the 4^2S ground state via the H and K lines, and also radiatively decay to the metastable 3^2D states via the so-called infrared triplet lines. To include this additional decay and radiative excitation channel provided by the infrared triplet lines, they divided each upper state into substates associated with each spectral line as seen in the observer's frame. This leads to statistical equilibrium equations for each of these substates and to a new redistribution function

$$R(\nu',\nu) = \gamma' R^x(\nu',\nu) + (1-\gamma')\phi(\nu')\phi(\nu) \quad , \qquad (7)$$

where $R^x(\nu',\nu)$, the cross-redistribution function, includes redistribution in the $i \rightarrow j$ transition via a third level k ($i \rightarrow j \rightarrow k \rightarrow j \rightarrow i$), and γ' is the branching ratio for this indirect process. This cross-redistribution provides additional noncoherence.

At this point I should mention that Hubený (1982) has introduced the concept of generalized redistribution functions p_i(i=1-5) in the atom's frame and P_i(i=1-5) in the observer's frame, which are analogous to the observer's frame redistribution functions R_i(i=1-5) previously described but differ from R_i(i=1-5) in that the final state may be different from the initial state. Thus R^x appears to be a special case of these generalized redistribution functions. Furthermore, Hubený, Oxenius, and Simonneau (1983a,b) have derived line profile coefficients for absorption and emission in multilevel atoms, taking into account generalized redistribution and the correlated re-emission that occurs when photoabsorption processes destroy the natural population of the excited state and elastic collisions tend to re-establish it.

In Paper V of their series, Shine, Milkey, and Mihalas (1975) applied the cross-redistribution formulation to a five-level, five-transition (H, K, 8498 Å, 8542 Å, 8662 Å) Ca II ion. The main purpose of the paper was to consider whether in principle PRD effects can account for the many discrepancies between line profiles calculated assuming CRD and the observations. For these calculations they assumed the HSRA solar model and several microturbulent velocity distributions with the following results:

(1) The PRD line profiles (see Fig. 4) show limb darkening at all wavelengths, consistent with the data, whereas the CRD profiles limb-brighten only at H_3 and K_3. Athay and Skumanich (1968b) could not account for the limb darkening of K_2 and H_2 assuming CRD.

(2) The decrease in intensity from K_2 to K_1 is abrupt in PRD but not CRD, in accord with observations. This answered Wilson's (1973) fundamental objections.

(3) The increase in line width is gradual toward the limb, as is observed.

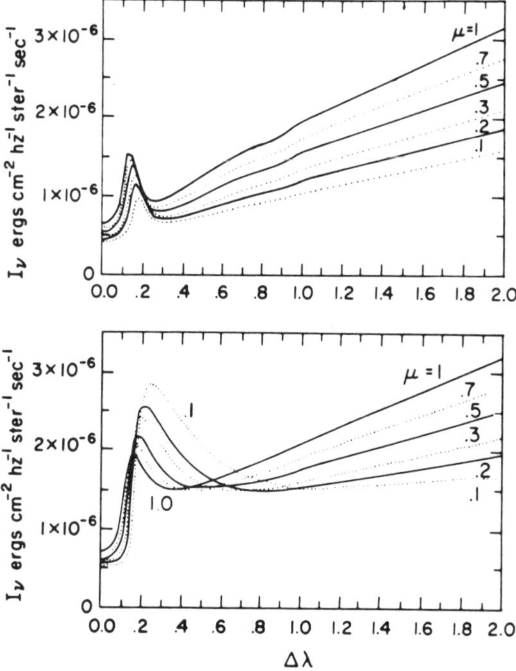

Fig. 4. Center-to-limb variation of the Ca II K line for a three-level atom with the HSRA model and the Linsky-Avrett velocity distribution. Top shows results for PRD; bottom shows results for CRD. Abscissa, wavelength separation from line center in Å; ordinate, specific intensity in absolute units. Curves are labeled with μ, the cosine of the angle from disk center (from Shine *et al.* 1975).

(4) $I(H_1)/I(K_1) = 1.17$, as is observed.

(5) The infrared triplet line PRD and CRD profiles are nearly identical, implying that cross-redistribution is extremely important for these lines as one would expect on the basis that the A values for depopulating of the 4^2P states via the resonance lines are far larger than the corresponding rates for the infrared triplet lines.

(6) Computed values of $I(H_1)$ and $I(K_1)$ are much smaller than observed when they use the HSRA model which has $T_{min} \approx 4170$ K, but are consistent with observations when they raise T_{min} to 4450 K.

(7) For a schematic solar plage model they found that K_2 brightens and $\Delta\lambda_{K_1}$ increases as is observed, and the infrared triplet lines develop double reversals in their cores as is also observed (Shine and Linsky 1972).

These successes provided strong evidence that the PRD approach includes much of the essential physics of the scattering process, however, even such approximate treatments of coherent scattering in the line wings as the partially coherent scattering model used by Vardavas and Cram (1974) can qualitatively account for points (1), (2), (3), and (6). Also there are alternative methods of writing the transfer equation that include the same physical processes but in some cases may be easier computationally (e.g. Heasley and Kneer 1976; Avrett and Loeser 1983).

I wish to mention briefly the next two papers in the series and postpone comments on the final paper until later. Shine, Milkey, and Mihalas (1975) considered the gravity dependence of the Ca II lines in the same way as Milkey, Ayres, and Shine (1975) did for the Mg II lines with qualitatively similar results. Milkey, Shine, and Mihalas (1975b) then included the full angle and frequency dependence of redistribution during the scattering processes in an <u>observer's frame</u> formulation. They found that angle-dependent effects are negligible for the Ca II lines in a homogeneous solar chromosphere model, but that these effects could be important for an inhomogeneous medium like the real solar chromosphere.

3.2. Models of the Solar Chromosphere Based on PRD Analyses of Resonance Emission Lines

I proceed now to several studies of the solar chromosphere in which atmospheric models were derived to match observed line profiles with profiles computed using the PRD formulation of the radiative transfer equation. In the first such study, Ayres and Linsky (1976) analyzed high resolution profiles of the Mg II h and k lines obtained by Kohl and Parkinson (1976) and Ca II H and K line profiles obtained by Brault and Testerman (1972). Their plane parallel hydrostatic equilibrium (HSE) models are parametrized by T_{min}, the column mass density at the temperature minimum $m(T_{min})$, the chromospheric temperature gradient $dT/d \log m$, and the chromospheric microturbulent velocity gradient $d\xi_T/d \log m$. For each model the electron density was computed from a three-level plus continuum representation for hydrogen, and LTE ionization of metals. They solved the transfer statistical equilibrium equations for a three-level (4^2S, $4^2P_{3/2}$, $3^2D_{5/2}$) Ca II ion and for a three-level $3^2S_{1/2}$, $3^2P_{1/2}$, $3^2P_{3/2}$) Mg II ion with the following results:

(1) There appears to be an inconsistency between the computed and observed shape of the Mg II line wings for the standard VAL (1973) and HSRA models no matter what values are assumed within factors of ±3 for the somewhat uncertain Mg abundances and van der Waals broadening parameters (see Fig. 5). Changing these parameters is equivalent to changing the percent coherency in the inner line wings and thus may mimic possible uncertainties produced by ignoring transitions to higher levels.

(2) Good fits to the inner wings of the Mg II lines were obtained by raising T_{min} to 4500 (+80, −110) K and to the wings of the Ca II H and K lines by raising T_{min} to 4450 ± 130 K, consistent with the previous study by Shine et al. (1975). These models are shown in Figure 6.

Ayres and Linsky (1976) pointed out that the Ca II and Mg II models are not really consistent in the sense that the Ca II model predicts Mg II wings that are too dark and the Mg II model predicts

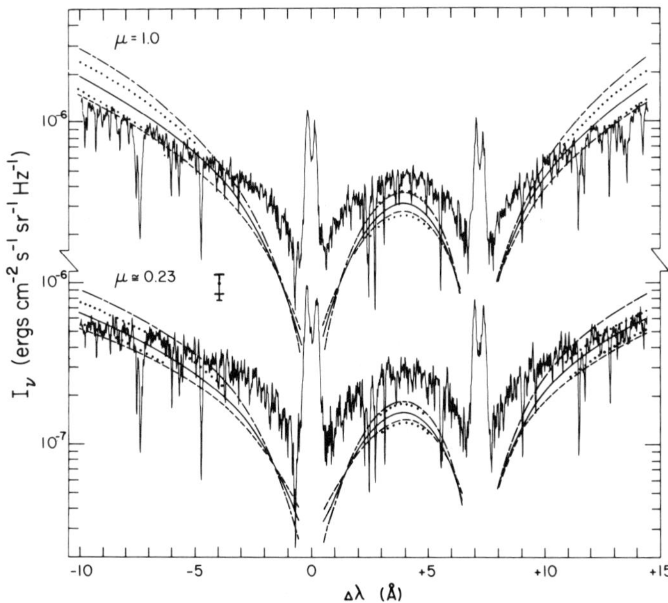

Fig. 5. Observed and synthesized Mg II h and k wings for the VAL model and a variety of γ_{VW} and A_{Mg} values: $\gamma_{VW} = \gamma_{VW}^{(0)} = 1.0 \times 10^{-8}$, $A_{Mg} = A_{Mg}^{(0)} = 3.89 \times 10^{-5}$ (———, adopted values); $\gamma_{VW} = 0.5 \times 10^{-8}$, $A_{Mg}^{(0)}$ (— - —); $\gamma_{VW} = 1.5 \times 10^{-8}$, $A_{Mg}^{(0)}$ (---); $A_{Mg} = 3.09 \times 10^{-5}$, $\gamma_{VW}^{(0)}$ (upper ••••); $A_{Mg} = 4.90 \times 10^{-5}$, $\gamma_{VW}^{(0)}$ (lower •••). The error bar refers to the estimated (+12, −20) percent accuracy of the absolute calibration at k_1 (from Ayres and Linsky 1976).

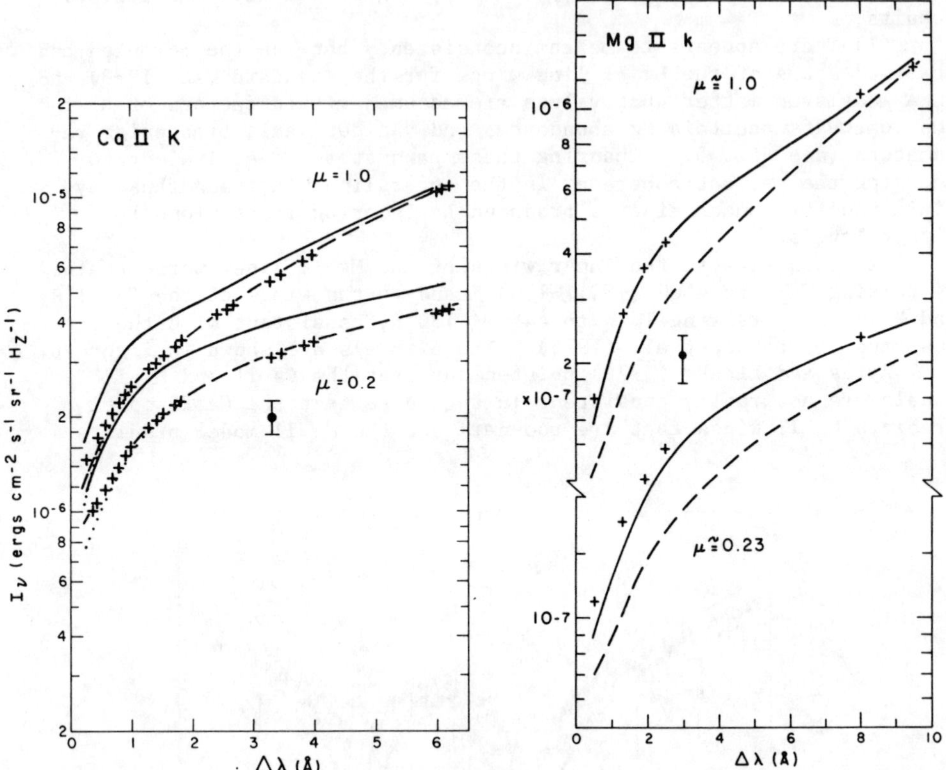

Fig. 6. (Left) Comparison of synthesized and measured (pluses) Ca II K line wings for the adopted "Ca II" model with T_{min} = 4450 K (dashed line) and the adopted "Mg II" model with T_{min} = 4500 K (solid line) at solar disk center (μ = 1.0) and near the limb (μ = 0.2). (Right) Same except pluses indicate measured Mg II line wings (from Ayres and Linsky 1976).

Ca II wings that are too bright (see Fig. 6). They were unable to decide whether the discrepancy should be explained as due to modeling an inhomogeneous atmosphere by a one component model, or whether it indicates that the degree of coherency in the line wings has been poorly estimated for either or both of the Ca II and Mg II lines. Vernazza, Avrett, and Loeser (1981) subsequently argued that the first explanation is likely incorrect since the Ca II and Mg II profiles computed with their mean atmospheric model (Model C) are nearly identical to those computed from an area-weighted average of the profiles compared with their six component atmospheric models.

Subsequently, Basri et al. (1979) presented profiles of the Lα emission core and wings for four components of the solar atmosphere (plage, network bright points, average quiet Sun, and dark points in supergranule cells) obtained with the NRL HRTS experiment. They then derived solar chromospheric models to match each Lα profile using the PRD formation and the line broadening parameters used by Milkey and Mihalas (1973a). For each assumed T(m) temperature distribution they computed the electron density selfconsistently with a five-level plus continuum representation for hydrogen and LTE ionization of the metals. They argued that the VAL 1973 model leads to a poor match to the Lα wing shape, a total line flux that is 35% too faint, and limb darkening of the line wings that is not observed. They therefore computed four new chromospheric models that match quite well each of the four Lα component line profiles. The Ca II and Mg II integrated line fluxes for their mean atmospheric model (computed to match the Lα line) are not, however, in satisfactory agreement with the data. Vernazza _et al._ (1981) pointed out that the discrepancy of the Lα wing flux between observations and the Basri _et al._ of calculations using the VAL 1973 model may be due to their assumption of LTE ionization equilibrium for C I, since the C I continuum overlaps the Lα line. Also Roussel-Dupré (1983) argued that the shape of the Lα wing may not be an accurate diagnostic with which to derive the chromospheric temperature structure, because the high degree of coherency in the wings produces thermalization lengths as large as the chromospheric thickness.

I now come to the solar chromospheric models of Vernazza, Avrett, and Loeser (1981) that have been the standard solar models for the last few years, but were not constructed to match the profiles of strong resonance lines. Instead they computed six atmospheric models (designated A-F) to match Skylab observations of the 1350-1680 Å flux consisting primarily of Si bound-free continua and to a lesser extent continua of Fe I. These models are for the darkest observed component (Model A), the mean component (Model C), the brightest component (Model F), and intermediate cases. For each component model the electron density was computed selfconsistently including nonLTE ionization of hydrogen and 17 metals. In addition, they computed profiles for the Lα, Lβ, Ca II and Mg II lines using a parametrized PRD formulation, but they did not use differences between the observed and computed line profiles to modify the chromospheric models themselves on the basis that the maximum wing coherency parameters may be uncertain. These computations differ from those of Ayres-Linsky and Milkey-Mihalas in that γ, the probability of coherent scattering in the atom's frame, is the smaller of that given by equation (6) and γ_{max}, an arbitrarily chosen parameter introduced to account for the effects of other possible redistribution channels, for example through higher atomic levels.

The computed PRD Lα line profile for Model C is close to the observed mean quiet Sun profile (assuming $\gamma_{max} = 0.98$), but Roussel-Dupré (1983) has shown that there is no physical basis for this value of γ_{max} and that computed Lα wings for Model C and values of γ calculated with the Yelnik _et al._ (1981) theory for degenerate levels lie a factor of 4 below the observed Lα wings. The Ca II and Mg II profiles computed with Model C and for a wide range of wing coherency parameters also do not match the observed wing fluxes or line shapes. In general, the computed inner wings are too dark, implying that to match these data the temperature minimum should be raised above the value

T_{min} = 4170 K determined for Model C. VAL pointed out that the Si I and Fe I ionization equilibria are far from LTE in the direction of overionization, and that if the ionization equilibrium should be closer to LTE by an error in either the assumed photoionization cross section or computed mean radiation field in the bound-free continua, then to match the ultraviolet continua they would have to raise T_{min} and thus achieve a better match to the Ca II and Mg II line wings.

Very recently Avrett, Kurucz, and Loeser (1984) have reassessed this particular point and computed a new six component model. This reassessment was stimulated by Kurucz's computed solar spectrum including a very large number of spectral lines in the 1350-1680 Å spectral region. This inclusion of line blanketing had the effect of drastically lowering the photoionizating radiation field that Si I and Fe I see at the base of the chromosphere (and thus the photoionization rates) and of increasing the "observed" ultraviolet continuum that the model must match. Both effects result in models for which the ionization equilibria of Si I and Fe I are closer to LTE and the temperature minimum must be raised to match the ultraviolet continuum. Their new mean atmospheric model (Model C') has $T_{min} \approx$ 4500 K and is very similar to the Ayres-Linsky (1976) model in the temperature minimum region. Model C' also matches well the ultraviolet and millimeter continua as well as the Ca II and Mg II lines, thereby presumably ending the long standing discrepancy between models based on analysis of solar continua and models based on matching PRD computations of the profiles of resonance lines with observations. Thus atmospheric models computed by matching the profiles of strong resonance lines should be reasonably valid in approximating the mean temperature-density structure, but one should never forget that the solar atmosphere is highly inhomogeneous and a mean atmospheric model is only a crude approximation to the real Sun.

3.3. Models of the Chromospheres of Late-Type Stars based on PRD Analyses of Resonance Emission Lines

The development of chromospheric models for stars based on the matching of observed and computed PRD profiles of resonance emission line profiles has proceeded in parallel with the solar modeling after the initial papers by Milkey, Mihalas, and their collaborators. Table 3 is a concise summary of the work to date. The first such model was constructed by Ayres and Linsky (1975) to match the Ca II K line of α Boo (Arcturus, K2 III). They first tried to fit the K line profile in a CRD calculation, but could not match the observed steep decrease in flux from K_2 to K_1 (see Fig. 7) as Wilson (1973) had previously argued was a major defect of the CRD calculations. Ayres and Linsky (1975) then found that a two-level atom PRD calculation with the same model atmosphere produces an inner wing profile of the correct shape but too low flux (see Fig. 7). They then obtained an excellent agreement to the inner wing shape and flux by raising T_{min} from 3050 K to 3200 K.

In subsequent papers, various authors have extended this work to derive atmospheric models for dwarf, giant, and supergiant stars by matching the Ca II and Mg II resonance lines with PRD calculations

Table 3. Summary of stellar chromospheric models computed using partial redistribution to fit observed line profiles.

Stars	Spectral Lines Fitted	Type of Model	Reference
Dwarfs			
α Cen A(G2 V), α Cen B (K1 V)	Ca II K	PP, static	Ayres *et al.* (1976)
70 Oph (K0 V), ε Eri (K2 V)	Ca II K, Mg II h, k	PP, static	Kelch (1978)
γ Vir N (F0 V), θ Boo (F7 IV-V) 59 Vir (F8 V), HD 76151 (G5 V) 61 UMa (G8 V), ξ Boo A (G8 V) EQ Vir (dK7e), 61 Cyg B (dM0)	Ca II K	PP, static	Kelch, Linsky and Worden (1979)
EQ Vir (dK7e), GL 616.2 (dM1.5e) YZ CMi (dM4.5e), GL 393 (dM2) GL 411 (dM2	Ca II K	PP, static	Giampapa, Worden and Linsky (1982)
Giants			
α Boo (K2 III)	Ca II K	PP, static	Ayres and Linsky (1975)
α Aur (G5 III), β Gem (K0 III) α Tau (K5 III)	Ca II K Mg II h, k	PP, static	Kelch *et al.* (1978)
λ And (G8 III-IV), α Aur (G5 III)	Ca II K, Mg II k	PP, static	Baliunas *et al.* (1979)
β Cet (G9.5 III)	Mg II h, k	PP, static	Eriksson, Linsky and Simon (1983)
α Boo (K2 III)	Mg II k	SS, wind	Drake (1985)
Bright Giants and Supergiants			
β Dra (G2 II-Ib), ε Gem (G8 Ib) α Ori (M2 Iab)	Ca II K, Mg II k	PP, static	Basri, Linsky and Eriksson (1981)
β Dra (G2 II-Ib)	Ca II K, Mg II k	PP, circulation pattern	Basri, Linsky and Eriksson (1981)
N-type Carbon Stars	Mg II k	PP, static	Avrett and Johnson (1984)

assuming three-level atoms or in a few cases full five-level atoms for Ca II. Until very recently, all of these atmospheric models were assumed to be plane-parallel, homogeneous, and static. Also, the atomic level broadening parameters were generally based on theoretical estimates rather then assuming *a priori* a range of values, although in many calculations the effect of the Ca II infrared triplet lines was not computed explicitly for a three-level or a five-level atom, but was rather simulated by including an additional incoherency fraction of 6% in a two-level atom PRD calculation.

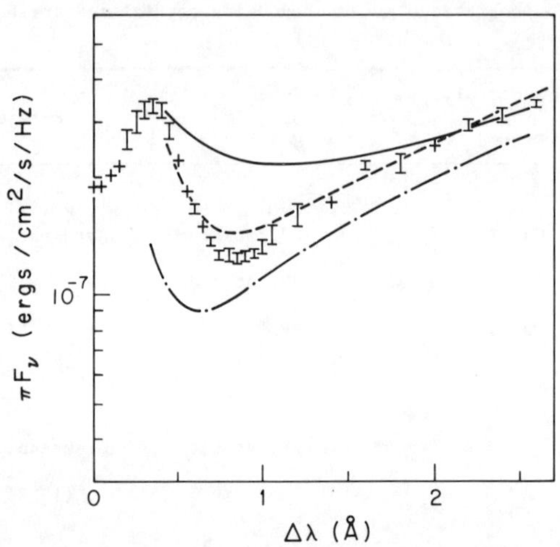

Fig. 7. Comparisons of observed (error bars) and computed fluxes for the Arcturus Ca II K line. The solid curve is the inner wing profile computed in CRD with an atmospheric model with T_{min} = 3050 K. The dot-dashed curve is the PRD wing profile for the same model, and the dashed curve is a PRD wing profile for a model in which T_{min} is raised to 3200 K (from Ayres and Linsky 1975).

In a thoughtful paper Basri (1980) assessed the effect of various processes that will substantially affect computed PRD profiles in the low density atmospheres that occur in low gravity stars (i.e. supergiants). For example, minimum features located outside the emission maxima near line center can be produced even for isothermal atmospheres as was previously noted by Jefferies and White (1960). However, the magnitude of the dip from K_2 to K_1 (using the Ca II notation) depends critically on the diffusion of photons in the line wing due to Doppler motions of the scattering atoms. Thus it is critical in these low density atmospheres to use the exact $R_{II}(x',x)$ function rather than the partially coherent scattering approximation (cf. Jefferies and White 1960; Kneer 1975) that in the line wings most scatterings are coherent in the observer's frame (eq. (2)). Basri (1980) also showed that Doppler diffusion of photons in the line wings (see Fig. 1) can substantially broaden the emission features and increase the redistribution from core to wings especially for intermediate scale turbulence, such as will occur when long wavelength waves are present in an atmosphere. He also argued that continuous opacity can change the shape of the line wings even when $r_0 \equiv \kappa_C/\kappa_L(\Delta\lambda = 0)$ is very small, because in the wings $r_{\Delta\lambda}$ is orders of magnitude larger and the continuum source function (generally the Planck function) can be orders of magnitude larger than the wing source function. Finally, he called attention to the role that transitions to other atomic levels can play in establishing the minimum incoherence fraction for a given transition in a low density atmosphere (cf. Hubený et al. 1983a,b).

3.4. The Roles Played by Velocity Fields and Atmospheric Extension

Until now I have discussed only the effects of PRD on computed line profiles for static, plane-parallel atmospheres, but real atmospheres expand and are often geometrically extended. In addition, wavelike motions, circulation patterns, and inhomogeneity must occur in many stars as in the Sun. In two early papers Vardavas (1974) and Cannon and Vardavas (1974) calculated PRD line profiles for an expanding atmosphere using the $R_I(x',x)$ redistribution function in the observer's frame. They found large differences between line profiles computed with CRD and PRD, which Magnan (1974) argued must be incorrect. Mihalas et al. (1976b) argued that in a moving atmosphere accurate source functions can be computed in the observer's frame only when the full angular and frequency coupling is taken into account for the frequency redistribution, which is computationally very expensive. However, calculations in the comoving frame of the fluid are greatly simplified because then the scattering atoms see a nearly isotropic radiation field of limited frequency range, so that angle-averaged static redistribution functions are a good approximation. They then formulated the PRD transfer problem in the comoving frame and showed that for $R_I(x',x)$ redistribution line profiles for CRD and PRD are nearly identical. Vardavas (1976) showed that accurate solutions to the PRD differentially expanding atmosphere problem can be obtained using the comoving frame formulation for R_I and R_{II} redistributions in an isothermal atmosphere, and Vardavas and Cannon (1976) did the same for a schematic chromosphere model with R_{II-A} redistribution.

In subsequent theoretical developments, solutions of the radiative transfer equation were generally formulated in the comoving frame for spherically symmetric geometries and they generally assumed CRD or, nearly equivalently, the Sobolev approximation, since the intended applications were for atmospheres of hot stars where the expansion velocities are much larger than the thermal and turbulent motions. To test the validity of the CRD approximation, Mihalas, Kunasz, and Hummer (1976a) solved the transfer equation for spherically symmetric flows with an angle-averaged redistribution function in the comoving frame using a variable-Eddington factor iterative scheme. They discussed the conditions for which the emergent line profiles for CRD and PRD can differ for $R_I(x',x)$ and $R_{II}(x',x)$ redistribution. Mihalas (1980) extended this work to include the full angle and frequency dependence of redistribution. Peraiah and Nagendra (1983) have used the $R_{II-A}(x',x)$ redistribution function to interpret line profiles formed in the moving atmospheres of hot stars.

Since winds in late-type giants and supergiants have expansion velocities of only a few times the combined thermal-turbulent halfwidth, the CRD and the Sobolev approximations should not be accurate. However, there is ample evidence in the asymmetric Mg II h and k lines (e.g. Stencel and Mullan 1980; Stencel et al. 1980) that significant expansion velocities occur in the chromospheres of these stars. To explore the range of velocities and chromospheric geometric extents that should be representative for the late-type giants, Drake and Linsky (1983) made a series of calculations using the comoving frame PRD code developed by Mihalas <u>et al.</u> (1976b). In their exploration of parameter space they found that with increasing flow speed and geometrical extension (see Fig. 8), the effect on the Mg II line profiles is to suppress the blue emission peak, shift the central minimum (k_3) to the blue, and enhance the red emission peak giving the Mg II lines a P Cygni character. For all cases in which the maximum flow speed is

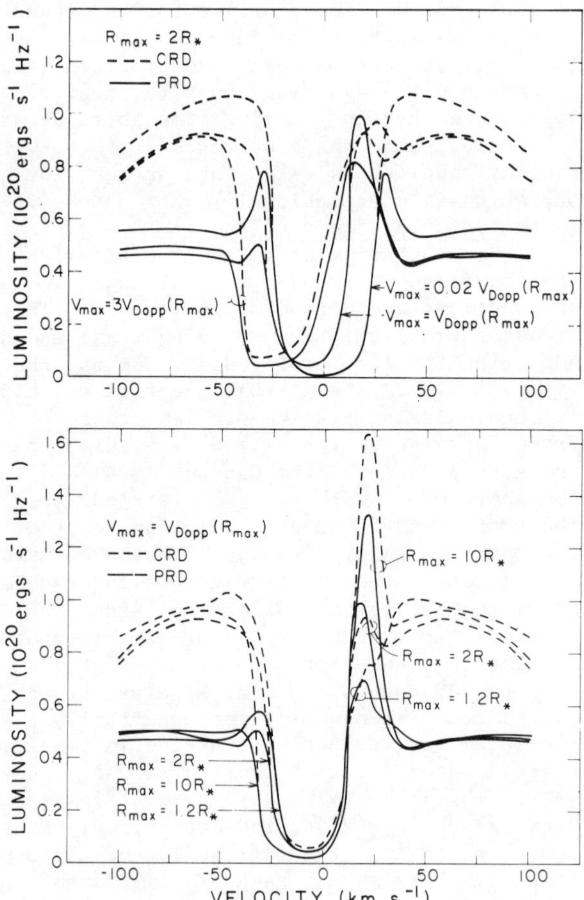

Fig. 8. (Top) Computed profiles of the Mg II k line for an early K giant star with $R_{max} = 2\ R_*$. CRD and PRD profiles are given for three different velocity laws. (Bottom) Computed profiles keeping the same expansion velocity law but changing the maximum size of the chromosphere from 1.2 to 10 times the photospheric radius (from Drake and Linsky 1983).

less than six times the Doppler width, the PRD profiles are very different from the CRD case. They also explored the effect of changing the chromospheric temperature gradient and the location of the chromosphere. Drake (1985) will report at this meeting on his calculations to match the observed Mg II line profiles of Arcturus.

Avrett and Loeser (1983) described an integral-equation method for solving the transfer equation in the comoving frame also including the effects of PRD, spherical geometry, and expansion. Using the partially coherent scattering formulation of $R_{II}(\nu',\nu)$ by Kneer (1975), they showed sample calculations for a constant property atmosphere and a velocity law with $v_{max}/\Delta v_D = 3$. Hartmann and Avrett (1984) used this formalism to compute resonant line profiles for their Alfven-wave driven wind model of α Ori (M2 Iab). Differences between computed and observed line profiles then led them to propose a more complex velocity field in the atmosphere of this star than initially computed. These calculations highlight the role that observations have and should continue to play in stimulating progress in spectral line formation theory.

4. CONCLUSIONS

As should be clear from the preceding discussion, the shape of a spectral line profile emergent from a stellar atmosphere depends in a complex way on the properties of the atmosphere (temperature-density structure, geometrical extent, inhomogeneity, velocity fields, and radiation fields in different lines and continua), as well as the radiative and collisional processes that lead to redistribution of photons during the scattering process. While the theory of PRD even for a multilevel atom is now becoming mature, we have no assurance that all of the important physical effects have been included properly. Nevertheless, the shape of spectral line profiles for optically thick resonance lines and perhaps also important subordinate lines depends critically on the proper inclusion of PRD, atmospheric extension, and expansion effects. Thus we cannot expect to derive the structure of a stellar atmosphere without properly including these effects, but at the same time we should not avoid using the powerful tool of spectral line profile analysis because of the mathematical complexity of the theory or possible uncertainties concerning its physical basis.

This work was supported in part by NASA grants NAG5-82 and NGL-06-003-057 through the University of Colorado. I would like to thank Drs. E. H. Avrett, T. R. Ayres, S. A. Drake, and D. G. Hummer for stimulating discussions. I would also like to thank the staff of the Osservatorio Astronomico di Trieste for their exceptional hospitality, and NATO for its support of the Workshop.

REFERENCES

Athay, R. 1972, Radiation Transport in Spectral Lines (Dordrecht, Holland: Reidel).

Athay, R. G. and Skumanich, A. 1968a, Solar Phys., **3**, 181.

Athay, R. G. and Skumanich, A. 1968b, Solar Phys., **4**, 176.

Avrett, E. H. 1965, SAO Special Report No. 174, p. 101.

Avrett, E. H. 1973, in Stellar Chromospheres, ed. S. D. Jordan and E. H. Avrett, NASA SP-317, p. 27.

Avrett, E. H. and Johnson, H. R. 1984, in Cool Stars, Stellar Systems, and the Sun, ed. S. L. Baliunas and L. Hartmann (Berlin: Springer-Verlag), p. 330.

Avrett, E. H., Kurucz, R. L., and Loeser, R. 1984, Bull. A. A. S., **16**, 450.

Avrett, E. H. and Loeser, R. 1983, in Methods in Radiative Transfer, ed. W. Kalkofen (Cambridge: Cambridge University Press), p. 341.

Ayres, T. R. 1979, Ap. J., **228**, 509.

Ayres, T. R. 1984, preprint.

Ayres, T. R. and Linsky, J. L. 1975, Ap. J., **200**, 660.

Ayres, T. R. and Linsky, J. L. 1976, Ap. J., **205**, 874.

Ayres, T. R., Linsky, J. L., Rodgers, A. W., and Kurucz, R. L. 1976, Ap. J., **210**, 199.

Ayres, T. R., Linsky, J. L., and Shine, R. A. 1975, Ap. J. (Letters), **195**, L121.

Baliunas, S. L., Avrett, E. H., Hartmann, L., and Dupree, A. K. 1979, Ap. J. (Letters), **233**, L129.

Basri, G. S. 1980, Ap. J., **242**, 1133.

Basri, G. S., Linsky, J. L., Bartoe, J.-D. F., Brueckner, G., and Van Hoosier, M. E. 1979, Ap. J., **230**, 924.

Basri, G. S., Linsky, J. L., and Eriksson, K. 1981, Ap. J., **251**, 162.

Brault, J. and Testerman, L. 1972, Preliminary Edition of the Kitt Peak Solar Atlas (Tucson: Kitt Peak National Observatory).

Cannon, C. J. and Vardavas, I. M. 1974, Astr. Ap., **32**, 85.

Drake, S. A. 1985, this volume.

Drake, S. A. and Linsky, J. L. 1983, Ap. J., **273**, 299.

Eriksson, K., Linsky, J. L., and Simon, T. 1983, Ap. J., **272**, 665.

Giampapa, M. S., Worden, S. P., and Linsky, J. L. 1982, Ap. J., **258**, 740.

Hartmann, L. and Avrett, E. H. 1984, Ap. J., in press.

Heasley, N. J. and Kneer, F. 1976, Ap. J., **203**, 660.

Heinzel, P. 1981, J. Quant. Spectrosc. Radiat. Transfer, **25**, 483.

Heinzel, P. and Hubený, I. 1982, J. Quant. Spectrosc. Radiat. Transfer, **27**, 1.

Heinzel, P. and Hubený, I. 1983, J. Quant. Spectrosc. Radiat. Transfer, **30**, 77.

Henyey, L. G. 1941, Proc. Nat. Acad. Sci., **26**, 50.

Holstein, T. 1947, Phys. Rev. **72**, 1212.

Hubený, I. 1982, J. Quant. Spectrosc. Radiat. Transfer, **27**, 593.

Hubený, I., Oxenius, J., and Simonneau, E. 1983a, J. Quant. Spectrosc. Radiat. Transfer, **29**, 477.

Hubený, I., Oxenius, J., and Simonneau, E. 1983b, J. Quant. Spectrosc. Radiat. Transfer, **29**, 495.
Hummer, D. G. 1962, M. N. R. A. S., **125**, 21.
Hummer, D. G. 1969, M. N. R. A. S., **145**, 95.
Jefferies, J. T. 1968, Spectral Line Formation (Waltham, Mass.: Blaisdell), Ch. 5.
Jefferies, J. T. and Morrison, N. D. 1973, in Stellar Chromospheres, ed. S. D. Jordan and E. H. Avrett, NASA SP-317, p. 3
Jefferies, J. T. and Thomas, R. N. 1959, Ap. J., **129**, 401.
Jefferies, J. T. and Thomas, R. N. 1960, Ap. J., **131**, 695.
Jefferies, J. T. and White, O. R. 1960, Ap. J., **132**, 767.
Kelch, W. L. 1978, Ap. J., **222**, 931.
Kelch, W. L., Linsky, J. L., Basri, G. S., Chiu, H.-Y. Chang, S. H., Maran, S. P., and Furenlid, I. 1978, Ap. J., **220**, 962.
Kelch, W. L., Linsky, J. L., and Worden, W. P. 1979, Ap. J., **229**, 700.
Kenty, C. 1932, Phys. Rev., **42**, 823.
Kneer, F. 1975, Ap. J., **200**, 367.
Kohl, J. L. and Parkinson, W. H. 1976, Ap. J., **205**, 599.
Linsky, J. L. 1968, SAO Special Report No. 274.
Linsky, J. L. and Avrett, E. H. 1970, P. A. S. P., **82**, 169.
Magnan, C. 1974, Astr. Ap., **35**, 233.
Mihalas, D. 1978, Stellar Atmospheres, Second Edition (San Francisco: W. H. Freeman), Ch. 13.
Mihalas, D. 1980, Ap. J., **238**, 1034.
Mihalas, D., Kunasz, P. B., and Hummer, D. G. 1976a, Ap. J., **210**, 419.
Mihalas, D., Shine, R. A., Kunasz, P. B., and Hummer, D. G. 1976b, Ap. J., **205**, 492.
Milkey, R. W., Ayres, T. R., and Shine, R. A. 1975, Ap. J., **197**, 143.
Milkey, R. W. and Mihalas, D. 1973a, Ap. J., **185**, 709.
Milkey, R. W. and Mihalas, D. 1973b, Solar Phys., **32**, 361.
Milkey, R. W. and Mihalas, D. 1974, Ap. J., **192**, 769.
Milkey, R. W., Shine, R. A., and Mihalas, D. 1975a, Ap. J., **199**, 718.
Milkey, R. W., Shine, R. A., and Mihalas, D. 1975b, Ap. J., **202**, 250.
Miyamoto, S. 1953, P. A. S. Japan, **5**, 142.
Miyamoto, S. 1954, P. A. S. Japan, **6**, 140.
Omont, A., Smith, E. W., and Cooper, J. 1972, Ap. J., **175**, 185.
Peraiah, A. and Nagendra, K. N. 1983, Astrophys. Space Sci., **90**, 237.
Roussel-Dupré, D. 1983, Ap. J., **272**, 723.
Shine, R. A. and Linsky, J. L. 1972, Solar Phys., **25**, 357.
Shine, R. A. and Linsky, J. L. 1974, Solar Phys., **39**, 49.
Shine, R. A., Milkey, R. W., and Mihalas, D. 1975, Ap. J., **199**, 724.
Sobolev, V. V. 1955, Vestnik Leningradskogo Gosudarstvennogo Universiteta No. 5, 85.
Stencel, R. E. and Mullan, D. J. 1980, Ap. J., **238**, 221.
Stencel, R. E., Mullan, D. J., Linsky, J. L., Basri, G. S., and Worden, S. P. 1980, Ap. J. Suppl., **44**, 383.
Thomas, R. N. 1957, Ap. J., **125**, 260.
Thomas, R. N. 1973, in Stellar Chromospheres, ed. S. D. Jordan and E. H. Avrett, NASA SP-317, p. 312.

Unno, W. 1952, P. A. S. Japan, **4**, 100.
Vardavas, I. M. 1974, J. Quant. Spectrosc. Radiat. Transfer, **14**, 909.
Vardavas, I. M. 1976, J. Quant. Spectrosc. Radiat. Transfer, **16**, 781.
Vardavas, I. M. and Cannon, C. J. 1976, Astr. Ap., **53**, 107.
Vardavas, I. M. and Cram, L. E. 1974, Solar Phys., **38**, 367.
Vernazza, J. E. 1972, Ph. D. Thesis, Harvard University.
Vernazza, J. E., Avrett, E. H., and Loeser, R. 1973, Ap. J., **184**, 605.
Vernazza, J. E., Avrett, E. H., and Loeser, R. 1981, Ap. J. Suppl., **45**, 635.
Warwick, J. W. 1955, Ap. J., **121**, 190.
Wilson, O. C. 1973, in Stellar Chromospheres, ed. S. D. Jordan and E. H. Avrett, NASA SP-317, p. 305.
Yelnik, B., Burnett, K., Cooper, J., Ballagh, R. J., and Voslamber, D. 1981, Ap. J., **248**, 705.

GENERAL ASPECTS OF PARTIAL REDISTRIBUTION AND ITS ASTROPHYSICAL IMPORTANCE

Ivan Hubeny
Astronomical Institute
Czechoslovak Academy of Sciences
251 65 Ondřejov
Czechoslovakia

1. INTRODUCTION

Let us start with a quotation from Section 5.5 of Jeffries' (1968) textbook which summarized the status of the astrophysical redistribution problem about 17 years ago:

> "Until better theoretical calculations and laboratory spectroscopy can give us a guide, we can probably do nothing but adopt the results given above. In general, we may suppose that the line cores, at least, of resonance lines are formed by a mechanism approaching that of complete redistribution, while the wings of the first resonance line in a series may have a significant degree of essentially coherent scattering; for the later resonance lines we shall find less coherence. If the line is collisionally broadened it seems legitimate to suppose it to be formed in complete redistribution in the core and wings".

This quotation provides us with a reasonable starting point for our subsequent discussion. We may ask: To what extent has our understanding of redistribution phenomena recently improved? Which aspects of the problem are now clear, and which aspects require further elaboration? Is the presently available theory capable of providing a reliable basis for spectroscopic diagnostics of astrophysical bodies? Are the refinements of the underlying theory really important for interpretational purposes, and, if so, under what circumstances? One of the purposes of this paper is to discuss the questions raised above.

The literature concerning this subject is abundant. Many important aspects of the problem can be found, and their historical development can be traced, in the reviews of Jefferies (1968), Hummer and Rybicki (1971), Athay (1972), Milkey (1976), and Mihalas (1978). However, no comprehensive critical summary exists that would cover the astrophysical results of the most recent period. Therefore, it seems worthwhile to attempt to fill in this gap.

Perhaps the most important point we would like to make here, is to emphasize the close connection between the two basic aspects of the redistribution problem, namely i) the underlying physical description, and ii) applications (in a wide sense) to astrophysical problems. We

J. E. Beckman and L. Crivellari (eds.), Progress in Stellar Spectral Line Formation Theory, 27–58.

hope it is fair to say that most of the studies have been focused more on the applications and, in particular, on computational problems, while many fundamental questions have remained unanswered. On the other hand, the problem of redistribution of line radiation has been studied extensively in "laboratory" physics, particularly due to the development of the laser. In a recent review paper, Hubeny (1984) has discussed the relation between the astrophysical and laboratory views on redistribution phenomena from the physical point of view. Here, we shall discuss both aspects of the redistribution problem, underlying physics and applications, with more emphasis on the astrophysical point of view.

Thus, the first part of the paper (Chapters 2 - 5) is intended as a summary of the basic physical concepts employed by current astrophysical approaches, and is meant, in some sense, as a rather pedagogical introduction to the subject. The second part (Chapter 6) reviews one of the most important topics of the theory, the redistribution function, which is the point where the fundamental and applicational aspects are particularly closely linked. The issue of the final part (Chapters 7 - 9) is a discussion of recent applications and numerical methods. Again, attention is devoted to the conceptual questions rather than to the results of the actual applications. We do not attempt to cover the whole range of recent astrophysical work; we shall primarily concentrate on the transfer of unpolarized radiation in planeparallel, static media, but we shall also briefly mention other models. These topics will be discussed in other sessions of this conference.

2. WHY DO WE STUDY PHOTON CORRELATION PHENOMENA?

The basic aim of spectroscopic diagnostics is to infer, from an observed spectrum, as much information about a distant medium as possible. Astrophysical spectroscopic diagnostics is complicated by radiation not only being the only probe of the physical properties of a studied medium, but usually also one of the dominant agents in establishing its overall energy balance.

Consequently, the radiative transfer equation, which may be viewed as the basic equation of the problem, must always be complemented by other equations (sometimes called constraints) that describe the coupling between the radiation and other structural parameters (temperature, density, atomic level populations, velocity distributions, etc.). This is the point which makes the topic of spectroscopic diagnostics so important from the practical point of view, and so interesting from the theoretical point of view.

Let us adopt, at the very outset, a view which will help us to understand many problems and uncertainties met in the past, as well as the contemporary trends of research: The interaction between radiation and matter is basically a quantum-mechanical (or, more precisely, a quantum-statistical) phenomenon. Thus any "classical" description (e.g. that based on a purely kinetic level of description) should be viewed as an approximation which may, but need not, be valid in actual applications. Moreover, a too literal classical interpretation of some concepts of the radiative transfer theory (e.g. stimulated emission) may thus be hazardous.

Fortunately, a typical astrophysical situation is characterized by a number of important simplifications. Amongst them, the most important ones are the assumption of a weak, incoherent radiation field, and that of media composed of randomly distributed particles separated by distances much larger than the wavelength of the radiation (for more detail, refer to Hubeny, 1984). It follows that a suitable description of astrophysical radiative transfer can be based on the kinetic (semi-classical) picture (i.e. on the usual radiative transfer + kinetic equations), provided that one carefully defines the concepts reflecting the quantum-mechanical nature of the problem (profile coefficients, generalized redistribution functions) - see Chapter 5.

Specifically, the time-independent transfer equation for unpolarized radiation in a spectral line that originates as a transition between a lower level l and an upper level u of a given atom (ion) may be written as (see, e.g., Mihalas, 1978)

$$\vec{n}.\nabla\, I(\nu,\vec{n}) = -\chi(\nu,\vec{n})\, I(\nu,\vec{n}) + \eta(\nu,\vec{n}) \ , \qquad (1)$$

where $I(\nu,\vec{n})$ is the specific intensity of radiation at frequency ν and $\vec{n}$ is the unit vector in the direction of propagation. The phenomenological (macroscopic) absorption coefficient χ and the emission coefficient η are conveniently expressed as

$$\chi(\nu,\vec{n}) = (h\nu/4\pi)\,[n_l B_{lu}\, \phi(\nu,\vec{n}) - n_u B_{ul}\, \psi(\nu,\vec{n})] \ , \qquad (2)$$

$$\eta(\nu,\vec{n}) = (h\nu/4\pi)\; n_u A_{ul}\, \psi(\nu,\vec{n}) \ , \qquad (3)$$

where A_{ul}, B_{ul} and B_{lu} are the Einstein coefficients for spontaneous emission, stimulated emission and absorption, respectively; n_l and n_u are the number densities of atoms in levels l and u (level populations). The quantities ϕ and ψ are the so-called line absorption profile coefficients for absorption and emission (often simply referred to as profiles). They represent the probability amplitude of absorption (emission) of a photon in the considered line, in the frequency range $(\nu,\ \nu + d\nu)$ and in an infinitesimal solid angle $d\Omega$ about the direction $\vec{n}$. They are normalized according to

$$\int\!\!\int \phi(\nu,\vec{n})\; d\nu d\Omega/4\pi = \int\!\!\int \psi(\nu,\vec{n})\; d\nu d\Omega/4\pi = 1 \ .$$

The profile coefficient for stimulated emission is taken to be equal to that for spontaneous emission, following Oxenius (1965); see also Cooper, Hubeny and Oxenius (1984).

In solving any specific transfer problem, one requires the knowledge of line profile coefficients as well as of the corresponding level populations. For local thermodynamic equilibrium (LTE), the problem is greatly simplified: the emission coefficient is given by the product of the absorption coefficient times the Planck function. In the absence of LTE, one has to determine the level populations and, in general, the line profile coefficients.

In the widely used approximation of complete redistribution (CRD), the profile coefficients for absorption and emission are supposed to be identical and a priori known. This assumption leads to the standard non-LTE line transfer problem (for a review, see Mihalas, 1978).

However, there is no a priori justification for assuming CRD. In order to be able to check the consistency of the CRD results, or to go beyond the approximation of CRD, one needs explicit expressions for the line profile coefficients. Physically, the assumption of CRD means that the probability of absorption (emission) of a given photon in a considered spectral line is independent of previous absorptions (emissions) of photons in the same or different spectral lines. In other words, the photons involved in consecutive radiative transitions of an atom are supposed to be uncorrelated. However, in reality, this is not generally true. Therefore, it is necessary to consider the photon correlation phenomena as a fundamental part of the line formation physics.

In astrophysics, the study of photon correlations is traditionally referred to as the domain of partial redistribution (PRD) to stress the conceptual differences from the usual complete redistribution approaches. We retain this name throughout the present paper, even if in some cases (e.g., a correlated two-photon absorption), it is a misnomer. The same also applies to terms like redistribution phenomena, redistribution function, etc.

3. HOW DO WE STUDY PHOTON CORRELATION PHENOMENA?

3.1. Two Aspects of the Problem

In the preceding chapter, we have clarified that one may think of the general line transfer problem in terms of two more or less distinct aspects, the kinetic and the quantum aspect. Let us take the profile coefficients as an example - Eqs (2) and (3): it is intuitively clear that the form of the macroscopic profiles ϕ and ψ should reflect two different physical phenomena viz., i) inherent atomic properties, i.e. probabilities of absorption (emission) in the atom's frame; photon correlations (quantum aspect); and ii) thermal, and possibly other, motions of atoms (kinetic aspect). Now the crucial question is the following: To what extent are both aspects linked? In other words, to what extent are the translational and internal degrees of freedom of the particles coupled?

A physically exact discussion of this problem is beyond the scope of the present paper. We shall thus only state that the common astrophysical description, the so-called semi-classical picture, is based on the conceptual decoupling of both aspects. The determination of the atom's frame quantities (profiles, redistribution functions) then belongs to the quantum aspect, while the determination of the macroscopic (laboratory-frame) quantities (e.g., the macroscopic profiles ϕ and ψ) belongs to the kinetic aspect.

The separation of both aspects may be formulated in terms of two basic assumptions of the semi-classical approach, namely i) an atom is supposed to be always in one of its energy eigenstates; and atomic transition, which is in fact linked with the time development of atomic wave functions, is viewed as an instantaneous process; and ii) distinct velocity distribution functions are ascribed to atoms in different energy eigenstates.

This is not to say that the physical quantities corresponding to both aspects are actually decoupled. On the contrary, this coupling may be quite substantial (e.g., velocity distributions and/or populations are affected by detailed microscopic mechanisms of photon correlations, etc.). By the term conceptual decoupling we mean that one is able to work in terms of three well-defined types of quantities: microscopic (atom's frame) quantities, macroscopic (laboratory-frame) quantities, and velocity distributions. Notice that in cases where such a decoupling is not justified, one should work in terms of the general density matrix.

From the point of view of the historical development of the subject, one may even set a boundary between "physical" and "astrophysical" contributions to the problem by putting the identity between the quantum aspect ≡ physics, and kinetic aspect ≡ astrophysics. Indeed, atomic physics is expected to supply the atom's frame quantities (profiles, redistribution functions), while astrophysics takes care of solving the coupled set of the radiative + rate (or, generally, kinetic) equations. This distinction is obviously only illustrative, yet it helps to understand certain features of the development of the subject.

3.2. Quantum Aspect

The quantum-mechanical studies of radiative transfer, which are focused on astrophysical problems, may be roughly divided into two categories: i) general studies that aim at deriving the relevant equations; and ii) "applied" studies, which provide the individual atom's frame quantities. The second group of studies thus represents the quantum aspect in the sense specified above.

The first category of approaches is discussed by Hubeny (1984); here, only a recent important paper by Cooper et al. (1982 - CBBH) is mentioned. They have derived the rate and kinetic equations, as well as the expressions for profile coefficients, starting with the general density matrix equation. In particular, their conclusions provide a justification for many semi-classical ideas, like the conceptual decoupling of the quantum and kinetic aspects, the meaning of the atom's frame and the laboratory-frame profile coefficients, etc., which were introduced and treated by astrophysics in a heuristic way. On the other hand, although specifically developed for multilevel atoms, their treatment is not a truly "multilevel" one, from the point of view of redistribution, because the only photon correlation process which is accounted for is an absorption + re-emission in the same line (see 5.4). This paper has also raised some questions concerning the treatment of stimulated emission; we shall return to this problem in 5.2 and 5.4.4.

In a subsequent paper, Cooper and Zoller (1984) extended the CBBH formalism to derive the radiative transfer equation. Their results indicate that the classical transfer equation is valid for practically all problems of traditional (visual) and ultraviolet astronomy.

The second category of quantum-mechanical studies will be discussed in Chapter 6.

3.3. Kinetic Aspect

This topic is comprehensively discussed by Oxenius (in these Proceedings) and will not be repeated here. For the purposes of subsequent discussion, we just mention that the basic equations of the problem are the radiative transfer equation (which may also be viewed as the kinetic equation for photons) and, generally, the kinetic equations for the distribution functions of atoms in individual energy states. One usually considers (explicitly or implicitly) the velocity distribution functions to be Maxwellian; thus the zero-order moment equations (also called statistical equilibrium or rate equations), determining atomic level populations, are sufficient. Other relevant points and some explicit formulas will be considered in Chapters 4 and 5.

4. STANDARD PARTIAL REDISTRIBUTION PROBLEM

Let us return to the quotation of Jeffries. To understand the state of affairs at that time, as well as to appreciate the subsequent development, let us examine Eqs (2) and (3) in some detail. The determination of the absorption (emission) coefficients was split into two entirely independent parts, namely a) the determination of the corresponding level populations (the standard non-LTE problem); and b) the determination of the emission profile (we emphasize that the absorption profile has been viewed as an a priori given quantity).

The determination of the emission profile has been found feasible under the following assumptions (however, not always spelt out explicitly), which specify a restricted problem we shall refer to as the <u>standard PRD problem</u>:

A) Photons in only one transition participate in a scattering process; i.e. the only correlated chain of radiative transitions considered is an absorption followed by a re-emission in the same line.
B) The stimulated emission is either negligible, or is treated as a negative absorption (i.e. with the absorption profile).
C) The velocity distributions of atoms in all energy states are assumed to be Maxwellian (more precisely, the velocity distribution of atoms in the lower level of a considered PRD transition is Maxwellian; the velocity distributions for other levels are irrelevant).
D) The atomic velocity is assumed unchanged during a scattering process.

Under these assumptions, the emission profile can be formulated entirely by means of the so-called <u>redistribution function</u> (RF), $R(\nu';\vec{n}',\nu,\vec{n})$, which represents the probability amplitude that a photon $(\nu',\vec{n}',d\nu',d\Omega')$ is absorbed from a radiation field of unit strength and a photon $(\nu,\vec{n},d\nu,d\Omega)$ is subsequently re-emitted in the same line, provided that no process destroying the excited state intervenes. $R(\nu',\vec{n}',d\nu,d\Omega)$ is the so-called laboratory-frame (velocity-averaged) redistribution function, and is given by (Hummer, 1962)

$$R(\nu',\vec{n}',\nu,\vec{n}) = \int r[\nu' - (\nu_0/c)\vec{n}'\cdot\vec{v},\ \nu - (\nu_0/c)\vec{n}\cdot\vec{v},\vec{n},\vec{n}'] \times$$

$$\times\ f_1(\vec{v})\ d^3\vec{v}\ , \tag{4}$$

where r is the atom's frame RF, and $f_1(v)$ is the velocity distribution of atoms in level 1 (taken here to be Maxwellian); ν_0 is the line-center frequency measured in the atom's rest frame.

Assuming a strict two-level atom, the emission profile is then given by (Oxenius, 1965)

$$\psi(\nu,\vec{n}) = [B_{lu}\bar{R}(\nu,\vec{n}) + C_{lu}\ \phi(\nu,\vec{n})]/(B_{lu}\bar{J} + C_{lu})\ , \tag{5}$$

where $\bar{R}(\nu,\vec{n})$ is the so-called redistribution integral,

$$\bar{R}(\nu,\vec{n}) = \iint I(\nu',\vec{n}')\ R(\nu',\vec{n}',\nu,\vec{n})\ d\nu'\ d\Omega'/4\pi\ , \tag{6}$$

and $\bar{J}$ is the frequency-averaged mean intensity of radiation,

$$\bar{J} = \iint I(\nu',\vec{n}')\ \phi(\nu',\vec{n}')\ d\nu'\ d\Omega'/4\pi\ . \tag{7}$$

Equation (5) has a simple physical interpretation. The first term in the square brackets corresponds to the emission following a previous photoexcitation (i.e. the true scattering event), while the second term describes the emission following a previous collisional excitation. The denominator accounts for normalization.

In the late sixties, when the above formulation had already been developed, there were basically three types of problems to be solved, namely i) the relevant form of the redistribution function, ii) efficient numerical methods for both evaluating the RF and solving the transfer equation; and iii) a whole complex of questions which can be summarized in one question: How does one go beyond the standard PRD problem?

A discussion of these questions will cover the rest of this paper. The third question, in a sense the fundamental one, will be analysed in the next chapter, whereas the first two problems, which are important from the practical point of view, will be discussed in Chapters 6 - 8.

5. GOING BEYOND THE STANDARD PROBLEM

5.1. Overview of Problems

Going beyond the standard PRD problem means relaxing some or all of its basic assumptions A - D. Let us stress immediately that a recent formulation of the semi-classical approach due to Hubeny, Oxenius and Simonneau (1983 - HOS) represents an internally consistent formalism in which all the assumptions A - D are relaxed. However, before describing its basic ideas, it seems worthwhile to analyse specific problems linked with relaxing the individual assumptions.

The main difficulty in treating a multilevel, multitransition, radiative coupling (item A) consists in the generally non-Markovian nature of consecutive radiative transitions of an atom. Unlike items B - D, A has not yet been sufficiently elaborated by laboratory physics. The current astrophysical treatment (see 5.4) is thus largely based on intuitive arguments.

The common feature of the three remaining items is that they may be sufficiently clarified within a simple model of a two-level atom, which is well understood by laboratory physics (see, e.g., Allen and Eberly, 1975; Cohen-Tannoudji, 1976). The laboratory (and, in particular, laser) physics may thus serve as a guide for developing relevant astrophysical approaches.

5.2. Two-level Atom; Stimulated Emission

An important step towards the understanding of line formation in a gas of two-level atoms, with relaxation of assumptions B and C, was made by Oxenius (1965). This paper clearly formulated the basic assumptions of the semi-classical approach (see Chapter 3). Further important assumptions energe when one considers the explicit form of the rate equations. For the excited state of a two-level atom, the rate equation reads

$$n_l(P_{lu} + C_{lu}) = n_u(P_{ul} + C_{ul}) \ , \tag{8}$$

where the C's are the corresponding rates of inelastic collisions, and the P's are the rates of radiative transitions. The latter are given by

$$P_{lu} = B_{lu} \iint I(\nu,\vec{n})\ \phi(\nu,\vec{n})\ d\nu\ d\Omega/4\pi \ , \tag{9}$$

and

$$P_{ul} = A_{ul} + B_{ul} \iint I(\nu,\vec{n})\ \omega(\nu,\vec{n})\ d\nu\ d\Omega/4\pi \ , \tag{10}$$

the first term of Eq. (10) corresponding to spontaneous emission and the second to stimulated emission.

The next basic assumption of the semi-classical approach is that the <u>same</u> profile coefficients are used in the radiative transfer equation (i.e. in the definition of the absorption and emission coefficients) <u>and</u> in the rate equations [i.e. the absorption profile in the absorption rate, Eq. (9), and the emission profile in both spontaneous and stimulated emission rates]. The spontaneous rate does not depend on the radiation field, so that the emission profile integrates to unity and thus does not appear explicitly in Eq. (10). This means, in particular, that $\omega(\nu,\vec{n}) = \psi(\nu,\vec{n})$.

The latter point has given rise to a drawn-out controversy in the astrophysical literature (Hubeny, 1981b; Baschek et al., 1981; Cooper et al., 1983). Although clarified to a large extent by Cooper et al. (1983), this question has not yet been completely resolved. The interested reader is referred to Hubeny (1984) where the present status of the problem is outlined. We just note that most conceptual difficulties concerning the stimulated emission rate (for a two-level atom) are due to this being the only term through which the non-Markovian character of consecutive radiative transitions enters the rate equations. Yet more fundamentally, in cases where the stimulated emission is expected to be important, and thus the actual form of the stimulated emission profile would really matter in the rate equations, the radiation field can no longer be viewed as a low-intensity field which is defined by the condition (Cooper et al., 1983):

$$B_{ul}\, \bar{J} \ll A_{ul} \ . \qquad (11)$$

Going beyond the strict limit of the low-intensity regime requires, in general, more fundamental changes of the formalism (see also 5.4.4).

We conclude, in agreement with Cooper et al. (1983), that for most astrophisically relevant cases, the use of the absorption profile in Eq. (10) (or rather the atomic absorption profile averaged over the velocity distribution of atoms in the excited state) instead of the emission profile serves as a quite satisfactory approximation.

5.3. Velocity-changing Collisions

The problems linked with relaxing assumptions C and D are, to a large extent, linked together, because the formulation of the kinetic equations requires an explicit consideration of the velocity-changing collisions. We do not intend to discuss this problem generally (see Oxenius, in these Proceedings); we shall rather concentrate on one important aspect of the problem, namely the description of photon correlations in the presence of velocity-changing collisions. In view of our preceding discussion, this is just the point where the kinetic and quantum aspects are mixed together most, and where exact treatment would thus require a more involved description than the semi-classical (see, e.g., Berman et al., 1982, and references cited therein).

The problem was first tackled in astrophysics by Oxenius (1965) who introduced two limiting approximations: i) the so-called correlated re-emission, which corresponds to the limit of no velocity-changing collisions; and ii) the non-correlated re-emission, which is the opposite limit; the velocity-changing collisions destroy any correlation between the absorbed and emitted photons, thus yielding complete redistribution in the atom's frame. Subsequent astrophysical work has practically always assumed the case of correlated re-emission (i.e. no velocity-changing collisions).

An attempt has been made at treating the intermediate case by Argyros and Mugglestone (1971) who formulated the problem by means of newly introduced generalized redistribution functions (which should not be confused with those introduced by Hubeny, 1982, and by HOS). However, this approach was correctly criticized by Hummer and Rybicki (1971) who argued that elastic collisions, which are able to change an atomic velocity, must also affect the re-emission of a photon (the effect neglected by Argyros and Mugglestone).

Recently, a quite satisfactory description of the effect of velocity-changing collisions has been developed both by the quantum-mechanical formulation (CBBH) and by the semi-classical approach (HOS). Both approaches are based on the so-called strong collision model which assumes that elastic collisions, according to the respective branching ratios, either redistribute the atom's vecloity to a Maxwellian distribution and, at the same time, destroy the correlations between photons, or else leave the velocity unchanged, and thus any photon correlation behaves as if velocity-changing collisions were absent. Some other aspects of velocity-changing collisions are discussed by Cooper and Hubeny (in preparation).

5.4. Multilevel Atoms - HOS Theory

Finally, we arrive at the point which was most uncertain during the development of the modern radiative transfer theory. Except for some simple cases (e.g., Milkey et al., 1975a) and attempts (Hubeny, 1981b), the first really multilevel formulation has been given by HOS, within the frame of the semi-classical picture. The extension of traditional approaches to multilevel atoms has been achieved by employing two important concepts which we shall discuss in turn.

5.4.1. Natural population. Although the notion of natural population has previously been used in astrophysics (see, e.g., Jefferies, 1968), its real importance emerges only when one considers the multilevel redistribution problems.

An atomic level is said to be naturally populated (HOS) if the probability of emitting (absorbing) a well-defined photon in a transition to a lower (higher) level, when averaged over an ensemble of identical atoms, is independent of the previous history of the ensemble, i.e. of the manner in which the level has been populated. Adopting the simple and intuitively appealing picture due to Weiskopf (1931) and Woolley (1931, 1938; see also Woolley and Stibbs, 1953) which views an atomic level as being broadened and composed of a continuous distribution of sublevels, as atomic level is naturally populated if it is populated by a process involving particles or photons whose energy spectrum is constant over the width of the level. Or, analogously, natural population corresponds to an uniform occupation of the sublevels of an atomic level.

Consequently, the processes that lead to natural population of the final level are all ionization and recombination processes, collisional excitation and de-excitation, and spontaneous emissions from upper states. In contrast, absorptions and stimulated emissions lead to deviations from natural population if the intensity of radiation varies within the line, or, equivalently, they populate the atomic level naturally only if the radiation intensity is constant over the frequency range of the line.

5.4.2. Generalized redistribution functions. The concept of natural population provides a well-defined starting point for considering the consecutive correlated radiative transitions of an atom. Consider an atom with bound levels i, j, k, l, ..., and consider consecutive radiative transitions $i \to j$, $i \to j \to k$, $i \to j \to k \to l$, ..., that start from a naturally populated level i. Any arrow may represent either an absorption ($i < j$), or an emission ($i > j$). To each individual chain is assigned the corresponding generalized redistribution function (GRF) $r_{ij}(\xi_{ij})$, $r_{ijk}(\xi_{ij},\xi_{jk})$, $r_{ijkl}(\xi_{ij},\xi_{jk},\xi_{kl})$, ..., (for convenience, we consider the frequency correlation only), which represents the probability amplitude that a photon(s) with frequencies in ranges (ξ_{ij}, ξ_{ij} + + $d\xi_{ij}$), etc., are involved in the given chain (ξ_{ij} is the frequency of a photon involved in the transition $i \to j$, measured in the atom's rest frame, etc.).

Note that the one-photon GRF r_{ij} corresponds to the commonly used atomic absorption profile; similarly, the GRF r_{iji} corresponds to the ordinary redistribution function for resonance fluorescence. The condition that the initial level i must be populated naturally is, however, rarely stated explicitly.

5.4.3. Profile coefficients. The atomic absorption (emission) profile for the transition $m \to n$ is then given by the sum of the contributions of all radiation chains ending with the transition $m \to n$; each term is given by the redistribution integral [GRF times the corresponding intensity(ies)] multiplied by the corresponding probability of occurrence of the process (for explicit formulas refer to HOS). The macroscopic (velocity-averaged) profile coefficients are then given by the relevant integrals of the atomic profile coefficients multiplied by the velocity distribution of atoms in the corresponding energy levels.

Let us emphasize several important features. Firstly, both the emission and absorption profiles may deviate from the natural excitation profile. Secondly, if stimulated emissions are neglected, only a finite number of GRF's has to be known to determine the profile coefficients. Specifically, for an n-level atom, the longest chain is the sequence of consecutive absorptions $1 \to 2 \to \ldots \to n$ followed by an emission $n \to i$. If stimulated emissions are taken into account, one has, in principle, to deal with infinite chains of repeated transitions, to and fro, like $\ldots \to i \to j \to i \to j \to i \to \ldots$, etc. (stimulated emissions may not populate a lower level naturally), which have to be truncated somewhere for practical applications.

This formalism clearly discriminates between the profile coefficients and the redistribution functions, thus avoiding uncertainties with the meaning of RF's met in the past. Atomic-frame GRF's are now the only quantities which are to be supplied by quantum mechanics; all the remaining quantities needed for the solution of the transfer problem are given by the theory. Also, effects of velocity-changing collisions are accounted for quite naturally here.

5.4.4. Stimulated emission once again. In view of the above results, we may discuss the problem of stimulated emission more exqlicitly. Consider a two-level atom. If the stimulated emission is important [i.e. condition (11) is not satisfied], one should take the deviations from the natural population of the ground level into account. Consequently, in addition to the ordinary RF r_{121} , which specifies the problem in the low-intensity limit completely, one should also consider multi-photon GRF's like r_{12121}, $r_{1212121}$, etc. Analogously, the velocity distribution of atoms in the ground state may differ from the Maxwellian. Anyway, the semi-classical formulation becomes very cumbersome.

However, this feature explains that i) an approximate form of Eq. (10), as discussed in 5.2, is indeed a quite satisfactory and, in some sense, consistent approximation, and ii) from the point of view of applications, it is thus not relevant to develop numerical approaches that would treat non-linearities induced by the use of the emission profile in Eq. (10), while considering only ordinary two-photon RF's.

6. REDISTRIBUTION FUNCTIONS

6.1. Terminology

Due to historical development, the term redistribution function appears in several different meanings; it is thus worthwhile to clarify them first. We have already defined the term generalized redistribution function, and the special case of ordinary redistribution functions. For either type, one should distinguish between the atomic-frame RF (ARF), which is defined in the atom's rest frame, and the laboratory-frame RF (LRF), which is defined as the corresponding ARF averaged over the velocity distribution of an atom in the initial level. It is convenient to designate ARF's as r, and LRF's as R. In the following text, we shall not specify the type of RF in detail, if there is no danger of confusion.

Although, from the point of view of application, an LRF is commonly viewed as an a priori given quantity, this need not, in general, be so, because the velocity distribution of an initial level may differ from the Maxwellian. In such cases, the LRF should be determined self-consistently with the global transfer solution. In order to distinguish this general case from the usual LRF's computed under the assumption of the Maxwellian distribution, we shall call the latter Maxwellian LRF (MLRF). We note immediately that practically all existing studies have used MLRF's.

Both ARF's and LRF's may be angle-averaged. In the atom's frame, an ARF is generally given by (see below)

$$r_{ijk..}(\xi_{ij},\xi_{jk},\ldots,\vec{n}_{ij},\vec{n}_{jk},\ldots) = \sum_K \alpha_K g_K(\vec{n}_{ij},\vec{n}_{jk},\ldots) \times \times \bar{r}_K(\xi_{ij},\xi_{jk},\ldots), \tag{12}$$

where $\vec{n}_{ij}$ is the unit vector in the direction of the photon involved in the transition $i \rightarrow j$, etc.; g is the phase function, and $\bar{r}$ is called the frequency redistribution function. To avoid confusion, we note that for ordinary RF's the latter term is sometimes used (e.g., Hummer, 1962; Mihalas, 1978) with a different meaning, namely that of our $\bar{r}$ divided by the natural-excitation absorption profile. α_K is a constant. In the absence of external fields (e.g., magnetic field) the phase function depends only on the respective angles between $\vec{n}_{ij}$, $\vec{n}_{jk}$, etc., and the angular and frequency dependences of the RF are separated as indicated by Eq. (12). A number of terms in the summation, as well as the actual form of α_K, g_K and $\bar{r}_K$, depend on the quantum properties of the relevant levels, particularly on their j and m quantum numbers.

The atom's frame angle-averaged ARF is then given by

$$r_{ijk..}(\xi_{ij},\xi_{jk},\ldots) = \sum_K \alpha_K r_K(\xi_{ij},\xi_{jk},\ldots) \ , \tag{13}$$

because the phase functions are normalized to unity.

The angular dependence of the LRF's is twofold: due to the intrinsic angular dependence of the corresponding ARF's, and an additional angular dependence is caused by the Doppler effect. Even if the corresponding ARF is angle-independent (or angle-averaged), the LRF depends on the angle(s) between the directions of the photons involved. This is basically due to the assumption that the atomic velocity remains unchanged during a scattering (photon-correlation) process. Moreover, the angular and frequency dependences are mixed together, so that the laboratory-frame angle-averaged LRF must be defined as the appropriate integral (given here, for convenience, for ordinary RF's only - see Hummer, 1962)

$$R(\nu',\nu) = 8\pi^2 \int_0^{\pi} R(\nu',\vec{n}',\nu,\vec{n}) \sin\gamma d\gamma \ , \tag{14}$$

where $\gamma = \vec{n}'.\vec{n}$.

Sometimes, one introduces so-called approximate forms of RF's (e.g., for R_{II}, R_{III} - see below). We stress that, from the physical as well as mathematical point of view, such approximations do not approximate the form of a redistribution function, but rather that of the corresponding scattering (redistribution) integral. Consequently, they will be discussed in Chapters 7 and 8.

For completeness, we shall briefly mention that all the above RF's are sometimes referred to as scalar RF's, in contrast to the more general redistribution matrices (Dumont et al., 1977; Rees and Saliba, 1982) which are introduced in order to study the transfer of polarized radiation (see 6.3.5).

Finally, it should be stressed that one uses the term RF in two different interpretations. Firstly, one speaks of RF having in mind some previously defined mathematical functions (e.g., R_I - R_V, P_I - P_V; see further on), which are not expected to describe an actual physical problem. Secondly, the physical RF is a function that describes an actual physical situation; it is usually given by the superposition of several mathematical RF's (e.g., a typical example of the OSC redistribution function - see 6.2.2).

6.2. Ordinary Redistribution Functions

6.2.1. Early treatments. In an important paper, Hummer (1962) summarized the earlier work and established an unified system of phenomenological redistribution functions R_I - R_{IV} . We recall briefly: R_I - an idealized case of zero line width; R_{II} - radiation damping in the upper state and coherence in the atom's frame; R_{III} - complete redistribution in the atom's frame; R_{IV} - resonance fluorescence between two broadened levels. The main attention has been devoted to deriving suitable expressions for MLRF's, while the ARF's were chosen to represent schematically the simplest physical cases. For each type I - IV, two possibilities of the phase function were considered: the isotropic (denoted A) and the dipole (or Rayleigh - denoted B). This system thus represents the mathematical RF's.

At the end of the sixties (refer to the quotation of Jefferies), two major problems were recognized concerning the application of RF's, namely i) the role of elastic collisions and the corresponding degree of incoherence (i.e. a relation between physical and mathematical RF's); and ii) the proper form of the ARF for resonance scattering in subordinate lines. As the second problem, there was a controversy between the quantum-mechanical result of Heitler (1954) - used by Hummer in deriving R_{IV} - and the semi-classical form of Woolley (1938; also see Woolley and Stibbs, 1953).

The effect of elastic collisions was treated, in a somewhat heuristic way, by several authors (Zanstra, 1941, 1946; Edmonds, 1955), who concluded that the RF for resonance lines should be given by a superposition of coherent scattering and complete redistribution (functions R_{II} and R_{III} in Hummer's notation), however, uncertainties existed as to the correct branching ratio between both contributions.

6.2.2. OSC redistribution function - resonance lines. Both the above problems have been resolved in a very important paper by Omont, Smith and Cooper (1972 - OSC) which established a firm basis for the subsequent astrophysical work in this area. They derived the quantum-mechanical form of the ARF for both resonance and subordinate lines, allowing for the effects of elastic and inelastic collisions. Their most important and often quoted result is the angle-averaged ARF for resonance lines (no lower state broadening),

$$r(\nu',\nu) = (1 - \Lambda)\, r_{II}(\nu',\nu) + \Lambda r_{III}(\nu',\nu) \ , \tag{15}$$

where the branching ratio (also called the incoherence function) is given by

$$\Lambda = Q_E/(\Gamma_R + \Gamma_I + Q_E) \ , \tag{16}$$

where Γ_R, Γ_I and Q_E are the spontaneous radiative rate, the inelastic collision rate, and the elastic collision rate, respectively. The corresponding LRF is also given by Eq. (15), replacing ARF's r_{II} and r_{III} by LRF's R_{II} and R_{III}.

6.2.3. Subordinate lines. OSC also resolved the long controversy concerning the ARF for subordinate lines by showing that Heitler's (1954) form is in error. However, an astrophysical study of subordinate lines was not undertaken until the eighties. This was partly due to the seemingly complicated form of the ARF even for the simplest case of pure radiative broadening [see OSC, Eq. (57)], partly due to the underlying physical problems. Indeed, a study of subordinate lines necessarily invokes departures from the standard two-level-atom view and, consequently it was not clear what the proper meaning of the corresponding LRF was until the HOS formalism had been developed.

By integrating OSC Eq. (57), using Henyey's (1946) identity, and applying the Fourier transform technique, Heinzel (1981) obtained a form of the LRF suitable for the case of purely natural broadening of both levels, and designated this function R_V. Function R_{IV}, which is based on an incorrect ARF, was retained in the system of (mathematical) RF's, in order to prevent possible confusion with the double meaning of R_{IV}. Heinzel (1981) also showed that function R_V mathematically contains all functions R_I - R_{IV} as corresponding limiting cases.

A function identical to R_V has also been calculated numerically by McKenna (1980) who used the original OSC expression (notice that he denoted this function R_{IV}).

Subsequently, Heinzel and Hubeny (1982) derived a form of ARF and LRF suitable for subordinate lines, starting from OSC and allowing for the radiative as well as collisional broadening of both levels involved. The resulting RF is quite analogous to that for resonance lines, viz.

$$r(\nu',\nu) = (1 - \Lambda)\ r_V(\nu',\nu) + \Lambda r_{III}(\nu',\nu)\ . \qquad (17)$$

The original OSC form (15) for resonance lines follows from Eq. (17) as a limiting case of negligible lower state broadening.

6.3. More Involved Ordinary Redistribution Functions

Although Eq. (15) [or Eq. (17)] is commonly viewed as the most important achievement of OSC, and indeed is mostly quite sufficient for applications, there are problems to which these simple forms can no longer be applied. This is the consequence of the assumptions used in deriving Eqs (15) and (17). Amongst them, the most important are the following: i) non-degenerate levels; ii) isolated lines; and iii) impact approximation. Although OSC have treated item i) (i.e. the case of spatial m-degeneracy of levels involved) in detail, the other two items represent an inherent limitation of the applicability of the OSC results. Relaxing each of the above items (assumptions) presents specific problems; we shall discuss them in turn.

6.3.1. Degenerate levels. The treatment of the atom's frame angular dependence of ARF's, as well as the study of polarized radiation, requires a formalism that is capable of considering the spatial degeneracy of the atomic levels. Compared to the non-degenerate case, the treatment of degeneracy presents but algebraic complications.

OSC have already treated this case; explicit expressions for a $j \to j+1 \to j$ transition have been derived by Ballagh and Cooper (1977). In the special case of an $s \to p \to s$ transition with no lower state interaction, the resulting RF contains a coherent term and two frequency-redistributed terms, one with the same angular dependence as the coherent term (described by the dipole phase function for unpolarized light), the other term being isotropic. [This result can be viewed as an actual example of our Eq. (12)]. An extension of the Ballagh and Cooper procedure to subordinate lines has been outlined by Heinzel and Hubeny (1982).

6.3.2. Inelastic collisions and the question of repopulation. The effect of inelastic collisions, which is linked with the assumption of isolated lines, has been a point of controversy for some time. OSC assumed that inelastic collisions depopulate the upper level (and then quench the line radiation), but they are not allowed to repopulate the upper level and thus give rise to a further contribution to the line radiation (consequently, the original OSC redistribution function is not normalized). This point has been criticized by Yelnik and Voslamber (1979; also see Cooper and Ballagh, 1978) who argued that inelastic collisions may well redistribute the line radiation.

The astrophysical significance of this question has been clarified by Heinzel and Hubeny (1982 - Appendix). They showed that the effects of repopulating the upper level are in fact accounted for by means of a specific additional term in the emission profile. Note also that the effects of repopulation are allowed for quite naturally in the HOS formalism. On the other hand, the results of Yelnik and Voslamber, and of Cooper and Ballagh offer, to some extent, a quantum-mechanical justification for the semi-classical HOS treatment of the profile coefficients (and even justify the very meaning of the normalization of the RF's !). Specifically, the fact that the inelastic collisions "redistribute" the radiation completely, which is incorporated in the HOS theory in an intuitive way, was derived by the above authors using an exact quantum-mechanical approach.

Furthermore, Cooper and Ballagh (1978) have also considered the effects of repopulating the lower level of a given transition. Their results were subsequently applied by Seitz et al. (1982) in calculating a corresponding MLRF aimed at taking the lower-state repopulation into account. However, this RF is not relevant to astrophysical problems as it is incompatible with the semi-classical picture.

This point is a good illustration of the fundamental limitations of the astrophysical semi-classical approach, and serves as an example of illegitimate extrapolation of the semi-classical concepts beyond the limits of their applicability; it is thus worth stressing here. The point is that Cooper and Ballagh have specifically considered an ensemble of atoms irradiated by monochromatic radiation. In the semi-classical picture, the frequency dependence of the incident radiation field is of secondary importance; the RF is viewed as a quantity independent of the actual behavior of the radiation field, and the corresponding redistribution integral, which gives the total response of an atom of the applied radiation field, is calculated from a general expression valid regardless of the frequency dependence of the radiation. However, from the fundamental, quantum-mechanical point of view, the frequency dependence of the applied radiation is a substantial property of the system under study as it determines the characteristic time for the development of the atomic density matrix due to coupling with the light beam. The inverse of the bandwidth of incident radiation gives the correlation time of the light beam; strictly monochromatic radiation thus has an infinite correlation time. The repopulation phenomena studied by Cooper and Ballagh are then a consequence of the infinite correlation time of the applied radiation. In contrast, in astrophysics, one always deals with radiation fields of a finite bandwidth (in fact, the infinite bandwidth limit - i.e. zero correlation time - is often quite a good approximation in astrophysics). In summing up, the semi-classical approach is capable of treating photon correlation processes in the presence of radiation fields of "short" (but not necessarily zero) correlation times, but is clearly incapable of dealing with phenomena of infinite correlation time, which are in conflict with its basic postulates (see Chapter 3).

6.3.3. Non-impact phenomena. The results of OSC are valid only within the domain of the impact approximation, i.e. their treatment of collisional redistribution is not applicable to the far wings of lines. It is, therefore, of considerable astrophysical interest to extent the OSC theory outside the impact regime.

On the other hand, this topic has recently been studied intensively in laboratory physics (see, e.g., reviews by Cooper, 1980; Burnett, 1981, 1983, and references cited therein). In fact, the theoretical work on a unified approach to collisional redistribution, valid both within and without the impact regime, was primarily stimulated by the pioneering experiments of Carlsten et al. (1977), and continues to be the subject of great interest because, by studying collisional redistribution, one learns more about the important details of atomic and molecular collision dynamics (Burnett, 1983).

Here, we shall only mention studies that are explicitly focused on astrophysical problems. Cooper (1979) presented an extension of the OSC theory and showed that the relevant OSC formula can be retained if one replaces the Lorentzian (impact) line shape (i.e. the natural-excitation absorption profile) by the actual absorption profile, and if one considers the elastic collision width Q_E (and thus the branching ratio Λ) to be frequency dependent. However, this simple procedure is only strictly valid for isolated (i.e., in particular, for non-hydrogenic) lines. Therefore, using the theory developed by Burnett et al. (1980) and Burnett and Cooper (1980), Yelnik et al. (1981) carried out a study of redistribution for lines originated between l-degenerated levels, with explicit attention to the hydrogen Lyman-alpha line.

The conclusions of both the above studies are particularly important to astrophysics, because they indicate that, in an excellent approximation, the study of the non-impact regime (i.e. the line wings) requires no additional information beyond that normally needed. In particular, the relevant RF can still be formulated by means of Hummer's RF's.

6.3.4. Other types of LRF's. This category comprises the RF's introduced by Argyros and Mugglestone (1981), which have already been mentioned in 5.3.

Another example is the redistribution functions introduced by Hummer (1968) that allow for the presence of a macroscopic velocity field. However, attention should be paid to the use of the angle-averaged RF's (i.e. those averaged in the observer's frame), for they may yield spurious results in transfer calculations (see 7.5).

Finally, in an interesting paper, Magnan (1975) has considered redistribution in the presence of turbulent velocities. Adopting a model of turbulent "bubbles", he introduced the RF that corresponds to complete redistribution in the rest frame of each bubble (velocity distributions of atoms in each bubble are supposed to be Maxwellian), averaged over the distribution of velocities of the bubbles (assumed Gaussian).

6.3.5. Redistribution matrices for polarized light. For the majority of cases of astrophysical interest, polarized radiation can be described by two Stokes parameters (Chandrasekhar, 1960); consequently, the redistribution of polarized radiation is described by means of a 2x2 redistribution matrix (Dumont et al., 1977; Rees and Saliba, 1982). The redistribution matrices considered so far are based on the simplifying assumption that frequency redistribution and polarization phenomena are not mixed together. This enables the frequency dependence to be treated by means of the usual (in practice Hummer's) RF's, while the only generalization consists in introducing phase matrices instead of (scalar) phase functions.

6.4. Generalized Redistribution Functions

The first function of this type that appeared in the astrophysical literature is the so-called cross-redistribution function R_X introduced by Milkey, Shine and Mihalas (1975a). This function describes the absorption of a photon in a line, with a sharp lower level, followed by a re-emission in another line, with the same upper level and a different (sharp) lower level; its introduction was motivated by the effort to handle the coupling of the Ca II H and K resonance lines with the infrared triplet.

A general case of arbitrary two-photon correlations (i.e. resonance fluorescence, resonance Raman and inverse Raman scattering, and resonance two-photon absorption and emission) has been considered by Hubeny (1982) who introduced a unified system of (mathematical) two-photon GRF's. This system, which is quite parallel to the system for ordinary RF's R_I - R_V , has been denoted by p_I - p_V for ARF's and P_I - P_V for MLRF's. The previously introduced function R_X appears quite naturally in this system and corresponds to function P_{II} . It was also shown that the general, collisional (i.e. physical) redistribution function for any type of resonance two-photon process is given by (schematically)

$$r_{ijk} = \alpha p_V + \beta p_{III} , \qquad (18)$$

in complete analogy with the case of resonance fluorescence (i.e. ordinary RF's). The derivation of Eq. (18) is based on a straightforward generalization of the OSC formalism (see also Nienhuis and Schuller, 1977, 1978), and rests on the same assumptions (impact approximation, isolated lines, non-degenerate levels).

Very recently, Magnan (1984) presented a classical derivation of the two- and three-photon GRF's by means of the sublevel formalism of Woolley and Stibbs (1953). This procedure is inherently limited to the case of pure radiative damping (i.e. to the case of no collisions; otherwise the sublevel formalism is not feasible). However, in calculating the corresponding LRF's, he assumed that the atomic velocities at the intermediate levels of consecutive radiative chains were completely uncorrelated. This approximation implicitly means that the atomic velocity changes during a scattering process, i.e. that elastic collisions are implicitly alloed to occur. Thus, an objection similar to that against the LRF's of Argyros and Mugglestone (1971) should be raised: if elastic collisions occur that change the atomic velocity, they must also affect (i.e. practically destroy) the correlation of the photons involved (or, in terms of the sublevel formalism, these collisions must also shift an atom from one sublevel to another sublevel within a given intermediate level). As explained in CBBH and HOS, the laboratory-frame complete redistribution is much more appropriate in such cases (see also 5.3).

6.5. Numerical Evaluation of Redistribution Functions

Since the ARF's are usually given by simple analytic functions, computational difficulties are only encountered in calculating the MLRF's and, in particular, angle-averaged MLRF's.

In the original paper, Hummer (1962) summarized the earlier numerical work and presented a numerical method of evaluating the angle-averaged R_{III}. Subsequently developed methods were usually based on suitable series expansions (see, e.g., Rees and Reichel, 1968; Yengibarian and Nicoghossian, 1973). A convenient numerical method of evaluating the angle-averaged R_{II} (which is, in fact, the most important LRF for actual applications - see further on) has been developed by Adams et al. (1971). Their computer code, albeit complicated, was widely used afterwards. However, the practical use of their formalism by an interested but non-specialist user is hampered by certain numerical constants (in their Table 1) being given with an insufficient number of significant figures.

Reichel and Vardavas (1975) have developed a method (the so-called error-correction method), originally devised for angle-dependent LRF's R_{III} and R_{IV}, which was later proved to be very efficient even in other cases (GRF's P_{III}, P_{IV}, and even three-photon GRF's). The angle-averaged R_{III} and R_{IV} were calculated by direct numerical integration using Eq. (14).

For completeness, it should also be mentioned that MLRF's R_I, R_{II} and R_{III}, both angle-dependent and angle-averaged, can be evaluated by the Monte Carlo method (Lee, 1974; 1977).

Finally, Heinzel (1981) employed the Fourier transform technique in devising the numerical method for calculating angle-dependent R_V; yet this method may also be applied to R_{III} and R_{IV} (and, with some improvements, to GRF's P_{III}, P_{IV} and P_V). The computer requirements of both methods, error-correction and Fourier transform, are practically equivalent. Heinzel and Hubeny (1983) presented comprehensive calculations and graphical plots of functions R_{II}, R_{III} and R_V, and showed that any type of angle-averaged LRF may be calculated using the definition Eq. (14), by the 8-point or 16-point Gaussian quadrature, provided one employs a sufficiently accurate method of evaluating the Voigt function. The methods of Matta and Reichel (1971), and of Hui et al. (1978) have been proved to satisfy this requirement.

As a by-product, this approach yields a very efficient and simple method of evaluating the angle-averaged R_{II} (also see Heinzel, in these Proceedings).

7. RADIATIVE TRANSFER SOLUTIONS

Comprehensive reviews of various PRD transfer solutions have been given by Mihalas (1978) and Milkey (1976). Here, we shall concentrate on work not covered by the above reviews and, in particular, on studies that contribute to a general understanding of the PRD problem.

7.1. Numerical Solutions for Idealized Models

The transfer solution of idealized models serves as the preliminary and /or complementary step to solutions of actual stellar atmospheric problems, as they can provide a good deal of insight into the nature of the enigma. Moreover, they often indicate a permissible degree of approximation and/or a quality of numerical methods to be used for more complex calculations.

In the case of partial redistribution, most attention has naturally been devoted to solutions using the simplest, i.e. mathematical, RF's. We emphasize that, to our knowledge, no work has yet been done on generalized redistribution functions, nor have some important questions of the kinetic asqect (e.g., departures from Maxwellian velocity distributions for atoms in the individual levels) been investigated. The idealized model calculations thus dealt with the standard PRD problem (see Chapter 4). Most of the studies have used the line source function of the form

$$S(\nu) = (1 - \varepsilon)\ \phi_\nu^{-1} \int_0^\infty R(\nu',\nu)\ J_{\nu'}\ d\nu' + \varepsilon B\ , \tag{19}$$

which employs the angle-averaged RF's - the so-called isotropic approximation (Avrett and Hummer, 1965); or, sometimes, the more general angle-dependent form

$$S(\nu,\vec{n}) = (1 - \varepsilon)[\phi_\nu(\vec{n})]^{-1} \iint R(\nu',\vec{n}';\nu,\vec{n})\ I(\nu',\vec{n}')\ d\nu'\ d\Omega'/4\pi + \varepsilon B\ . \tag{20}$$

For R, one substitutes one of the functions $R_I - R_V$. In Eqs (19) and (20), we use the standard notation; for an explanation of symbols and further discussion, refer to Mihalas (1978).

The most important results of this type of studies are the following: For the source function (19), the departures of the solutions using R_I (Hummer, 1969) and R_{III} (Finn, 1967; Vardavas, 1976b) from those assuming complete redistribution are small; the CRD may thus serve as a good approximation. On the other hand, solutions using R_{II} differ from both CRD and (laboratory-frame) coherent scattering (Hummer, 1969).

Important work has been carried out by Milkey et al. (1975b; also seeVardavas, 1976a), who tested the validity of the isotropic approximation by comparing results using both forms of the source function, Eqs (19) and (20). They proved that there were practically negligible differences between both solutions, at least for media where neither pronounced inhomogeneous structures, nor significant differential motions are present.

The above results indicate that the most important redistribution function is the angle-averaged R_{II}. Therefore, a number of approximations have been devised that enable this problem to be treated in a more economical way (see 7.3). Further, since R_{II} is relevant to treating radiative transfer in gaseous nebulae, some studies have thus been specifically devoted to inferring the scaling laws for finite slabs by means of accurate numerical solutions (Hummer and Kunasz, 1980; Bonilha et al., 1979).

Idealized model solutions have recently been obtained for the angle-averaged R_V (Hubeny and Heinzel, 1984; Mohan Rao et al., 1984). They have shown that the solutions using R_V lie between those for R_{II} and CRD; the larger the ratio of the damping parameters for the upper and the lower levels, a_u/a_l, the greater the departures from CRD. On the other hand, for damping parameters of a comparable value, $a_u \approx a_l$, the departures from CRD are small; CRD may thus again serve as a good approximation.

7.2. Asymptotic Analytical Solutions

An asymptotic analysis of PRD line transfer is very useful in deducing the large-scale behavior of the solutions and in extracting the scaling laws; i.e. the global way in which the basic characteristics of the problem depend on the appropriate parameters. Like the idealized model examples, the analytical solutions provide a physical insight into the nature of the solutions in more complex cases.

In an important paper, Frisch (1980a) summarized the earlier work and thoroughly discussed the large-scale properties of the source function in the presence of PRD effects. Using an elegant analytical method, she confirmed the conclusions following from the individual idealized model solutions, and established the scaling laws for the behavior of the source functions (19) and (20), with $R = R_I$, R_{II}, R_{III}, R_{IV}, for small values of the inelastic destruction probability ε. In subsequent papers (Frisch, 1980b; Frisch and Bardos, 1981), this procedure was extended to treat specifically the case of R_{II} in the presence of continuous absorption and a more general source term.

7.3. Partial Coherent Scattering Approximation and Doppler Diffusion Effects

The results of both the above types of studies indicate that, at least for exploratory applications, CRD serves as a satisfactory approximation for all ordinary RF's, except R_{II} (and, to some extent, R_V). (Note, however, that this conclusion concerns the standard PRD problem; more general cases remain to be clarified). Since in many applications the contribution of R_{II} is dominant [see Eq. (15)], and since function R_{II} is somewhat costly to compute, a good deal of effort has been expended to find suitable approximate descriptions.

The most natural, and most widely used, approximation is the so-called partial coherent scattering (PCS) approach. Roughly speaking, it consists in replacing R_{II} by complete redistribution in the line core and by (laboratory-frame) coherent scattering in the line wings. [Notice that this replacement only has sense in evaluating the redistribution integral - the first term on the r.h.s. of Eq. (19)]. This idea was originally suggested by Jefferies and White (1960) who examined the graphical plots of numerically evaluated function R_{II} , and subsequently modified by Kneer (1975). The latter form correctly satisfies the requirements of normalization and symmetry, and has later often been used in applications. However, this approach has been criticized because it does not allow for photon diffusion in the line wings (also called Doppler diffusion - see Milkey, 1976; Basri, 1980). The name Doppler diffusion comes from the fact that a photon frequency measured in the laboratory frame changes, on the average, by one Doppler width during one individual act of scattering (i.e. the exact coherence in the atom's rest frame is modified by the Doppler shift due to atomic thermal motion). Since the photon frequency is more likely to be shifted towards the line center, the whole process may be thought of as photon diffusion in frequency space. Such phenomena are a priori neglected by the PCS approximation.

Recently, the latter inadequacy of the PCS approximation has been overcome by introducing the concept of depth-dependent division frequency, i.e. the frequency separating the CRD core region from the coherent wing region (Hubeny, 1985b; also see Hubeny - in these Proceedings). From the general point of view, this approach is very interesting, because it enables most PRD transfer problems with the ordinary (and, to some extent, even with generalized) RF's to be solved in a good approximation without having to evaluate the redistribution functions explicitly.

7.4. Solar and Stellar Applications

Most applications have been focused on analysing the physical properties of the solar and stellar chromospheres. In fact, a study of stellar chromospheres motivated the general progress in astrophysical research into partial redistribution and, consequently, is the domain where the PRD theory finds the largest field of application. The reason for this is obvious: the important probes into chromospheric conditions are the observed profiles of strong resonance lines, like the hydrogen Lyman-alpha and beta lines, the Mg II and Ca II resonance lines. Simultaneously, the density of chromospheric material is comparatively low, hence the effect of collisions is rather small and the PRD effects are important.

In stellar chromospheres, an approximate PRD analysis of the above mentioned lines is now an almost routine procedure used in constructing semi-empirical models, following the original treatments of Ayres (1975) and Ayres and Linsky (1976).

For stellar photospheres, the effects of PRD are generally smaller due to the higher density in those regions. Yet there are some specific cases in which at least an approximate treatment of PRD is necessary, for example, the interpretation of the broad Lyman-alpha wings in A-type and late B-type stars (Hubeny, 1980, 1981a); or the analysis of the Mg II resonance lines in A-type stars (Freire-Ferrero et al., 1983), etc.

Finally, the main area of PRD applications is the solar chromosphere. After the pioneering studies carried out during the seventies (Vernazza et al., 1973; Milkey and Mihalas, 1973a, 1973b, 1974; Shine et al., 1975, Vardavas and Cram, 1974; in general, see reviews by Milkey, 1976; and Rybicki, 1976), attention has been devoted to constructing detailed semi-empirical models, using the PRD treatment for all the above mentioned lines (Vernazza et al., 1981), and, in particular, to allow for non-impact phenomena in calculating the theoretical Lyman-alpha profile (Basri et al., 1979; Roussel-Dupré, 1983). Basri et al. used the modification of the OSC redistribution function suggested by Cooper (1979), while Roussel-Dupré employed the recent non-impact RF of Yelnik et al. (1981).

However, the results of no recent study have shown agreement between the observed and calculated profiles, particularly for the Lyman-alpha and Lyman-beta lines. There are several likely reasons for this discrepancy, which have already been discussed by the above authors (influence of inhomogeneities, microscopic velocity fields, etc.). Nevertheless, we stress that even within the simple homogeneous, static model, there are the follwing important points which may, at least partly, explain this discrepancy. Firstly, the above authors (Vernazza et al., Basri et al., Roussel-Dupré) have used the (depth-independent) PCS approximation, which is not quite adequate,particularly for Lyman-alpha. Secondly, for Lyman-beta, there is an additional and more fundamental reason (note that the agreement between the observed and theoretical profiles of Lyman-beta is particularly poor - Vernazza et al., 1981), namely the inadequacy of the simple two-level-atom picture of redistribution and the consequent use of only ordinary RF's (also see Heinzel and Hubeny - in these Proceedings).

7.5. Moving Atmospheres

Finally, we shall briefly mention several studies of PRD effects in moving atmospheres. Practically all studies of this type belong to the category of idealized models; attention was primarily devoted to assessing the importance of departures from CRD.

Vardavas (1974) and Cannon and Vardavas (1974), using the angle-averaged RF's (actually R_I) in the observer's frame, found large differences between the CRD and PRD results for a planeparallel, low-velocity case. However, Magnan (1974) and Mihalas et al. (1976) showed that these effects are in fact, spurious due to the inadequacy of the usage of the angle-average LRF's in the observer's frame. In contrast, using a comoving-frame method, where the usual static form of (even angle-averaged) LRF's may be employed, Mihalas et al. (1976) demonstrated that for R_I there is no substantial difference between the results of CRD and PRD.

The calculations for expanding spherical atmospheres have been carried out by Mihalas, Kunasz and Hummer (1976) for R_I and R_{II} by using the comoving-frame formalism; and by Peraiah (1978, 1979) who worked in the observer's frame. Recently, some more realistic (yet still idealized) model calculations have been performed by Drake and Linsky (1983; also see Linsky - in these Proceedings). They used the OSC redistribution function (without the PCS approximation) with realistic (depth-dependent) damping parameters and incoherence fraction. They proved the PRD effects to be important and generally decreasing with increasing expansion velocity.

8. NUMERICAL TECHNIQUES

In discussing numerical methods of solving the PRD line transfer problems, one should distinguish between two types of approaches, namely i) methods of solving the radiative transfer equation with the frequency- (and, in general, angle-) dependent line source function, and ii) global approaches, devised to solve a coupled set of radiative transfer + constraint equations. As constraints, one usually considers the equations of statistical equilibrium, but one may also consider other equations (e.g., kinetic) for more complex studies.

8.1. Solution of the Linear Transfer Equation

Methods of the first type may be characterized by stressing that they solve the radiative transfer equation with the source function given as a linear functional of the specific intensity of radiation. This feature then limits the region of their applicability: they may thus be used either for intrinsically linear problems and/or for idealized models, or as formal-solution steps in (usually iterative) solutions of non-linear problems. Specifically, one solves the transfer equation with the source function given by Eq. (19) or (20), where ε, ϕ_ν, and B represent prespecified functions of position (and, generally, frequency).

Due to the specific nature of the PRD transfer problems, a number of numerical methods have been developed which may roughly be divided into the following three categories: a) direct methods, b) perturbation methods, and c) inherently approximate approaches.

Category a) comprises all standard methods capable of treating a frequency-dependent line source function. In astrophysics, one uses either the difference-equation methods (Feautrier, 1964; Auer, 1976), or the integral-equation methods (for a review refer to Athay, 1972; Avrett and Loesser, 1983). Another possibility is the application of the Monte Carlo methods (Auer, 1968; Avery and House, 1968; Magnan, 1968; Meier and Lee, 1978, 1981; Lee and Meier, 1980); or discrete-space-theory methods (Peraiah, 1978).

The perturbation methods (Category b) are based on the assumption that departures from CRD are not very large, and may thus be treated perturbatively. The simplest approach of this type is the method of the frequency-iterated source function (Avrett and Hummer, 1965). More sophisticated appraoches have been developed by Cannon et al. (1975 - the redistribution perturbation technique; also see Cannon, 1976), and recently by Scharmer (1983). Generally, as discussed, e.g., by Frisch (1980a), these methods are well suited for all RF's but R_{II}, whose large-scale properties differ substantially from those of CRD.

The methods of Categories a) and b) (if the latter converge) may yield, in principle, solutions of arbitrary accuracy. On the other hand, the methods of Category c) are, from the outset, devised to be approximate. This category includes, for example, the PRD core-saturation method (Stenholm and Wehrse, 1984), or the modified Rybicki method (Hubeny, 1985b; also see Hubeny - in these Proceedings). The latter method is very advantageous as it retains the basic computational advantages of the original Rybicki (1971) method, does not require an explicit evaluation of the RF's, and yet yields very good agreement with exact solutions. Moreover, it can easily be applied to the perturbation approaches in which it can replace the less suitable CRD step.

8.2. Non-linear Coupled Problems

From the physical point of view, one should distinguish among several kinds of non-linear coupling involved in PRD transfer. The first one is that already encountered in the standard non-LTE multilevel CRD problem (Mihalas, 1978) which consists in the coupling between different atomic transitions via the response of the level population to the radiative rates. The second kind is the non-linearity induced by the deviation of the profile of stimulated emission from the absorption profile; this non-linearity applies even to a two-level atom (notice that in CRD the two-level-atom problem is exactly linear). Other possible non-linear couplings, not however, treated numerically yet (with the exception of Milkey et al., 1975a), are the so-called redistribution interlocking (several atomic transitions are coupled by photon correlation effects, described by means of GRF's); or the even more complicated coupling if the departures from Maxwellian velocity distributions are taken into account. In the following, we shall only discuss the first two kinds of non-linearities.

In devising suitable global numerical methods, experience with the standard multilevel CRD problems may serve as a guide. Thus, one can use either the powerful complete linearization technique (for a review refer to Mihalas, 1978); or the less powerful, but sometimes advantageous formulation based on the equivalent two-level-atom (ETA) approach (Mihalas, 1978). The latter method treats the overall coupling iteratively; consequently, it may fail to converge in some cases.

The feature of double non-linearity, discussed above, complicates the solutions of PRD problems. Milkey and Mihalas (1973a) have developed a variant of the complete linearization method that is, in principle, able to treat both the above non-linearities simultaneously. Subsequently, this method was extended to include some simple cases of redistribution interlocking (the Ca II problem - see Milkey et al., 1975a). The basic idea of both methods is to divide the upper state of a given PRD transition into a set of substates; the population of each substate is directly linked to the discretized laboratory-frame emission profile. The radiative transfer equation is then solved together with a set of "rate equations" for the individual substates.

This method is very effective, but relatively very expensive computationally. Moreover, any generalization to treat more PRD transitions would require a considerable increase of computational effort. Therefore, the following idea is very attractive: one retains the treatment of the first non-linearity (i.e. the standard multilevel one) using the complete linearization method (or some other suitable global multilevel technique), but treats the second one (i.e. that induced by stimulated emission) by means of the ETA approach. We shall refer to such methods as being of the <u>semi-ETA type</u>. This idea was explicitly formulated by Heasley and Kneer (1976). Other existing approaches are, more or less, based on the same idea (e.g., Ayres and Linsky, 1976; Vernazza et al., 1981; Avrett and Loesser, 1983). The authors last mentioned, moreover, have used the ETA-type method even for the first (multilevel) part of the problem.

However, the explicit formulas of Milkey and Mihalas (1973a) and of Heasley and Kneer (1976) contain an error in the stimulated emission term. This error has been pointed out and corrected by Baschek, Mihalas and Oxenius (1981; also see Hubeny, 1981b) using the strictly semi-classical picture, i.e. assuming the emission profile in the stimulated emission rate (see 5.2). As explained in 5.4.4, even this approach is not entirely consistent. From this point of view, we may conclude that the basic objection to the semi-ETA-type global methods, as expressed, e.g., by Mihalas (1978, p. 438), that: "... when stimulated emissions are very important one must use the full linearization technique described above" (i.e. the substate formalism of Milkey and Mihalas) is not quite relevant, because the stimulated emission is not treated properly anyway.

In summing up, the semi-ETA-type methods appear to be the most advantageous global methods of solving multilevel PRD problems, at least for the usual astrophysically important cases. An explicit formulation and further discussion of this problem is presented by Hubeny (1985a). Obviously, in studying cases with substantial redistribution interlocking, it will be necessary to develop more sophisticated techniques, most probably based on the idea of complete linearization.

8.3. Numerical Methods for Moving Atmospheres

As above, we shall consider separately two types of approaches. As far as the global multilevel problem is concerned, the situation is the same, or even more favorable, than for static atmospheres: Mihalas and Kunasz (1978) have demonstrated that the ETA-type methods are, in fact, better suited to moving than to static atmospheres.

The solution of the linear transfer equation can be obtained using either the observer's-frame or the comoving-frame formalisms. Practically any method mentioned in 8.1 can be suitably modified to treat PRD phenomena in moving atmospheres using the observer's-frame formalism. However, as discussed in 7.4, one should be extremely careful in using angle-averaged RF's. The basic disadvantage of these methods consists in substantially higher computer requirements (large number of frequency and angle points) as compared to the static case. Nevertheless, as pointed out by Mihalas (1980b), they can be used to an advantage on computers with vector-processing capabilities (e.g., Cray 1) and, moreover, these methods offer practically the only possible way of treating radiative transfer in the presence of non-monotonic velocity fields.

On the other hand, the comoving-frame methods are computationally very advantageous. In particular, the usual isotropic approximation is as good here as in the static problem. Specific methods of treating the angle-averaged redistribution in spherical expanding atmospheres have been developed by Mihalas, Kunasz and Hummer (1976), and extended to angle-dependent redistribution by Mihalas (1980a). Recently, Kunasz (1984) formulated a similar approach for moving cylinders.

All the above comoving-frame methods are based on the Feautrier (1964) elimination technique. As stated explicitly by the above authors, the use of the more economical Rybicki (1971) elimination is not possble here due to the intrinsic frequency dependence of the line source function. Thus, the modified Rybicki method (Hubeny, 1985b) offers an attractive possibility of solving, at least in a good approximation, PRD transfer problems in the comoving frame.

9. SUMMARY OF UNRESOLVED PROBLEMS

Finally, we shall briefly summarize some important problems to be solved, and outline possible perspectives of the future development of the astrophysical redistribution theory.

From the fundamental point of view, there is a whole complex of questions connected with the validity and quantum-mechanical justification of the semi-classical approach. This question is also important from the physical point of view, because the semi-classical approach can provide a suitable description for some laboratory situations which are not yet sufficiently understood. Nevertheless, the present studies strongly indicate that the semi-classical picture will be approved by quantum mechanics as a consistent description, or at least as a good approximation.

A large amount of work remains to be done on the theoretical level within the scope of the semi-classical picture: evaluation of three- and multiphoton GRF's; formulation of the two-level-atom problem in the presence of the important stimulated emission; clarification of the effects of velocity-changing collisions; developing a practical formalism for treating departures from Maxwellian velocity distributions, and others. It would also be interesting to elaborate Magnan's (1975) treatment of redistribution in the presence of turbulent motions.

At the level of numerical applications, attention should be devoted both to idealized model calculations and to actual spectroscopic diagnostics. Future idealized models could give an insight into both: i) the problem of redistribution interlocking, or, generally, non-standard PRD problems, under otherwise standard conditions (planeparallel stratification, constant-property media); and ii) the standard PRD problem under non-standard conditions (extended and moving atmospheres, turbulence, multidimensional media, time-dependent transfer, polarization, magnetic fields, etc., and combinations thereof).

Problem i) can be treated by taking an archetype model of a three-level atom and solving idealized transfer problems (in the spirit of Hummer, 1969; Hubeny and Heinzel, 1984), by adding consecutively more and more GRF's: R_{121}, R_{131}, R_{232}, R_{132}, R_{231}, R_{123}, R_{1231} . As the next step, one might consider departures from the Maxwellian velocity distributions. This set of idealized models should also enable possible numerical methods of solving highly non-standard PRD problems to be tested.

The complexity of actual future applications will depend, to a large extent, on conclusions drawn from the results of the above idealized models. The solar atmosphere will most likely remain the most straightforward area of application of the more involved PRD theory. As regards the solar chromosphere and transition region, progress in the PRD description is in fact one of the substantial prerequisites of the further improvement of their models. One of the crucial tests of the theoretical description of PRD phenomena will indeed be our ability to reproduce the observed solar Lyman-alpha and, in particular, Lyman-beta profiles. The same applies to studying some specific solar atmospheric features, like prominences, spicules, etc.

For stellar applications, the importance of the PRD description will certanly increase with increasing quality of stellar spectroscopic observations. In the near future, an interpretation of high-resolution spectra, taken e.g. by the Space Telescope, will require the developing of a standard methodology for describing, at least approximately, PRD phenomena (their effect on determining chemical abundances; combined effects of velocity fields and redistribution, etc.).

10. CONCLUSIONS

We have demonstrated that the progress achieved during the last 17 years has indeed been considerable. As expected, laboratory physics (represented here by the quantum-mechanical derivation of the OSC redistribution function) has stimulated enormous progress in understanding solar and stellar spectral line formation. The apparent success of the astrophysical partial redistribution theory has convinced astronomers that photon correlation phenomena must be considered as a quite natural and substantial part of line formation physics.

On the other hand, this increased level of our understanding, together with the improved quality of astronomical spectroscopic observations, have given rise to new demands on the fundamental theoretical description. Thus, the present effort may be characterized by the endeavour to study multilevel, multitransition photon correlation phenomena and effects linked with the correlation between radiative and collisional processes. From the point of view of spectroscopic diagnostics, attention is devoted to the study of the combined effects of redistribution and velocity fields, geometrical structure and other physical phenomena.

Some of these questions have already been resolved, but much work remains to be done before a completely satisfactory understanding of spectral line formation in the presence of partial redistribution effects is achieved.

REFERENCES

Adams, T.F., Hummer, D.G., Rybicki, G.B.: 1971, J. Quant. Spectrosc. Radiat. Transfer 11, 1365

Allen, L., Eberley, J.H.: 1975, Optical Resonances and Two-Level Atoms, Wiley, New York

Argyros, J.D., Mugglestone, D.: 1971, J. Quant. Spectrosc. Radiat. Transfer 11, 1621 and 1633

Athay, R.: 1972, Radiation Transport in Spectral Lines, Reidel, Dordrecht
Auer, L.H.: 1968, Astrophys. J. 153, 783
Auer, L.H.: 1976, J. Quant. Spectrosc. Radiat. Transfer 16, 931
Avery, L.W., House, L.L.: 1968, Astrophys. J. 152, 493
Avrett, E.H., Hummer, D.G.: 1965, Monthly Notices Roy. Astron. Soc. 130, 315
Avrett, E.H., Loesser, R.: 1983, in Methods in Radiative Transfer, W. Kalkofen, Ed., Cambridge University Press
Ayres, T.R.: 1975, Astrophys. J. 201, 799
Ayres, T.R., Linsky, J.L.: 1976, Astrophys. J. 205, 874
Ballagh, R.J., Cooper, J.: 1977, Astrophys. J. 213, 479
Baschek, B., Mihalas, D., Oxenius, J.: 1981, Astron. Astrophys. 97, 43
Basri, G.S.: 1980, Astrophys. J. 242, 1133
Basri, G.S., Linsky, J.L., Bartoe, J.-D., Brueckner, G.E., Van Hoosier, M.E.: 1979, Astrophys. J. 230, 924
Berman, P.R., Mossberg, T.W., Hartmann, S.R.: 1982, Phys. Rev. A 25, 2550
Bonilha, J.R., Ferch, R.L., Salpeter, E.E., Slater, G., Noerdlinger, P.D.: 1979, Astrophys. J. 233, 649
Burnett, K.: 1981, in Spectral Line Shapes, Proc. 5th Int. Conf. on Spectral Line Shapes, B. Wende, Ed., Wlater de Gruyter, Berlin - New York
Burnett, K.: 1983, Comments At. Mol. Phys. 13, 179
Burnett, K., Cooper, J.: 1980, Phys. Rev. A 22, 2027 and 2044
Burnett, K., Cooper, J., Ballagh, R.J., Smith, E.W.: 1980, Phys. Rev. A 22, 2005
Cannon, C.J.: 1976, Astron. Astrophys. 52, 337
Cannon, C.J., Vardavas, I.M.: 1974, Astron Astrophys. 32, 85
Cannon, C.J., Lopert, P.B., Magnan, C.: 1975, Astron. Astrophys. 42, 347
Carlsten, J.L., Szöke, A., Raymer, M.G.: 1977, Phys. Rev. A 15, 1029
Chandrasekhar, S.: 1960, Radiative Transfer, Dover, New York
Cohen-Tannoudji, C.: 1976, Frontiers in Laser Spectroscopy, Les Houches Summer School 1975, Session 27, Eds R. Balian, S. Haroche, and S. Liberman, Amsterdam: North-Holland, p. 3
Cooper, J.: 1979, Astrophys. J. 228, 339
Cooper, J.: 1980, in Laser Physics, Eds D.F. Walls and J.D. Harvey, Academic, Sydney, p. 241
Cooper, J., Ballagh, R.J.: 1978, Phys. Rev. A 18, 1302
Cooper, J., Ballagh, R.J., Burnett, K., Hummer, D.G.: 1982, Astrophys. J. 260, 299 - CBBH
Cooper, J., Hubeny, I., Oxenius, J.: 1983, Astron. Astrophys. 127, 224
Cooper, J., Zoller, P.: 1984, Astrophys. J. 277, 813
Drake, S.A., Linsky, J.L.: 1983, Astrophys. J. 273, 299
Dumont, S., Omont, A., Pecker, J.C., Rees, D.: 1977, Astron. Astrophys. 54, 675
Edmonds, F.N.: 1955, Astrophys. J. 121, 418
Feautrier, P.: 1964, C.R. Acad. Sci. Paris 258, 3189
Finn, G.D.: 1967, Astrophys. J. 147, 1085

Freire Ferrero, R., Gouttebroze, P., Kondo, Y.: 1983, Astron. Astrophys. 121, 59
Frisch, H.: 1980a, Astron. Astrophys. 83, 166
Frisch, H.: 1980b, Astron. Astrophys. 87, 357
Frisch, H., Bardos, C.: 1981, J. Quant. Spectrosc. Radiat. Transfer 26, 119
Heasley, J.N., Kneer, F.: 1976, Astrophys. J. 203, 660
Heinzel, P.: 1981, J. Quant. Spectrosc. Radiat. Transfer 25, 483
Heinzel, P., Hubeny, I.: 1982, J. Quant. Spectrosc. Radiat. Transfer 27, 1
Heinzel, P., Hubeny, I.: 1983, J. Quant. Spectrosc. Radiat. Transfer 30, 77
Heitler, W.: 1954, Quantum Theory of Radiation, 3rd Ed., Clarendon Press, Oxford
Henyey, L.G.: 1946, Astrophys. J. 103, 332
Hubeny, I.: 1980, Astron. Astrophys. 86, 225
Hubeny, I.: 1981a, Astron. Astrophys. 98, 96
Hubeny, I.: 1981b, Bull. Astron. Inst. Czechosl. 32, 271
Hubeny, I.: 1982, J. Quant. Spectrosc. Radiat. Transfer 27, 593
Hubeny, I.: 1984, in Spectral Line Shapes, Proc. 7th Int. Conf. on Spectral Line Shapes (in press)
Hubeny, I.: 1985a, Bull. Astron. Inst. Czechosl. (in press)
Hubeny, I.: 1985b, Astron, Astrophys. (submitted)
Hubeny, I., Heinzel, O.: 1984, J. Quant. Spectrosc. Radiat. Transfer 32,
Hubeny, I., Oxenius, J., Simonneau, E.: 1983, J. Quant. Spectrosc. Radiat. Transfer 29, 477 and 495 - HOS
Hui, A.K., Armstrong, B.H., Wray, A.A.: 1978, J. Quant. Spectrosc. Radiat. Transfer 19, 509
Hummer, D.G.: 1962, Monthly Notices Roy. Astron. Soc. 125, 21
Hummer, D.G.: 1968, Monthly Notices Roy. Astron. Soc. 141, 479
Hummer, D.G.: 1969, Monthly Notices Roy. Astron. Soc. 145, 95
Hummer, D.G., Kunasz, P.B.: 1980, Astrophys, J. 236, 609
Hummer, D.G., Rybicki, G.B.: 1971, Ann. Rev. Astron. Astrophys. 9, 237
Jefferies, J.T.: 1968, Spectral Line Formation, Blaisdell, Walthamm, Mass.
Jefferies, J.T., White, O.R.: 1960, Astrophys. J. 132, 767
Kneer, F.: 1975, Astrophys. J. 200, 367
Kunasz, P.B.: 1984, Astrophys. J. 276, 677
Lee, J.-S.: 1974, Astrophys. J. 192, 465
Lee, J.-S.: 1977, Astrophys. J. 218, 857
Lee, J.-S., Meier, R.R.: 1980, Astrophys. J. 240, 185
Magnan, C.: 1968, Astrophys. Letters 2, 213
Magnan, C.: 1974, Astron. Astrophys. 35, 233
Magnan, C.: 1975, J. Quant. Spectrosc. Radiat. Transfer 15, 979
Magnan, C.: 1984, Astron. Astrophys. (in press)
Matta, F., Reichel, A.: 1971, Math. Comput. 25, 339
McKenna, S.J.: 1980, Astrophys. J. 242, 283
Meier, R.R., Lee, J.-S.: 1978, Astrophys. J. 219, 262
Meier, R.R., Lee, J.-S.: 1981, Astrophys. J. 250, 376
Mihalas, D.: 1978, Stellar Atmospheres, 2nd Ed., Freeman, San Francisco

Mihalas, D.: 1980a, Astrophys. J. 238, 1034
Mihalas, D.: 1980b, Astrophys. J. 238, 1042
Mihalas, D., Kunasz, P.B.: 1978, Astrophys. J. 219, 635
Mihalas, D., Kunasz, P.B., Hummer, D.G.: 1976, Astrophys. J. 210, 419
Mihalas, D., Shine, R.A., Kunasz, P.B., Hummer, D.G.: 1976, Astrophys. J. 205, 492
Milkey, R.W.: 1976, in Interpretation of Atmospheric Structure in the Presence of Inhomogeneities, Commission 12, IAU General Assembly 1976, Ed. C.J. Cannon, Publ. Univ. Sydney, p. 55
Milkey, R.W., Mihalas, D.: 1973a, Astrophys. J. 185, 709
Milkey, R.W., Mihalas, D.: 1973b, Solar Physics 32, 361
Milkey, R.W., Mihalas, D.: 1974, Astrophys. J. 192, 769
Milkey, R.W., Shine, R.A., Mihalas, D.: 1975a, Astrophys. J. 199, 718
Milkey, R.W., Shine, R.A., Mihalas, D.: 1975b, Astrophys. J. 202, 250
Mohan Rao, D., Rangarajan, K.E., Peraiah, A.: 1984, J. Astrophys. Astron. 5, 169
Nienhuis, G., Schuller, F.: 1977, Physica 92C, 397
Nienhuis, G., Schuller, F.: 1978, Physica 94C, 394
Omont, A., Smith, E.W., Cooper, J.: 1972, Astrophys. J. 175, 185 - OSC
Oxenius, J.: 1965, J. Quant. Spectrosc. Radiat. Transfer 5, 771
Peraiah, A.: 1978, Kodaikanal Obs. Bull. Ser. A 2, 115
Peraiah, A.: 1979, Kodaikanal Ons. Bull. Ser. A 2, 203
Rees, D., Reichel, A.: 1968, J. Quant. Spectrosc. Radiat. Transfer 8, 1795
Rees, D., Saliba, G.J.: 1982, Astron. Astrophys. 115, 1
Reichel, A., Vardavas, I.M.: 1975, J. Quant. Spectrosc. Radiat. Transfer 15, 929
Roussel-Dupré, D.: 1983, Astrophys. J. 272, 723
Rybicki, G.B.: 1971, J. Quant. Spectrosc. Radiat. Transfer 11, 589
Rybicki, G.B.: 1976, in The Energy Balance and Hydrodynamics of the Solar Chromosphere and Corona, IAU Coll. No. 36, Ed. R.-M. Bonnet and P. Delache, p. 191
Scharmer, G.B.: 1983, Astron. Astrophys. 117, 83
Seitz, M., Baschek, B., Wehrse, R.: 1982, Astron. Astrophys. 109, 10
Shine, R.A., Milkey, R.W., Mihalas, D.: 1975, Astrophys. J. 199, 724
Stenholm, L.G., Wehrse, R.: 1984, Astron. Astrophys. 131, 399
Vardavas, I.M.: 1974, J. Quant. Spectrosc. Radiat. Transfer 14, 909
Vardavas, I.M.: 1976a, J. Quant. Spectrosc. Radiat. Transfer 16, 1
Vardavas, I.M.: 1976b, J. Quant. Spectrosc. Radiat. Transfer 16, 715
Vardavas, I.M., Cram, L.E.: 1974, Solar Physics 38, 367
Vernazza, J.E., Avrett, E.H., Loesser, R.: 1973, Astrophys. J. 184, 605
Vernazza, J.E., Avrett, E.H., Loesser, R.: 1981, Astrophys. J. Suppl. 45, 635
Weisskopf, V.: 1931, Ann. d. Physik 9, 23
Woolley, R.v.d.R.: 1931, Monthly Notices Roy. Astron. Soc. 91, 977
Woolley, R.v.d.R.: 1938, Monthly Notices Roy. Astron. Soc. 98, 624
Woolley, R.v.d.R., Stibbs, D.W.N.: 1953, The Outer Layers of a Star, Clarendon Press, Oxford
Yelnik, J.-B., Voslamber, D.: 1979, Astrophys. J. 230, 184
Yelnik, J.-B., Burnett, K., Cooper, J., Ballagh, R.J., Voslamber, D.: 1981, Astrophys. J. 248, 705

Zanstra, H.: 1941, Monthly Notices Roy. Astron. Soc. **101**, 273
Zanstra, H.: 1946, Monthly Notices Roy. Astron. Soc. **106**, 225

KINETIC ASPECTS OF REDISTRIBUTION IN SPECTRAL LINES

J.Oxenius
Université Libre de Bruxelles
Campus Plaine U.L.B., C.P.231
B-1050 Bruxelles
Belgium

Association Euratom-Etat Belge

1. INTRODUCTION

The classical problem of photon redistribution in a spectral line due to resonance fluorescence has two distinct aspects: the atomic one and the kinetic one. The atomic aspect concerns the redistribution of photon frequency (and photon direction) in the rest frame of the atom, while the kinetic aspect concerns the velocity distribution of the atoms, which affects via the Doppler effect the redistribution of photon frequencies as observed in the laboratory frame.

Here we consider the somewhat more general question of how atomic velocity distributions affect the line profile coefficients of the radiative transfer equation. To this end, we shall discuss in turn the two-level atom and the three-level atom. In particular, in the discussion of the two-level atom it is pointed out that in deriving the profile coefficient for emission, the streaming of excited atoms must be taken into account, and in the discussion of the three-level atom it is shown that atomic velocity distributions not only enter the laboratory profile coefficients in the usual way, but even show up in the atomic profile coefficients.

In the following we adopt the semi-classical picture according to which an atom is always in one of its energy eigenstates, and its motion can be described by a classical trajectory. It follows that, in this picture, transitions between atomic levels are considered as instantaneous processes, and that a well-defined velocity distribution can be ascribed to the atoms being in a given energy level.

This paper complements in some respect Hubený's review article in these Proceedings. We also refer to Oxenius (1985) for a somewhat more detailed discussion.

J. E. Beckman and L. Crivellari (eds.), Progress in Stellar Spectral Line Formation Theory, 59–71.

2. TWO-LEVEL ATOM

Consider spectral line formation by two-level atoms (ground state 1, excited state 2, photon energy $h\nu_0 = E_2 - E_1$). The time-independent tranfer equation for the radiation intensity $I_\nu(\underline{n})$ of frequency ν and direction $\underline{n}$ ($|\underline{n}| = 1$) is

$$\underline{n}\cdot\underline{\nabla} I_\nu(\underline{n}) = \kappa_\nu(\underline{n})\left[S_\nu(\underline{n}) - I_\nu(\underline{n})\right] \quad . \tag{1}$$

In the absence of stimulated emissions, the line absorption coefficient κ_ν and the line source function S_ν are given by

$$\kappa_\nu(\underline{n}) = (h\nu_0/4\pi)\; n_1 B_{12}\; \varphi_\nu(\underline{n}) \quad , \tag{2}$$

$$S_\nu(\underline{n}) = n_2 A_{21}\; \psi_\nu(\underline{n})/n_1 B_{12}\; \varphi_\nu(\underline{n}) \quad , \tag{3}$$

in terms of Einstein A- and B-coefficients and atomic number densities n_1, n_2. The normalized line profile coefficients φ_ν and ψ_ν,

$$\iint \varphi_\nu(\underline{n}) d\nu d\Omega/4\pi = \iint \psi_\nu(\underline{n}) d\nu d\Omega/4\pi = 1 \quad , \tag{4}$$

describe respectively the probabilities of absorbing and emitting photons $(\nu,\underline{n})$ by a volume element of the gas of two-level atoms.

The profile coefficients φ_ν and ψ_ν are related to the corresponding atomic profile coefficients α_{12} and η_{21} through

$$\varphi_\nu(\underline{n}) = \int d^3v\; f_1(\underline{v})\, \alpha_{12}(\xi) \quad , \tag{5}$$

$$\psi_\nu(\underline{n}) = \int d^3v\; f_2(\underline{v})\, \eta_{21}(\xi,\underline{n}) \quad , \tag{6}$$

where $f_1(\underline{v})$ and $f_2(\underline{v})$ are the velocity distributions of nonexcited and excited two-level atoms, and ξ denotes the frequency of the photon $(\nu,\underline{n})$ in the rest frame of an atom of velocity $\underline{v}$,

$$\xi = \nu - (\nu_0/c)\; \underline{n}\cdot\underline{v} \quad . \tag{7}$$

The normalizations are

$$\int \alpha_{12}(\xi)\; d\xi = \iint \eta_{21}(\xi,\underline{n})\; d\xi d\Omega/4\pi = 1 \quad , \tag{8}$$

$$\int f_i(\underline{v})\; d^3v = 1 \quad , \tag{9}$$

where $i = 1, 2$.

In order to know the profiles φ_ν and ψ_ν one must know the atomic profiles α_{12} and η_{21}, on the one hand,

and the velocity distributions $f_1(\underline{v})$ and $f_2(\underline{v})$, on the other.

Complete redistribution in the laboratory frame, $\psi_\nu = \varphi_\nu$, requires simultaneously $\eta_{21} = \alpha_{12}$ and $f_2(\underline{v}) = f_1(\underline{v})$.

The atomic absorption profile is the subject matter of the so-called "line broadening" calculations (see e.g. Griem, 1964). In simple cases, it may be a Lorentzian profile

$$\alpha_{12}(\xi) = (\delta/\pi)/[(\xi - \xi_0)^2 + \delta^2] \, . \tag{10}$$

To this "line broadening" contribute the interaction of the atoms with particles, including electromagnetic microfields ("phase-changing elastic collisions"), on the one hand, and the finite lifetime of the atomic levels owing to spontaneous emissions and inelastic collisions, on the other.

The atomic emission profile is a more involved quantity as it depends in general on the previous history of the atom. Consider excited two-level atoms that spontaneously emit photons. These atoms have previously been excited from the ground state either by the absorption of a photon ("radiation-excited atoms") or by an excitation collision ("collision-excited atoms"). Quite independent of this distinction between the two excitation mechanisms, an emitting atom has either suffered a velocity-changing elastic collision during the time it was excited, or it has not.

Consider first the limiting case where all emitting atoms have suffered velocity-changing collisions. The atomic velocities at the instances of excitation and emission are then uncorrelated, and as a result of the collisions, the spectrum of emitted photons in the atomic rest frame corresponds to the atomic absorption profile,

$$\eta_{21}(\xi, \underline{n}) = \alpha_{12}(\xi) \quad . \tag{11}$$

In the opposite limiting case, no elastic collisions occur during the lifetime of the excited level. Hence $\underline{v}$ = const for any atom during the time where it is excited. If velocity changes owing to inelastic collisions (with electrons) and radiative interactions are neglected, then an emitting atom has the same velocity it had at the instance of its previous excitation.

In the absence of collisions, the atomic emission profile is composed of contributions from radiation- and collision-excited atoms, respectively, which in general differ from each other. Let the probabilities for exciting an atom of velocity $\underline{v}$ by absorption and collisional excitation be prob($1 \Rightarrow 2$) and prob($1 \rightarrow 2$), respectively.

Then

$$\eta_{21}(\xi,\underline{n}) = \text{prob}(1\Rightarrow 2)\ j(\xi,\underline{n}) + \text{prob}(1\rightarrow 2)\,\alpha_{12}(\xi), \quad (12)$$

where we have used the fact that collision-excited atoms emit according to complete redistribution. The emission profile of radiation-excited atoms, on the other hand, can be written as

$$j(\xi,\underline{n}) = (1/I_{12}) \iint I_{\xi'}(\underline{n}')\ r(\xi',\underline{n}';\xi,\underline{n})\ d\xi' d\Omega'/4\pi\ . \quad (13)$$

Here the atomic redistribution function r describes the joint probability of absorbing a photon $(\xi',\underline{n}')$ and re-emitting a photon $(\xi,\underline{n})$, with the normalization

$$\iiiint r(\xi',\underline{n}';\xi,\underline{n})\ d\xi' d\Omega' d\xi d\Omega/(4\pi)^2 = 1\ . \quad (14)$$

The quantity I_{12} is defined by

$$I_{12} = \iint I_{\xi'}(\underline{n}')\,\alpha_{12}(\xi')\ d\xi' d\Omega'/4\pi \quad (15)$$

so as to normalize the quantity j,

$$\iint j(\xi,\underline{n})\ d\xi d\Omega/4\pi = 1\ . \quad (16)$$

In many cases (see e.g. Omont et al., 1972) the atomic redistribution function corresponds to a superposition of coherent re-emission and complete redistribution,

$$r(\xi',\underline{n}';\xi,\underline{n}) = \gamma(\underline{n}',\underline{n})\,\alpha_{12}(\xi')[\beta\,\delta(\xi-\xi') + (1-\beta)\,\alpha_{12}(\xi)]\ . \quad (17)$$

The frequencies ξ', ξ in the atomic rest frame are connected with the laboratory frequencies ν', ν through

$$\xi' = \nu' - (\nu_0/c)\ \underline{n}'\cdot\underline{v}\ , \quad \xi = \nu - (\nu_0/c)\ \underline{n}\cdot\underline{v} \quad (18)$$

with the <u>same</u> atomic velocity $\underline{v}$.

The two probabilities entering the emission profile (12) are given by the corresponding excitation rates of atoms of velocities $\underline{v}$, up to a normalization factor N. Denoting by

$$F_i(\underline{v}) = n_i f_i(\underline{v})\ , \quad \int F_i(\underline{v})\ d^3v = n_i \quad (19)$$

the distribution function of atoms in level i, one has

$$\text{prob}(1\Rightarrow 2) = NB_{12}\,I_{12}(\underline{v})F_1(\underline{v})\ , \quad (20)$$

$$\text{prob}(1\rightarrow 2) = NS_{12}\,F_1(\underline{v})\ , \quad (21)$$

where S_{12} is the rate coefficient for electronic excitation collisions, and [cf.Eq.(15)]

$$I_{12}(\underline{v}) = \iint I_\nu(\underline{n})\alpha_{12}(\xi)\, d\nu d\Omega/4\pi \quad . \tag{22}$$

In Eqs.(20) and (21), use has been made of the approximation that neglects changes of the atomic velocity owing to absorptions and excitation collisions. Hence finally

$$\text{prob}(1\Rightarrow 2) = B_{12}I_{12}(\underline{v})/[B_{12}I_{12}(\underline{v}) + S_{12}] \quad , \tag{23}$$

$$\text{prob}(1\rightarrow 2) = S_{12}/[B_{12}I_{12}(\underline{v}) + S_{12}] \quad . \tag{24}$$

Note that the distribution function $F_1(\underline{v})$ has dropped out in Eqs.(23) and (24).

In the general case, intermediate between the two limiting cases discussed so far, one can again write (Hubený et al., 1983)

$$\eta_{21}(\xi,\underline{n}) = \text{prob}(1\Rightarrow 2)\ j(\xi,\underline{n}) + \text{prob}(\rightarrow 2^*)\alpha_{12}(\xi), \tag{25}$$

where now $\text{prob}(\rightarrow 2^*)$ refers to all processes that create excited atoms of velocity $\underline{v}$ in a naturally populated level 2^*, that is, atoms whose emission spectrum corresponds to α_{12}. Writing the elastic collision term for the distribution function $F_2(\underline{v})$ in the form

$$(\delta F_2/\delta t)_{el} = \Gamma_2^+(\underline{v}) - \Gamma_2^-(\underline{v}) \tag{26}$$

where Γ_2^+ and Γ_2^- describe respectively the creation and destruction of excited atoms of velocity $\underline{v}$ by elastic collisions, one has

$$\text{prob}(1\Rightarrow 2) = (1/N)\ B_{12}I_{12}(\underline{v})\ F_1(\underline{v}) \quad , \tag{27}$$

$$\text{prob}(\rightarrow 2^*) = (1/N)\left[S_{12}F_1(\underline{v}) + \Gamma_2^+(\underline{v})\right] \quad , \tag{28}$$

$$N = [B_{12}I_{12}(\underline{v}) + S_{12}]F_1(\underline{v}) + \Gamma_2^+(\underline{v}) \quad . \tag{29}$$

Note that in the probabilities prob(...) , and hence in the emission profile (25), the distribution function $F_1(\underline{v})$ does not drop out owing to the presence of the term Γ_2^+.

To sum up, the atomic absorption profile is given by Eq.(10), and the atomic emission profile by Eq.(25) together with Eqs.(13) and (27) - (29).

In order to know the profiles φ_ν and ψ_ν entering the transfer equation, one must also know the velocity distributions $f_1(\underline{v})$ and $f_2(\underline{v})$ [see Eqs.(5) and (6)]. In principle, these have to be determined from the kinetic equations for the distribution functions $F_1(\underline{v})$ and $F_2(\underline{v})$.

In the simplest model of spectral line formation one has to consider four distribution functions: the distribution functions $F_1(\underline{v})$ and $F_2(\underline{v})$ of nonexcited and excited two-level atoms, the electron distribution function $F_e(\underline{v})$, and the photon distribution function $I_\nu(\underline{n})$. Correspondingly, one has four kinetic equations which are coupled.

One usually assumes that the distribution functions $F_1(\underline{v}) = n_1 f^M(v)$ and $F_e(\underline{v}) = n_e f^M(v)$ are Maxwell distributions of known temperature T and with known densities n_1 and n_e. The line absorption profile is then known, too. Note that an a priori specification of $F_1(\underline{v})$ makes sense only when the number of excited atoms is low: $n_2 \ll n_1$ or, more precisely, $F_2(\underline{v}) \ll F_1(\underline{v})$ for all velocities $\underline{v}$.

Hence, one is left with two kinetic equations for the two unknown distribution functions $F_2(\underline{v})$ and $I_\nu(\underline{n})$.

In the absence of force fields, the time-independent kinetic equation for the excited atoms is

$$\underline{v}\cdot\underline{\nabla} F_2 = (\delta F_2/\delta t)_{rad} + (\delta F_2/\delta t)_{inel} + (\delta F_2/\delta t)_{el}, \quad (30)$$

where the collision terms are due to radiative interactions, inelastic collisions, and elastic collisions, respectively. Neglecting again recoil effects due to radiative interactions and inelastic collisions, one has

$$(\delta F_2/\delta t)_{rad} = B_{12} I_{12}(\underline{v})\, F_1(\underline{v}) - A_{21} F_2(\underline{v}) \quad , \quad (31)$$

$$(\delta F_2/\delta t)_{inel} = S_{12} F_1(\underline{v}) - S_{21} F_2(\underline{v}) \quad , \quad (32)$$

$$(\delta F_2/\delta t)_{el} = \gamma_2 n_2 [f^M(v) - f_2(\underline{v})] \quad . \quad (33)$$

Collision terms (31) and (32) describe the creation and destruction of excited atoms of velocity $\underline{v}$ by radiative and inelastic collision processes. On the other hand, the elastic collision term has been approximated by the simple relaxation term (33), with γ_2 being a constant collision frequency and $f^M(v)$ a normalized Maxwell distribution.

The usual treatment of partial redistribution tacitly makes the following two assumptions:

(1) $\gamma_2 = 0$, no elastic collisions with excited atoms;

(2) $\underline{v}\cdot\underline{\nabla} F_2 = 0$, no streaming of the excited atoms.

The kinetic equation (30) then reduces to

$$(\delta F_2/\delta t)_{rad} + (\delta F_2/\delta t)_{inel} = 0 \quad . \quad (34)$$

Hence

$$F_2(\underline{v}) = \{[B_{12} I_{12}(\underline{v}) + S_{12}]/[A_{21} + S_{21}]\} F_1(\underline{v}) \quad , \quad (35)$$

$$n_2 = [(B_{12} J_{12} + S_{12})/(A_{21} + S_{21})] n_1 \quad , \quad (36)$$

$$f_2(\underline{v}) = \{[B_{12}I_{12}(\underline{v}) + S_{12}]/[B_{12}J_{12} + S_{12}]\} f^M(v) \quad . \qquad (37)$$

Here the relation $F_1 = n_1 f^M$ has been used, and

$$J_{12} = \int d^3v \; f_1(\underline{v}) I_{12}(\underline{v}) = \iint I_\nu(\underline{n}) \varphi_\nu d\nu d\Omega/4\pi \quad , \qquad (38)$$

$I_{12}(\underline{v})$ being given by Eq.(22).

Equation (37) shows that partial redistribution calculations implicitly take into account that the velocity distribution of excited atoms differs in general from a Maxwellian.

Without these simplifications, the kinetic equation (30) takes the form

$$\underline{v}\cdot\underline{\nabla}(n_2 f_2) = B_{12} I_{12} n_1 f^M - A_{21} n_2 f_2$$
$$+ S_{12} n_1 f^M - S_{21} n_2 f_2 + \gamma_2 n_2 (f^M - f_2), \qquad (39)$$

or, in dimensionless form,

$$\eta [\underline{v}\cdot\underline{\nabla}][n_2 f_2(\underline{v})] = [\varepsilon + (1-\varepsilon) I_{12}(\underline{v})] f^M(v)$$
$$- n_2 f_2(\underline{v}) + \xi n_2 [f^M(v) - f_2(\underline{v})] \, . \qquad (40)$$

In Eq.(40), all quantities are dimensionless. For example, the velocity v is expressed in terms of the thermal velocity of the atoms, the frequency in terms of the Doppler width, the number density n_2 in terms of the corresponding Boltzmann density, etc.

The three dimensionless parameters appearing in Eq. (40) are defined as follows:

$$\varepsilon = \varepsilon_0/(1+\varepsilon_0) \; , \quad \xi = \xi_0/(1+\varepsilon_0) \; , \quad \eta = \eta_0/(1+\varepsilon_0) \qquad (41)$$

where

$$\varepsilon_0 = S_{21}/A_{21} \; , \quad \xi_0 = \gamma_2/A_{21} \; , \quad \eta_0 = \kappa w/A_{21} \quad . \qquad (42)$$

Here w is the thermal velocity of the atoms,

$$w = (2kT/M)^{1/2} \quad , \qquad (42)$$

and κ is the mean line absorption coefficient

$$\kappa = (h\nu_0/4\pi)(n_1 B_{12}/\Delta\nu_D) \quad , \quad \Delta\nu_D = (\nu_0/c)\, w \quad . \qquad (44)$$

Thus, in contrast to the usual partial redistribution calculations which involve only the single parameter ε, in the complete model appear three parameters ε, ξ, η.

The radiation intensity I_ν enters the kinetic equation (40) through the term I_{12}. Equation (40) must therefore be solved simultaneously with the transfer equation, which can be written in dimensionless form as

$$\underline{n}\cdot\underline{\nabla}I_\nu(\underline{n}) = \varphi_\nu[S_\nu(\underline{n}) - I_\nu(\underline{n})] \quad , \tag{45}$$

with the line source function

$$\begin{aligned} S_\nu(\underline{n}) &= n_2 \psi_\nu(\underline{n})/\varphi_\nu \\ &= (1/\varphi_\nu) \int d^3v\ n_2 f_2(\underline{v})\, \eta_{21}(\xi,\underline{n}) \end{aligned} \tag{46}$$

using Eq.(6), which contains the distribution function $F_2 = n_2 f_2$.

Solutions of the coupled equations (40) and (45) for some simple cases have been obtained by Cipolla and Morse (1979), and by Borsenberger et al. We also refer to Simmoneau's paper in these Proceedings.

3. THREE-LEVEL ATOM

The three-level atom (ground state 1, excited states 2 and 3; level energies $E_1 < E_2 < E_3$) is the prototype of a multilevel atom: no new features are encountered when taking more bound levels and the continuum into consideration.

In the semi-classical picture, the laboratory profile coefficients are the averages of the atomic profile coefficients over the velocity distribution of atoms in the initial state of the considered transition. For instance, the absorption profile for the transition $2\rightarrow3$ and the emission profile for the transition $3\rightarrow1$ can be written in an obvious shorthand notation as

$$\varphi_{23} = \int d^3v\ f_2\alpha_{23} \quad , \quad \psi_{31} = \int d^3v\ f_3\eta_{31} \quad . \tag{47}$$

Notice that the absorption profiles of subordinate lines, like φ_{23} , are to be determined using velocity distributions of excited atoms which may differ from Maxwellian owing to non-LTE line transfer, as discussed in the preceding section [see e.g. Eq.(37)].

In the following we restrict our discussion to the atomic profile coefficients, neglecting again stimulated emissions.

The velocity of an atom will be assumed to be constant during the whole time when it is excited, no matter if it remains in just one excited level or makes transitions between several excited levels before returning to the ground state. Thus, velocity-changing elastic collisions with excited atoms, as well as recoil due to radiative interactions and inelastic collisions (with electrons) are neglected. As a consequence, in a given volume element of the gas, the ensemble of three-level atoms with velocities in a given range $(\underline{v}, d^3v)$ can be considered independently from all the other atoms having velocities outside this range.

The determination of atomic profile coefficients is difficult because of the correlations between photons involved in consecutive radiative transitions of the atom. For a two-level atom, the only case in question is resonance fluorescence: The probability of re-emitting a photon of frequency ξ in the transition $2 \rightarrow 1$ depends in general on the frequency ξ' of the photon previously absorbed in the transition $1 \rightarrow 2$ [see Eq.(17)]. For a three-level atom, there are still other possibilities of photon correlations. For example, the probability of absorbing a photon of frequency ξ in the transition $2 \rightarrow 3$ depends in general on the frequency ξ' of the photon previously absorbed in the transition $1 \rightarrow 2$.

The formal solution of the problem of determining atomic profile coefficients, due to Hubený et al.(1983), is based on the concepts of natural population, on the one hand, and generalized redistribution functions, on the other, which will be discussed in turn.

Processes involving particles or photons whose energy spectrum is constant over the atomic level width, as well as spontaneous emissions populate an atomic level "naturally". Thus, excitation and de-excitation collisions and spontaneous emission in spectral lines lead to natural population of the final atomic level, as do all ionization and recombination processes. By contrast, absorptions (and stimulated emissions) in spectral lines populate an atomic level naturally only when the radiation intensity is constant over the frequency width of the spectral line, while they lead to deviations from natural population if the radiation intensity in the spectral line varies with frequency.

The importance of the concept of natural population derives from the fact that for an <u>ensemble</u> of atoms in a given naturally populated level, the probability of emitting or absorbing a well-defined photon $(\xi, \underline{n})$ is independent of the previous history of the ensemble. In other words, atoms in a naturally populated level behave (emit or absorb) in the same way, if the term "atom" is interpreted as meaning "ensemble of atoms". So the emission spectrum in the line $2 \rightarrow 1$ of an ensemble of atoms in the naturally populated level 2 is always the same, no matter if 20% of the atoms were produced by collisional excitation $1 \rightarrow 2$ and 80% by spontaneous emissions $3 \rightarrow 2$, or if 100% of them were produced by de-excitation collisions $3 \rightarrow 2$.

The fact that spontaneous emissions lead to natural population of the final level does not mean that the photons in a radiative cascade are uncorrelated. To make this point clear, consider a three-level atom with a sharp ground state and two naturally broadened excited levels. Let the atoms be excited from the ground state 1 into level 3 by optical pumping with monochromatic light of frequency ξ_{13}, and consider the radiative cascade $3 \rightarrow 2 \rightarrow 1$. In the absence of collision broadening, energy conservation requires $\xi_{32} + \xi_{21} = \xi_{13}$, so that the frequency ξ_{21} of the photon in the transition $2 \rightarrow 1$ is not only correlated with, but even uniquely determined by the frequency ξ_{32} of the photon previously emitted in the transition $3 \rightarrow 2$. Nevertheless, the spontaneous emissions $3 \rightarrow 2$ populate the level 2 naturally. Indeed, taking the ensemble average over the atoms here corresponds to integrating over all photons ξ_{32} spontaneously emitted in transitions $3 \rightarrow 2$, which leads to a distribution of spontaneously emitted photons ξ_{21} that is identical with the emission spectrum of atoms produced by excitation collisions $1 \rightarrow 2$ say.

We note in passing that, in the absence of stimulated emissions, the ground state of an atom is always naturally populated.

The (tacit) assumption of natural population of atomic levels is ubiquitous in the literature on plasma spectroscopy and the theory of spectral line formation.

The fact that spontaneous emissions populate the final level of an atomic transition naturally has the important consequence that, as far as line profile coefficients are concerned, one needs to consider the non-Markovian character of consecutive radiative transitions of an atom only up to the point where a spontaneous emission takes place. Therefore, if stimulated emissions are neglected, the determination of atomic profile coefficients requires the knowledge of only a <u>finite number</u> of different types of non-Markovian radiative processes. Indeed, for an n-level atom, non-Markovian radiative processes relevant to line profile coefficients comprise n radiative transitions at most, the longest chains being provided by the $n - 1$ consecutive absorptions $1 \rightarrow 2 \rightarrow \ldots \rightarrow n$ followed by a spontaneous emission $n \rightarrow i$.

One therefore defines a "generalized redistribution function" for each finite sequence of radiative transitions that starts from a naturally populated atomic level. These functions, which describe photon correlations in these sequences of radiative transitions, are straightforward generalizations of the usual redistribution function which describes the correlation between absorbed and reemitted photons in the case of resonance fluorescence.

Let us from now on suppress in the formulas the photon direction $\underline{n}$. Furthermore, we denote by ξ_{ij} the frequency of a photon involved in the radiative transition $i \rightarrow j$ of the atom; for example, ξ_{31} denotes an emitted, and ξ_{13} an absorbed photon in the spectral line that corresponds to the atomic transition $1 \leftrightarrow 3$.

Generalized atomic redistribution functions r_{ij}, r_{ijk}, r_{ijkl}, ... are defined in the following way. Imagine the atom to be immersed in an isotropic radiation field of frequency-independent intensity, and consider consecutive radiative transitions $i \to j$, $i \to j \to k$, $i \to j \to k \to l$, ... <u>that start from a naturally populated initial level i</u>. Then

$$r_{ij}(\xi_{ij})\, d\xi_{ij}$$

is the probability that a photon in the frequency range $(\xi_{ij}, d\xi_{ij})$ is involved (i.e. emitted if $i > j$, or absorbed if $i < j$) in the radiative transition $i \to j$;

$$r_{ijk}(\xi_{ij}, \xi_{jk})\, d\xi_{ij} d\xi_{jk}$$

is the probability that two photons in the frequency ranges $(\xi_{ij}, d\xi_{ij})$ and $(\xi_{jk}, d\xi_{jk})$, respectively, are involved in the consecutive radiative transitions $i \to j \to k$, etc.

For example, r_{23} is the absorption profile α_{23} corresponding to a naturally populated level 2; r_{121} is the ordinary redistribution function for the lowest resonance transition $1 \leftrightarrow 2$; r_{1232} describes two successive absorptions $1 \to 2 \to 3$ followed by a spontaneos emission $3 \to 2$; etc.

The atomic line profile coefficients are readily formulated in terms of generalized redistribution functions. In order to determine a particular absorption or emission profile, one must first look for all independent sequences of radiative transitions that start from a naturally populated level and terminate with the particular radiative transition under consideration, and one must then add the contributions of all these sequences, weighted with their respective probabilities of occurrence.

Consider a particular atomic level i. We denote by $\to i^*$ the ensemble of all processes that create atoms with a naturally populated level i^* (the asterisk standing for natural population); by contrast, as they may give rise to deviations from natural population, absorptions from lower levels $j < i$ are denoted by $j \Rightarrow i$.

So, in order to determine the atomic absorption profile α_{23} of a three-level atom one has to consider the following processes, each of them starting from a naturally populated level,

$$\to 2^* \Rightarrow 3 \quad ,$$

$$1^* \Rightarrow 2 \Rightarrow 3 \quad ,$$

which occur with relative probabilities prob($\rightarrow 2^*$) and prob($1^* \Rightarrow 2$), respectively. Here the ground state has been denoted by 1^* as it is naturally populated. Or, for the emission profile η_{31} one has to consider the processes (interpreting the final transition $3 \rightarrow 1$ as a spontaneous emissions),

$$\rightarrow 3^* \rightarrow 1 \ ,$$

$$1^* \Rightarrow 3 \rightarrow 1 \ ,$$

$$\rightarrow 2^* \Rightarrow 3 \rightarrow 1 \ ,$$

$$1^* \Rightarrow 2 \Rightarrow 3 \rightarrow 1 \ ,$$

which occur with relative probabilities prob($\rightarrow 3^*$), prob($1^* \Rightarrow 3$), prob($\rightarrow 2^* \Rightarrow 3$), and prob($1^* \Rightarrow 2 \Rightarrow 3$), respectively.

In terms of generalized redistribution functions one can therefore write

$$\alpha_{23}(\xi_{23}) = \text{prob}(\rightarrow 2^*)r_{23}(\xi_{23}) + \text{prob}(1^* \Rightarrow 2)j_{123}(\xi_{23}) \ , \tag{48}$$

$$\eta_{31}(\xi_{31}) = \text{prob}(\rightarrow 3^*)r_{31}(\xi_{31}) + \text{prob}(1^* \Rightarrow 3)j_{131}(\xi_{31}) + \text{prob}(\rightarrow 2^* \Rightarrow 3)j_{231}(\xi_{31}) + \text{prob}(1^* \Rightarrow 2 \Rightarrow 3)j_{1231}(\xi_{31}) \ , \tag{49}$$

where, up to normalization factors N,

$$j_{ijk}(\xi_{jk}) = N_{ijk}\int I(\xi_{ij})r_{ijk}(\xi_{ij}, \xi_{jk})d\xi_{ij} \ , \tag{50}$$

$$j_{ijkl}(\xi_{kl}) = N_{ijkl}\iint I(\xi_{ij})I(\xi_{jk})r_{ijkl}(\xi_{ij}, \xi_{jk}, \xi_{kl})d\xi_{ij}\,d\xi_{jk} \ . \tag{51}$$

Notice that, owing to the term j_{123}, the absorption profile (48) differs from the function r_{23} which would have been expected from a naive point of view; likewise, owing to the terms j_{231} and j_{1231}, the emission profile (49) differs from that of a two-level atom made up of the levels 1 and 3 [cf.Eq.(25)], observing that $r_{31} = r_{13} = \alpha_{13}$.

It remains to determine the probabilities prob(...). Recalling that recoil is neglected, one has up to normalization factors N,

$$\text{prob}(\rightarrow 2^*) = N_2[S_{12}F_1(\underline{v}) + (A_{32} + S_{32})F_3(\underline{v})] \ , \tag{52}$$

$$\text{prob}(1^* \Rightarrow 2) = N_2[B_{12}I_{12}(\underline{v})F_1(\underline{v})] \ , \tag{53}$$

$$\text{prob}(\rightarrow 3^*) = N_3[S_{13}F_1(\underline{v}) + S_{23}F_2(\underline{v})] \quad , \tag{54}$$

$$\text{prob}(1^* \Rightarrow 3) = N_3[B_{13}I_{13}(\underline{v})F_1(\underline{v})] \quad , \tag{55}$$

$$\text{prob}(\rightarrow 2^* \Rightarrow 3) = N_3[B_{23}I_{23}(\underline{v})F_2(\underline{v})] \times [N_2\{S_{12}F_1(\underline{v}) + (A_{32} + S_{32})F_3(\underline{v})\}] , \tag{56}$$

$$\text{prob}(1^* \Rightarrow 2 \Rightarrow 3) = N_3[B_{23}I_{23}(\underline{v})F_2(\underline{v})] \; N_2[B_{12}I_{12}(\underline{v})F_1(\underline{v})] , \tag{57}$$

where

$$I_{ij}(\underline{v}) = \int I(\nu_{ij})\alpha_{ij}(\xi_{ij})d\nu_{ij} \qquad (i < j) \quad . \tag{58}$$

Hence, apart from their dependence on the radiation intensity $I_\nu(\underline{n})$, the atomic line profile coefficients depend in general on the distribution functions $F_i(\underline{v})$ of the atoms in the various energy levels through the probabilities prob(...).

REFERENCES

Borsenberger,J., Oxenius,J., Simonneau,E.: to be published

Cipolla,J.W., Morse,T.F.: 1979, J.Quant.Spectrosc.Radiat. Transfer 22 , 365

Griem,H.R.: 1964, Plasma Spectroscopy, McGraw-Hill, New York

Hubený,I., Oxenius,J., Simonneau,E.: 1983, J.Quant.Spectrosc.Radiat.Transfer 29 , 477 and 495

Omont,A., Smith,E.W., Cooper,J.: 1972, Astrophys.J. 175 , 185

Oxenius,J.: 1985, Kinetic Theory of Particles and Photons, Springer-Verlag, Berlin, Heidelberg, New York

NON LOCAL EFFECTS ON THE REDISTRIBUTION OF RESONANT SCATTERED PHOTONS

Eduardo SIMONNEAU
Laboratoire d'Astrophysique Theorique
du College de France
Institut d'Astrophysique du C.N.R.S.
98bis, boulevard Arago
75014 Paris
France.

1. INTRODUCTION

The preceeding paper by Oxenius leads to the conclusion that photon redistribution during spectral line formation presents two problems : the first is the obtention of the atomic absorption and emission coefficients and the second is the relation between these atomic profile coefficients and the corresponding laboratory profiles. The classical or <u>local</u> theory of redistribution simplifies considerably this second problem because it do not consider the streaming of the excited atoms, that is, it takes the position of the atoms at the emission moment as the same as the position at the absorption moment. But the motion of absorbed and emitted atoms in the basic element is the description of the absorption and emission laboratory profile coefficients, for any given atomic profiles, so the study of this motion must be important in the analysis of the shape of a spectral line.

Let us focus our attention on a gas composed of two-level atoms and electrons under stationary, homogeneous and isothermal conditions, and we shall study here the transfer of photons with frequencies corresponding to the transition between these two atomic levels. We suppose that the gas is self-excited, that is, that the energy of the emergent spectral line is provided only by electronic excitation of these two-level atoms. We also suppose that the electron temperature is small compared to the atomic excitation energy so that the density of excited atoms is much smaller than that of the non-excited atoms. In this low-temperature limit, stimulated emissions can be neglected. Moreover, under these conditions the velocity distribution function of non-excited atoms is not affected by collisional and radiative transitions between the two atomic levels ; we can therefore assume this distribution to be maxwellian with the same kinetic temperature as the electron gas.

J. E. Beckman and L. Crivellari (eds.), Progress in Stellar Spectral Line Formation Theory, 73–86.

On the contrary, the velocity distribution function of excited atoms is not more maxwellian. On one hand, because of the selectivity of the absorption processes, this velocity distribution depends on the radiation field's direction and frequency distribution : this effect of the radiation field on the density and velocity distribution of the excited atoms is considered to be local, at least from the kinetic point of view. The classical, local, redistribution theory considers only this effect. But on the other hand, the non-LTE line transfer gives rise to a density gradient of the excited atoms, which in turn will lead to a transport of these excited atoms and thus to a modification of its velocity distribution function. We must take this modification into account in any self-consistent treatment of the spectral line formation. A non-local description of the redistribution phenomena becomes necessary. This new description requires a kinetic equation for the velocity distribution function of the excited atoms, where the source and sink terms depend on the radiative excitations and de-excitations. So the transfer equation is coupled with this kinetic equation for the excited atoms distribution, requiring that both equation be solved simultaneously.

Since our aim here is to concentrate on the non-local, kinetic aspects of photon frequency redistribution, the atomic description will be reduced as much as possible. To do this we choose an atomic model in which the two levels are perfectly sharp. In this case, the atomic absorption and emission profiles are the same and both are the Dirac $\delta(\xi-\xi_o)$function. Of course, with this model, effects of the local laboratory redistribution are practically non existent (Mihalas, 1978). But, on the contrary, this property will allow us to put in clear only the non-local aspects of the redistribution.

Although the preceeding Oxenius' paper described this kinetic formalism this present paper will concentrate on presenting and discussing results for cases of astrophysical and laboratory interest. These results correspond to a more larger work in progress by J. Borsenberger, J. Oxenius and E. Simonneau.

2. THE NON-LOCAL EQUATIONS AND LIMITING CASES

Under the conditions described above, the transfer equation for a gas composed of atoms with two perfectly sharp levels (see Oxenius preceeding paper), becomes

$$\left[\bar{n}.\nabla_{\nu}\right] I_{\nu,\bar{n}}(\tau) = -\phi_{\nu}\left[I_{\nu,\bar{n}}(\tau) - S_{\nu,\bar{n}}(\tau)\right] , \qquad (1)$$

where τ is the optical distance,

$$d\tau = k\ dr \quad ,$$

k the line mean opacity,

$$k = \frac{h\nu_o}{4\pi}\ \frac{1}{\Delta\nu_D}\ n_1 B_{12} \quad , \tag{2}$$

and ϕ_ν the normalized aborption profile, which takes a Gaussian shape for the chosen atomic model. The source function is

$$S_{\nu,\bar{n}}(\tau) = n_2\ \frac{\Psi_{\nu,\bar{n}}}{\phi_\nu} \quad , \tag{3}$$

where n_2 is the excited atoms density and $\Psi_{\nu,\bar{n}}$ the emission laboratoy profile

$$\Psi_{\nu,\bar{n}} \equiv \iiint d^3v f_2(\bar{v},\tau)\ \eta_{21}(\xi,\bar{n}) \quad , \tag{4}$$

where $f_2(\bar{v},\tau)$ is the normalized velocity distribution function of the excited atoms and $\eta_{21}(\xi,\bar{n})$ is the normalized atomic emission profile at frequency ξ and direction $\bar{n}$. The atomic emission frequency ξ , the laboratory frequency ν , the emission direction $\bar{n}$ and the atomic velocity $\bar{v}$ are related by the Doppler effect

$$\xi = \nu.(1 - \frac{\bar{n}.\bar{v}}{c}) \ . \tag{5}$$

The excited atoms are described by a velocity distribution function

$$F_2\ (\bar{v},\tau) \equiv n_2(\tau)\ f_2(\bar{v},\tau) \quad , \tag{6}$$

with

$$\iiint d^3v f_2(\bar{v},\tau) = 1.$$

Under the above conditions, the kinetic equation described by this distribution function is

$$\eta\left[\bar{v}.\nabla_\tau\right]\ \left[n_2(\tau) f_2(\bar{v},\tau)\right] = \left[\varepsilon + (1-\varepsilon)\ I_{12}(\bar{v})\right] f_M(\bar{v}) - n_2(\tau) f_2(\bar{v},\tau) + \zeta n_2(\tau)\left[f_M(\bar{v}) - f_2(\bar{v},\tau)\right] \tag{7}$$

with

$$I_{12}(\bar{v}) = \int\!\!\oint d\nu\ \frac{d\Omega}{4\pi}\ I_{\nu,\bar{n}}(\tau)\ \alpha_{12}(\xi) \quad , \tag{8}$$

where we must take into account the Doppler relation

$$\xi = \nu.(1 - \frac{\bar{n}.\bar{v}}{c}).$$

$f_M(\bar{v})$ is the normalized Maxwell velocity distribution at the temperature of the gas, and $\alpha_{12}(\xi)$ is the normalized atomic absorption profile.

We now have three characteristic non-dimensional parameters: ε, η, and ζ. ε is the classic non-LTE parameter

$$\varepsilon = \frac{ne\, C_{21}}{A_{21} + ne\, C_{21}} , \tag{9}$$

ζ measures the importance of the elastic collisions

$$\zeta = \frac{\lambda_{exc}}{\lambda_{e\ell}} , \tag{10}$$

and η measures the importance of the streaming

$$\eta = \frac{\lambda_{exc}}{\lambda_{ph}} . \tag{11}$$

λ_{ph} is the mean free path of the photons, such that

$$\lambda_{ph} = \frac{1}{k} \tag{12}$$

$\lambda_{e\ell}$ is the mean free path between two elastic collisions

$$\lambda_{e\ell} = \frac{1}{n_1 \sigma} , \tag{13}$$

where n_1 is the non-excited atoms density and σ is a mean cross section for elastic atom-atom collisions.
λ_{exc} is the mean free path of the excited atoms

$$\lambda_{exc} = \frac{w}{A_{21} + ne\, C_{21}} \tag{14}$$

that is the mean path during the time of excitation. w is the mean thermal velocity.

Now we have two kinetic equations coupled by their respective source terms, that must be solved them for some values of the characteristic quantities ε, η and ζ. As we shall show later, the influence of elastic collisions will not introduce any new important physical modification; in a first approximation its influence is only a scale factor for the characteristic lengths of the solution. For ε we shall take the two values: $\varepsilon = 10^{-2}$ and 10^{-4}, which describe two typical non-LTE situations; and for η we analyse several situations between $\eta \ll 1$ and $\eta \gg 1$.

To solve these equations we choose the customary plane-parallel semi-infinite geometry. For this we need boundary conditions for the distribution function of the excited atoms $f_2(\bar{v},\tau)$, as well as for the photon distribution. These boundary conditions needed for $f_2(\bar{v},\tau)$ is an important difference between the non-local and the classic local description of the excited atoms in the line transfer problem.

It is normal to assume that the incident intensity for the photons is zero, a condition we can choose for the excited atoms velocity distribution as well. It is the case where the excited atoms are "destroyed" when they arrive at the surface. This can occur as well by adsorption, by de-excitation or even by free escape. But another boundary condition physically illustrative is the "reflexion" condition. It is the case where the excited atoms undergo elastic reflexions on the boundary surface.

So, if we separate the two distribution functions $I_{\nu,\bar{n}}(\tau)$ and $f_2(\bar{v},\tau)$ in two parts

outgoing $I^+_{\nu,\bar{n}}(\tau)$, $f_2^+(\bar{v},\tau)$,

incoming $I^-_{\nu\bar{n}}(\tau)$, $f_2^-(\bar{v},\tau)$

the boundary conditions are

$$I^-_{\nu,\bar{n}}(0) = 0 ,$$

and

$$f_2^-(\bar{v},0) = 0 \qquad \text{for DESTRUCTION}$$

$$f_2^-(\bar{v},0) = f_2^+(\bar{v},0) \qquad \text{for REFLEXION}$$

Once we have the transfer and kinetic equation we can now study some extreme cases that allow us to better understand the redistribution processes.

If the elastic collisions dominate all other processes, $\zeta \to \infty$ then

$$f_2(\bar{v},\tau) = f_M(\bar{v}) ,$$

and the emission profile takes the same equilibrium form ϕ_ν as the absorption profile. The excited atoms density becomes

$$n_2(\tau) = \varepsilon + (1-\varepsilon)\, J_{12}(\tau) , \tag{15}$$

where

$$J_{12}(\tau) = \iint d\nu \frac{d\Omega}{4\pi}\, \phi_\nu\, I_{\nu,\bar{n}}(\tau) . \tag{16}$$

This is the wellknown case of the laboratory complete redistribution. In a first approximation, only one scale length, we have

$$n_2(\tau) = 1-(1-\sqrt{\varepsilon})\, e^{-\tau/L_T} , \tag{17}$$

where

$$L_T \sim 1/\varepsilon , \tag{18}$$

is the classic thermalization length. If the elastic collisions are large, but are unable to cancel the streaming completely, the kinetic equation can be studied using the diffusion approximation, a quasi-local equation is obtained for the source function (Duchs and Oxenius, 1977).

On opposite extreme, one finds the obvious case where the two equations are not coupled. This is the electronic collision dominated case, $\varepsilon = 1$, that is, the LTE problem where the homogeneity of the excited atoms distribution is only broken by the boundary surface. Here the density of the excited atoms is given only by kinetic considerations and is independent of the radiation field, leaving to a characteristic length, the streaming length

$$L_S \quad \frac{\eta}{\sqrt{1+\zeta}} . \tag{19}$$

Since both local and non-local redistribution effects are only important when there are few elastic collisions, we can limit future discussions to kinetic and transfer equations where we exclude these elastic collisions, and will only return briefly to the elastic collision effects in the final discussions.

3. RESULTS

We have solved the preceeding transfer (1) and kinetic (7) equations for a plane-parallel layer with a total optical depth $\tau_M = 10^6$; the thermalization is then reached in the middle of the layer and all the characteristic lengths will be exhibited in the numerical solutions.

We have employed the traditional discrete ordinated method for the frequencies, velocities and directions in both distribution functions. For the directions, because of the azimuthal symmetry, we have used only light discrete values for the cosine of the zenithal angles. For the absolute values of the velocity we have taken 16 discrete points, and for the intensity frequencies 32 values. We then have 256 specific intensities $I_{\nu_i \mu_j}(\tau)$ and 128 specific distribution functions $f(v_k, \mu_\ell, \tau)$ and therefore an equal number of transfer and kinetic equations.

These specific distribution functions $f(v_k, \mu_\ell, \tau)$ are now coupled non-trivially in the source terms of the transfer equations for each intensity $I_{\nu_i \mu_j}(\tau)$, according to Expression (4). The specific intensities $I_{\nu_i \mu_j}(\tau)$ are likewise coupled in the source terms of the kinetic equations for each function $f(v_k, \mu_\ell, \tau)$, in accordance with the Expression (8). As a consequence these source terms are two matrix operators whose product is the redistribution operator R_I, but to describe the non-local effects it is necessary to use them separately.

Once we have the two vectorial equations, transfer and kinetic, coupled by their respective source terms, the solution also becomes non trivial. We followed an iteration method, using the Eddington variable factor technique on the equation for the first moments of the two distributions $I_{\nu\mu}(\tau)$ and $f(v, \tau)$, as habitual for transfer problems.

We obtained results for several values of ε and η , and presented some of them here. For exemple, for $\varepsilon=10^{-4}$ which represents a typical non-LTE case and for the characteristic values $\eta=10^{-2}$, $\eta=1$, $\eta=10^{2}$. The $\eta=10^{-2}$ case recovers the classic, local redistribution, but the other two show clearly non-local effects.

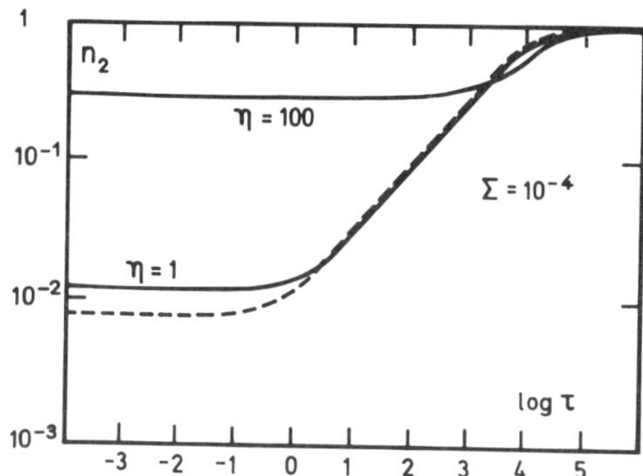

Figure 1. Density of excited atoms, for reflexion boundary conditions; $\varepsilon = 10^{-4}$ and the quoted values of η. For $\eta \ll 1$ this density reachs the values for the classic, static redistribution (dashed lines).

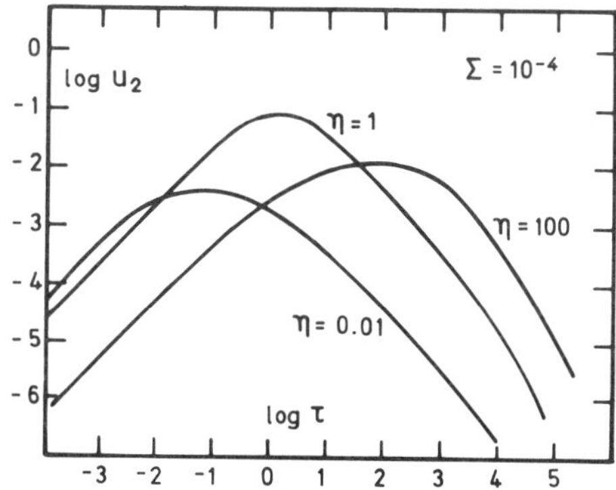

Figure 2. Mean velocity of the excited atoms for reflexion boundary conditions; $\varepsilon = 10^{-4}$, and the quoted value of η.

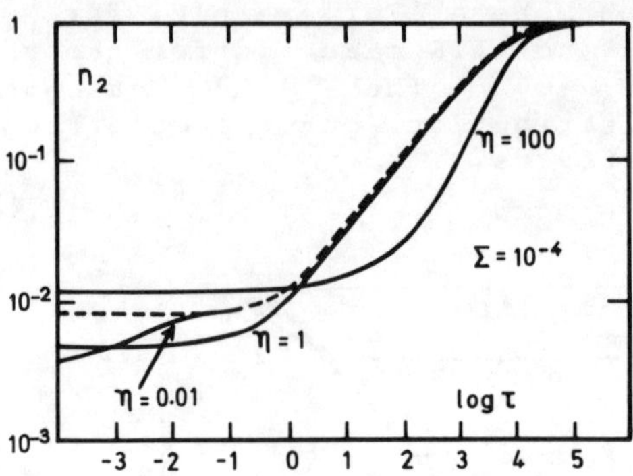

Figure 3. The same as figure 1, but for destruction boundary conditions.

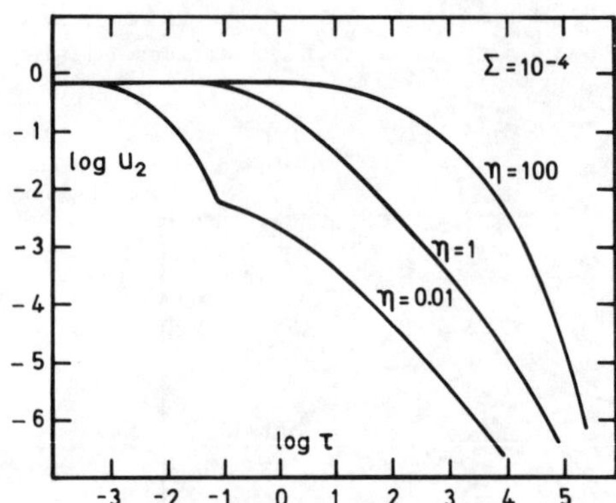

Figure 4. The same as figure 2, but for destruction boundary conditions.

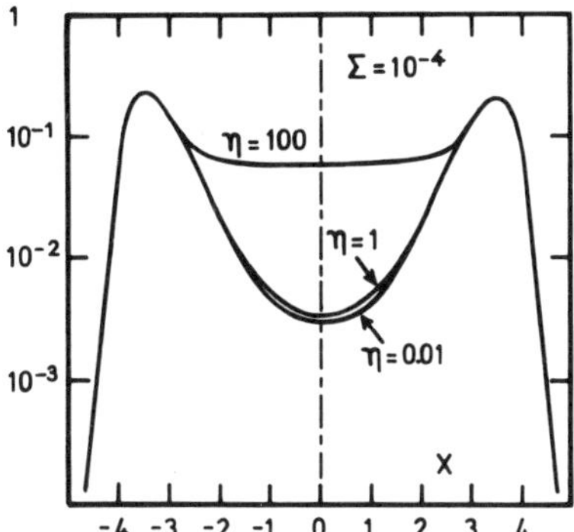

Figure 5. Emergent Eddington flux for the reflexion boundary conditions; $\varepsilon = 10^{-4}$ and the quoted values of η. Although it is not visible on this figure, the curves for $\eta = 1$ and $\eta = 100$ are slightly asymmetric.

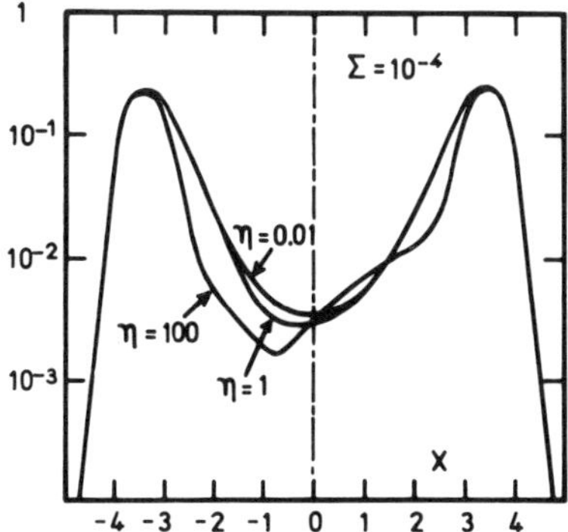

Figure 6. The same as figure 5, but for the destruction boundary conditions.

Before giving an interpretation of these results, we want to point out the importance of boundary conditions for the velocity distribution function of the excited atoms. It is a consequence of the non-local treatment. For the "reflexion" conditions, the surface density and therefore the total density distribution of the excited atoms n_2 (τ), are greater than in the local case, because excited atoms are transported from the deeper regions (fig.1). In the "destruction" case there are a competition between the atoms which come from the deeper regions and the atoms which are destroyed on the surface (fig.3).

Changes in boundary conditions can also cause wide differences in the mean velocity of excited atoms (fig.2 and 4). In the case of "reflexion", this velocity is zero both at the surface and at great optical depth where the two distributions are thermalized, and is highest at optical depths greater than the corresponding streaming length; although this latter value is not very high: less than 0.1 of the thermal velocity. In contrast mean velocity is great at the surface in the case of "destruction" with a value near $1/\sqrt{2}$. This velocity decreases first with a scale length, the streaming length, and afterwards with a different scale, the thermalization length.

Since the mean velocity is not very great with reflexion boundary conditions, there will not be important asymmetries on the emergent line; only the density effects (the emergent intensity is greater than in the local case) are visible (fig.5). In the "destruction" case, the highest value of the velocity is the same for all values of η. But only when these highest velocity values stay within the regions of line formation, the effects of asymmetry will be large on the emergent line, as is the case in figure 4. For $\eta = 0.01$, even when the velocity is on the order of $1/\sqrt{2}$ at $\tau = 0$, it will decrease too quickly and at the line formation depth, this velocity is too small to produce an asymmetry on the emergent spectral line (fig.6). For $\eta = 1$, and even more so for $\eta = 100$, the mean velocity at line formation depth is great, leading therefore to the greater effects of asymmetry, as well as of density, on the emergent line.

4. INTERPRETATION OF RESULTS

Due to the anisotropy and frequency-dependance of the radiation field, the excited atoms take a velocity distribution function which, in a sense, is an image of this radiation field. For some values of ε and η , this velocity distribution can vary quite far from the maxwellian. In such cases a flux of excited atoms exists which is very important in relation to the density of the excited atoms so that the

$n_2(\tau)$ distribution can be seriously perturbed. Likewise we have seen that these non-local, streaming effects, can strongly modify the laboratory emissions profile and, therefore, the shape of the emergent spectral line.

To obtain a first analytic approximation to describe these non local effects. We begin with the custumary two-fluid model for each distribution function. The two distribution functions $I_{\nu,\bar{n}}(\tau)$ and $f_2(\bar{v},\tau)$ are each separated into two parts outgoing $I^+_{\nu,\bar{n}}(\tau)$ and $f^+_2(\bar{v},\tau)$ and incoming $I^-_{\nu,\bar{n}}(\tau)$ and $f^-_2(\bar{v},\tau)$ and then we take a local equilibrium form for each one of these four partial distribution functions. The next step is to write the four kinetic equations for these functions and to realize that in the corresponding source terms these four partial distribution functions are coupled in a simple linear way. After some algebraic transformations, the four equations can be reduced to two equations of second order for the excited atoms density $n_2(\tau)$ and for the frequency integrated mean intensity $J_{12}(\tau)$. These equations are:

$$\frac{1}{3\varepsilon}\frac{d^2}{d\tau^2}J_{12}(\tau) = J_{12}(\tau) - n_2(\tau)\ , \tag{20}$$

$$\frac{\eta^2}{2}\frac{d^2}{d\tau^2}n_2(\tau) = (1+\zeta)\left[n_2(\tau)-\varepsilon-(1-\varepsilon)J_{12}(\tau)\right]\ , \tag{21}$$

with the boundary conditions at $\tau = 0$,

$$\frac{1}{\sqrt{3\varepsilon}}\left[\frac{d}{d}J_{12}(\tau)\right]_o = J_{12}(0) \tag{22}$$

for the photon's density, and

$$\left[\frac{d}{d\tau}n_2(\tau)\right]_o = 0 \qquad \text{REFLEXION} \tag{23}$$

$$\frac{\eta}{2}\left[\frac{d}{d\tau}n_2(\tau)\right]_o = \frac{(1+\zeta)}{\sqrt{2}}n_2(0) \qquad \text{DESTRUCTION} \tag{24}$$

for the excited atoms density.
This two fluid model leads to a double exponential solution for the density of the photons and of the excited atoms; that is, the solutions are described by two-scale lengths L_T, L_S, in the form

$$n_2(\tau) = 1-\left[A_1 e^{-\tau/L_S} + A_2 e^{-\tau/L_T}\right], \tag{25}$$

$$J_{12}(\tau) = 1 - \left[\frac{A_1 3\varepsilon L_S^2}{3\varepsilon L_S^2 - 1} e^{-\tau/L_S} + \frac{A_2 3\varepsilon L_T^2}{3\varepsilon L_T^2 - 1} e^{-\tau/L_T}\right], \tag{26}$$

with the corresponding mean velocity

$$U_2(\tau) \equiv \frac{\Phi(\tau)}{n_2(\tau)} = \frac{1}{n_2(\tau)}\ \frac{\eta}{2}\left[\frac{A_1}{L_S}e^{-\tau/L_S} + \frac{A_2}{L_T}e^{-\tau/L_T}\right], \tag{27}$$

where $\Phi(\tau)$ is the excited atom's flux. The scale lengths

L_T and L_S solve the characteristic equation

$$\left(1 - \frac{Q}{3\varepsilon L^2}\right)\left(1 - \frac{1}{3\varepsilon L^2}\right) = (1-\varepsilon) , \tag{28}$$

where

$$Q = \frac{3}{2}\ \frac{\varepsilon\eta^2}{1+\zeta} \tag{29}$$

is a non-dimensional quantity which determine the importance of the non-local effects. The streaming influence only becomes important when $Q \gg 1$. The A_1 and A_2 coefficients are given by the boundary conditions.

Because the four partial distribution functions are assumed to be in local equilibrium, this two-fluid model can't explain the local redistribution effects. But this model is, on one hand, sufficient enough to describe the non-local streaming properties in a first approximation. And on the other hand the local redistribution effects are not great because of the simplified atomic model we have chosen, with its two infinitely sharp levels.

Now, when $\varepsilon \ll 1$, we have for the two scale-lengths

$$L_T = \frac{\sqrt{1+Q}}{\sqrt{3}}\ \frac{1}{\varepsilon} \quad , \quad L_S = \frac{1}{\sqrt{2}}\ \frac{1}{\sqrt{1+Q}}\ \eta \quad . \tag{30}$$

Thermalization length is greater here than in the local description, which is a natural consequence of the streaming, because the excited atom's density decreases at great optical depths due to the outgoing flux of these atoms. In contrast, in relation to the case where the excited atom's distribution and the photon's distribution are not coupled, the radiation field acts on the excited atom's velocity distribution in such a way that this function becomes, at great optical depths, more isotrope and the streaming length is therefore less.

Boundary conditions will allow us to obtain the values of the coefficients A_1 and A_2, and we have:
For the reflexion boundary conditions, the excited atom's surface density takes the value

$$n_2(0) = \frac{\sqrt{\varepsilon}}{\sqrt{1+2\sqrt{\varepsilon}\sqrt{Q}+Q}-\sqrt{Q}} \quad , \tag{31}$$

and, since $A_2 \gg A_1$, in a first approximation it will be

$$n_2(\tau) = 1 - \left[1-n_2(0)\right] e^{-\tau/L_T} \quad . \tag{32}$$

As we have seen already in figure 1, in this case, one exponential decay is sufficient to describe the $n_2(\tau)$ distribution. In contrast, the mean velocity of the excited atoms takes the approximate form

$$U_2(\tau) = U_M \left[e^{-\tau/L_T} - e^{-\tau/L_S} \right], \tag{33}$$

with

$$U_M = \frac{1}{\sqrt{2}} \frac{\sqrt{Q}}{1+\sqrt{Q}} \frac{1-\varepsilon}{1+\sqrt{\varepsilon}+2\sqrt{Q}} . \tag{34}$$

$U_2(\tau)$ reaches the maximum value U_M when $L_s < \tau < L_T$. Its highest value is not very great and it depends on the values of Q and ε. In any case U_M is less than 0.1, that is less than the value that U_M takes for $Q=1$ (fig.2). For $Q > 1$, the velocity is small because η is small; for $Q < 1$, the velocity is small because the density gradient is small. The effects of this velocity $U_2(\tau)$ on the emergent line profile will at any rate not be very great.

For the destruction boundary conditions, the excited atom's density surface takes the value

$$n_2(0) = \frac{\sqrt{\varepsilon}}{\sqrt{1+\zeta} + \sqrt{1+2\sqrt{\varepsilon}\sqrt{Q} + Q} - \sqrt{Q}}, \tag{35}$$

and the $n_2(\tau)$ distribution becomes

$$n_2(\tau) = 1 - \left[1-\sqrt{\varepsilon}\right] e^{-\tau/L_T} - \left[\sqrt{\varepsilon} - n_2(0)\right] e^{-\tau/L_S}, \tag{36}$$

in a first approximation. In this case, owing to the destruction of the excited atoms which arrive at the surface, the streaming length is clearly visible. When $\eta \gg 1$, the two-scales lengths L_s and L_T are of the same order of magnitude, so that the solution can be expressed as a single exponential decay (fig.3). The excited atoms mean velocity now reaches the value $1/\sqrt{2}$ at the surface; and

$$U_2(\tau) \simeq \frac{1}{\sqrt{2}} \left[1-(1-\sqrt{\varepsilon})\sqrt{Q}\right] e^{-\tau/L_S} + \frac{1}{\sqrt{2}} (1-\sqrt{\varepsilon})\sqrt{Q}\, e^{-\tau/L_T}, \tag{37}$$

if $Q \to 0$.

$$U_2(\tau) \simeq \frac{1}{\sqrt{2}} e^{-\tau/L_T} \tag{38}$$

if $Q \to \infty$.
as we show in figure 4.
In this case, the effects of velocity on the emergent line profile can be verylarge (fig.6).

5. CONCLUSIONS

In some astrophysical situations: resonance lines of the most abundant elements, the non-LTE radiative transfer can induce a convective transport of excited atoms; see numerical values of the characteristic lengths in Oxenius numerical values of the characteristic lengths in Oxenius (1979). This transport can strongly modify the excited atom's density and therefore the intensity of the spectral line and, under some conditions, can also alter the shape of the line profile. From a conceptual point of view, and even if the described model is, to some extent, academic, we have developed here the first truly self-consistent non-LTE solution to the two-level atom transfer problem.

REFERENCES

Mihalas, D., 1978, Stellar atmospheres, 2nd ed., Freeman, San Francisco.
Duchs, D.F., Oxenius, J., 1977, Z. Naturforsch., **32a**, 156.
Oxenius, J., 1979, Astron. Astrophys., **76**, 312.

ASYMPTOTIC PROPERTIES OF COMPLETE AND PARTIAL FREQUENCY REDISTRIBUTION

H. FRISCH
Observatoire de Nice, C.N.R.S.
P.O. Box 139
06003 Nice Cedex
France

ABSTRACT. Radiative transfer problems with frequency redistribution may be investigated by asymptotic methods when the mean number of scatterings undergone by photons is very large. These methods, which rely on the existence of a small parameter ε, which can be the probability of collisional destruction, provide scaling laws for characteristic parameters of the line radiation field, such as the thermalization length, the thermalization frequency or the saturation intensity of the radiation field. These methods also provide asymptotic transfer equations which describe the large scale behaviour of the radiation field away from boundaries. Boundary layers equations valid close to surfaces can also be obtained. Complete redistribution and the four standard types of partial redistribution will be discussed. It will be shown that resonance line partial redistribution (R_{II}) is a scattering process of the diffusive type, the diffusion being not only in space but also in frequency, while the other partial redistribution mechanisms are of the complete redistribution type and not diffusive. The origin of these drastically different behaviours will be explained in physical terms. Implications for numerical calculations will be briefly discussed.

1. INTRODUCTION

The formation of spectral lines in dilute media such as atmospheres or planetary nebulae is essentially a problem of multiple scatterings. Photons are preferentially created at line center, for instance by collisional excitation of the lower level. They then undergo several scatterings with changes in direction and frequency until they escape the medium or are destroyed by some process, such as deexcitation of the upper level. For strong lines formed in dilute media, the probability of collisional deexcitation ε may be very small, 10^{-4} or less, and the line optical thickness quite large, say 10^6 or more. The mean number of scatterings undergone by line photons is then very large, as large as ε^{-1} if the line has reached thermalization. The formation of such lines can be analyzed by asymptotic techniques, using ε as small expansion parameter.

J. E. Beckman and L. Crivellari (eds.), Progress in Stellar Spectral Line Formation Theory, 87–100.

When photons undergo a very large number of scatterings, the mean distance they can travel between creation and destruction is much larger than the mean free path at line center. This distance, known as the thermalization length, determines the characteristic scale of variation of the radiation field in space. The idea of the asymptotic analysis is to obtain equations which describe the variation of the radiation field on this large scale. These equations are to the usual transfer equation exactly what the hydrodynamic equations are to Boltzman equations.

Systematic asymptotic analyses of transport equations were first developed to investigate the scattering of neutrons (say in a nuclear reactor) (Larsen and Keller, 1974). Neutrons being scattered coherently, i.e. with a negligible change in energy, the large scale transport equation is simply a diffusion equation. For spectral lines the asymptotic analysis leads to a variety of large scale transport equations, not all of the diffusive type, the structure of which depends on the frequency redistribution mechanism. These asymptotic transfer equations depend only on the behaviour of the absorption profile and of the redistribution function at large frequencies. They are always simpler than the starting transfer equation (e.g. no angular variable anymore). The asymptotic analysis provides also scaling laws for characteristic parameters such as the thermalization frequency, or the saturation value of the radiation field.

For transport problem in bounded media, stars or nuclear reactors, the asymptotic transfer equation will be valid only inside the medium. Near the boundaries the radiation field varies rapidly and is solution to a boundary-layer problem, which usually reduces to a one-dimensional transfer equation in a semi-infinite medium. An approximation of the radiation field, valid everywhere in the medium, can be constructed by matching asymptotically the interior and the boundary-layer solution.

My purpose now is to illustrate these very general ideas with various frequency redistribution mechanisms. I'll consider complete redistribution and the four basic type of partial redistribution introduced by Hummer (1962), R_I to R_{IV}. For R_{IV} it is of course the correct form, first established by Woolley and Stibbs (1953), which is being used. A detailed analysis of these simple processes should help to understand more realistic situations such as resonance line redistribution, which is linear combination of R_{II} and R_{III}. The line formation problem considered in this paper is the standard one : a two-level atom without continum in a stationnary and static medium. Photons are being created and destroyed only by inelastic collisions with electrons. The transfer equation may then be written

$$\underline{\Omega}.\underline{\nabla} = k(\underline{r})\phi(x)\,|I-S| \,, \tag{1.1}$$

with the source function $S(\underline{r},x,\underline{\Omega})$ defined by

$$S = \varepsilon B + \frac{1-\varepsilon}{\phi(x)}\int\frac{d\omega}{4\pi}\int_{-\infty}^{+\infty} R(x,\underline{\Omega},x',\underline{\Omega}')I(\underline{r},x',\underline{\Omega}')dx' \quad . \tag{1.2}$$

Here $\underline{r}$, x and $\underline{\Omega}$ are the space, frequency and direction variables. The frequency is measured in Doppler units and is equal to zero at line center. $I(\underline{r},x,\underline{\Omega})$ is the specific intensity, $B(\underline{r})$ the Planck function, $k(\underline{r})$ the absorption coefficient and $\phi(x)$ the absorption profile normalized to unity. $R(x,\underline{\Omega},x',\underline{\Omega}')$ is the joint probability of absorbing a photon $(x',\underline{\Omega}')$ and emitting a photon $(x,\underline{\Omega})$. ε is the probability of collisional destruction per scattering. It is the small parameter of the problem and the asymptotic analysis will be performed in the limit $\varepsilon\to 0$.

There are of course several other transfer problems which are amenable to an asymptotic analysis. For instance conservative scattering ($\varepsilon=0$) in a bounded medium having a large optical thickness T. The small parameter is then the inverse of T. Or conservative scattering in a medium where there is a small amount of a continuous absorption produced by dust particles. The small parameter is now the ratio β of the continuous opacity to the line opacity. If the situation is more complex, for instance non conservative scattering in a bounded, absorbing medium, the large scale behaviour will be controlled by the process with the smallest characteristic scale.

The paper is organized as follows. Section 2 is devoted to complete redistribution, Sect. 3 to the two types of partial redistribution R_I and R_{III}, Section 4 to R_{IV} and Sect. 5 to R_{II}. The mechanisms R_I, R_{III} and R_{IV} are studied by a perturbation technique around complete redistribution. For R_{IV} this procedure provides a proof that the R_{IV} redistribution function can be approximated by complete redistribution near line center and coherent scattering in the wings. The scaling laws for R_{II} are recovered with simple random walk arguments, using the space and frequency diffusive behaviour. Some boundary layers problems are discussed in Sect. 6. Implications of asymptotic results on numerical methods are examined in Sect. 7.

2. COMPLETE REDISTRIBUTION.

When there is complete frequency redistribution, the frequencies of the absorbed and emitted photon are totally decorrelated and the redistribution function is equal to the product $\phi(x)\phi(x')$. Photons with large frequencies are thus most likely to be reemitted in the line core. Complete redistribution occurs when atoms are strongly perturbed by collisions during the scattering process. The idea of looking for a large scale description of the radiation field has been introduced by Ivanov (1973, pp 390-392) for conservative scattering in a finite slab of large optical thickness and, independently, by Frisch and Frisch (1977) for non conservative scattering in infinite and semi-infinite one dimensional media. The techniques are similar and lead to a singular integral equation for the leading term in an asymptotic expansion of the source function. We give now the main lines of the expansion technique.

First one defines a rescaled transfer problem by introducing a rescaled optical depth,

$$\tilde{\tau} = \tau/\tau_c(\varepsilon) \quad , \tag{2.1}$$

and by assuming that the source function S is a function $S^{\varepsilon}(\tilde{\tau})$ of the slow variable . Then one assumes for $S^{\varepsilon}(\tilde{\tau})$ an expansion of the form :

$$S^{\varepsilon}(\tilde{\tau}) = t(\varepsilon)\left[\tilde{S}(\tilde{\tau}) + \text{terms of higher order in } \varepsilon\right] . \quad (2.2)$$

The scaling factor $\tau_c(\varepsilon)$ appears thus as the characteristic scale of variation of the radiation field. It will be determined, together with $t(\varepsilon)$, by the asymptotic analysis. These new variables are then introduced in the integral equation for the source function, which has first been set into a form appropriate to the asymptotic analysis by making use of the normalization of the kernel. The condition that all terms in the equation be of the same order as $\varepsilon \to 0$ uniquely determines the two scaling factors. The usual thermalization lengths are recovered, namely $(\varepsilon\sqrt{-\ell n\varepsilon})^{-1}$ and $a\varepsilon^{-2}$ for Doppler and Voigt profiles, respectively. The factor $t(\varepsilon)$ scales like the primary creation term divided by ε. Hence, for the model described by Eq.(1.2) $t(\varepsilon)$ is of order unity.

For a semi-infinite medium, and assuming Doppler complete redistribution, the singular integral equation may be written as

$$\tilde{S}(\tilde{\tau}) - \tilde{B}(\tilde{\tau}) = P\int_0^{\infty} \frac{\tilde{S}(\tilde{\tau})-\tilde{S}(\tilde{\tau}')}{4(\tilde{\tau}-\tilde{\tau}')^2}\, d\tilde{\tau}' - \frac{\tilde{S}(\tilde{\tau})}{4\tilde{\tau}} . \quad (2.3)$$

P means Cauchy principal value. $\tilde{B}(\tilde{\tau})$ is the leading term in an expansion of the thermal source. The second term in the r.h.s of (2.3) takes into account the escape of photons through the boundary. It is zero in the case of an infinite medium. The kernel of the singular integral is the asymptotic form of the kernel $K_1(\tau)$ (cf. Avrett and Hummer, 1965), but without the logarithmic correction which is included in the scaling factor τ_c. Similarly, the factor $1/4\tilde{\tau}$ is one-half of the asymptotic form of $K_2(\tau)$. In the case of Voigt (= Lorentz) complete redistribution the corresponding quantities are $1/6|\tilde{\tau}-\tilde{\tau}'|^{3/2}$ and $1/3\tilde{\tau}^{1/2}$.

The singular equation (2.3) provides a better approximation for the radiation field at large optical depths than the first or even second-order escape probability approximation (Hummer and Rybicki, 1982 ; Rybicki, 1984). For instance, for a semi-infinite medium with a chromospheric temperature rise the second order escape probability approximation predicts a source function which lies significantly below the exact value at depths or order $\tau_c(\varepsilon)$ (cf. Fig. 3 in Frisch and Frisch, 1975). The large scale equation on the contrary gives an excellent fit to the exact values because it properly takes into account backscatterings (Fig. 1e in Frisch and Froeschlé, 1977).

3. R_I AND R_{III} PARTIAL REDISTRIBUTION.

The physical picture for R_I is an infinitely sharp line. Redistribution is thus purely coherent in the atom's frame. R_{III} describes the frequency redistribution of a line having a sharp lower level and an upper level, of width a, broadened essentially by collisions. In the atom's frame there is thus complete frequency redistribution with a Lorentz profile. Profiles and redistribution functions in the laboratory frame are obtained by folding the atom's frame corresponding quantities with the velocity distribution of the atoms, usually taken to be a Maxwellian. Also the velocity of the atoms is assumed constant during the scattering process.

For frequencies larger than a few Doppler widths, which are the only one relevant to the asymptotic analysis, $R_{III}(x,\underline{\Omega},x',\underline{\Omega}')$ reduces to the atom's frame redistribution function, hence to complete redistribution (in the convolution product of a Lorentzian and a Gaussian, the asymptotic behaviour is controlled by the Lorentzian). The situation is not as simple for R_I, but it turns out that R_I behaves asymptotically also like complete redistribution, with Doppler characteristics of course, since the absorption profile is a Gaussian. This result, first shown for purely time-dependent transfer equations (e.g. Field, 1959), was then extended to space dependent ones (Frisch, 1980a).

The main lines of the asymptotic technique, which proceeds by perturbation around complete redistribution, are the following. First the frequency and angle dependent source function $S(\tau,x,\underline{\Omega})$ is decomposed into a "mean value", which depends only on optical depth, and a "fluctuating" part, viz.

$$S(\tau,x,\underline{\Omega}) = \overline{S}(\tau) + S_1(\tau,x,\underline{\Omega}) \quad , \tag{3.1}$$

where

$$\overline{S}(\tau) = \int \frac{d\omega}{4\pi} \int dx\ \phi(x)\ S(\tau,x,\underline{\Omega}) \quad . \tag{3.2}$$

A coupled set of equations for $\overline{S}$ and S_1 is deduced from the integral equation for the source function. These equations are of the form

$$\overline{S} - B = \overline{L}(\overline{S}) + \overline{L}_1(S_1) \quad , \tag{3.3}$$

$$S_1 = L(\overline{S}) + L(S_1) \quad , \tag{3.4}$$

where $\overline{L}$, $\overline{L}_1$, L are linear integral operators. The idea of the perturbation method is to assume that the terms with S_1 in the r.h.s of (3.3) and (3.4) are of higher order than the terms with $\overline{S}$ as $\varepsilon \to 0$, and then to check the validity of this assumption. When the term with S_1, is dropped, (3.3)reduces simply to the integral equation for complete redistribution and (3.4) provides S_1 in terms of $\overline{S}$. The asymptotic analysis is carried out by introducing a rescaled optical depth $\mathcal{T}=\tau/\tau_c(\varepsilon)$ and a rescaled profile $\mathfrak{F}= \phi\tau_c(\varepsilon)$. The frequency variable needs to be rescaled because S_1 is frequency dependent. The major outcome of the analysis is that the leading term of $\overline{S}(\tau)$ satisfies the singular integral equation (2.3) for complete redistribution with Doppler profile and that the fluctuation S_1 scales like $1/(-\ell n\varepsilon)$. In the case of R_{III} the fluctuation S_1 scales like ε and the leading term in $\overline{S}(\tau)$ satisfies the Voigt version of (2.3). This differences in the behaviour of the fluctuating terms shows clearly why complete redistribution is a very good approximation for R_{III} but not for R_I.

The most likely reason for which R_I behaves like complete redistribution is that a photon with a large frequency x' is most likely to be reemitted with a frequency x having a smaller absolute value (for a given x', R(x,x') is constant for $|x|<|x'|$ and decreases exponentially for $|x|>x'$). Hence it is most probable that after a few scatterings, the photon will be back to line center.

4. SUBORDINATE LINE REDISTRIBUTION : R_{IV}.

The scattering process described by R_{IV} is a transition from a naturally broadened lower state (with width a_ℓ) to a naturally broadened upper state (with width a_u) followed by a radiative decay to the lower state. It holds for subordinate lines in the limit where both levels are broadened by natural damping only. The atom's frame redistribution function r_{IV} consists in products and sums of Lorentzian functions (Woolley and Stibbs, 1953). The laboratory (R_{IV}) and the atom's frame (r_{IV}) redistribution functions are thus asymptotically equal at large frequencies, as was the case for R_{III}. For frequencies larger than a few Doppler widths, $r_{IV}(x,x')$ shows two well separated Lorentzian peaks, one centered close to the line center and the other one close to the absorption frequency. Their areas are approximately equal to $a_\ell/(a_\ell+a_u)$ and $a_u/(a_u+a_\ell)$, respectively. This property has lead Frisch (1980a) and Heinzel and Hubeny (1983) to independently suggest that R_{IV} could be approximated by a superposition of coherent scattering in the wings and complete redistribution in the line core. The validity of this approximation has been checked numerically by Hubeny and Heinzel (1984) and analytically by Frisch (1980a), by means of an asymptotic analysis of the transfer equation in infinite space. The main steps of this analysis are as follows.

First the redistribution function is rewritten as

$$r_{IV} = \frac{a_\ell}{a_\ell+a_u}\,\phi(x)\phi(x') + \frac{a_u}{a_\ell+a_u}\,\phi(x)\delta(x'-x)+\phi(x)C(x,x') \;, \tag{4.1}$$

where the third term $\phi(x)\ C(x,x')$ contains all the contributions which have been neglected, in particular the difference between the Lorentzian peak at $x=x'$ and the δ-function. The asymptotic analysis of the transfer equation can then be carried out on the Fourier transform S(k,x) of the source function (as r_{IV} is angle independent S is isotropic). By regrouping in the integral equation for S(k,x) the two terms which are not integrated over frequencies (one of them comes from the coherent scattering part of r_{IV}), one obtains an integral euqation of the form,

$$\Sigma(k,x) = \varepsilon B(k) + (1-\varepsilon)\left[M_1(\Sigma) + M_2(\Sigma)\right] . \tag{4.2}$$

The auxiliary function $\Sigma(k,x)$ is related to the source function S by

$$\Sigma(k,x) = S(k,x)\ T(k,x) \quad . \tag{4.3}$$

T(k,x) has a simple analytic expression in terms of k and x. The function Σ describes essentially the emission of photons about the line core. The operator M_1 resembles closely a complete redistribution operator and the operator M_2 contains the correction term C(x,x'). The asymptotic analysis of (4.2) is then carried out by perturbation around complete redistribution, with the term $M_2(\Sigma)$ playing the part of the perturbation (cf. Sect. 3). One finds that the thermalization length is of order $(a_\ell+a_u)\varepsilon^{-2}$ and that the leading term in the asymptotic expansion of $\Sigma(k,x)$ is frequency independent and obeys the singular integral equation (2.3) in its infinite space and Lorentz version. The R_{IV} source function itself is frequency dependent at leading order, the deviation from complete redistribution being proportionnal to the ratio a_u/a_ℓ. However, for depths of order the thermalization length, it will stay essentially flat up to a thermalization frequency $x_c \sim \phi^{-1}[\tau_c(\varepsilon)]$.

The scaling laws obtained with an infinite space analysis hold presumably also for a semi-infinite medium. Complete redistribution, R_I, R_{III} and R_{II}, for which various geometries have been considered, show that the scaling laws depend on the scattering mechanism but not on boundary conditions. The reason for which r_{IV} behaves as complete redistribution is presumably that photons are not likely to stay in the wings for many scatterings. The probability per scattering to stay in the wing is $a_u/(a_\ell+a_u)$. Hence, if a_u and a_ℓ are of the same order, this probability will be only around one-half.

When elastic collisions contribute to the broadening of the upper and lower levels of the transition, the redistribution function is of the form (Heinzel and Hubeny, 1981).

$$r = (1-\bar{\alpha})r_{IV} + \bar{\alpha}r_{III} \tag{4.4}$$

As r_{III} behaves asymptotically like complete redistribution, the effect of collisional broadening on the large scale behaviour is simply to modify the respective parts of complete and coherent redistribution.

5. RESONANCE LINE REDISTRIBUTION : R_{II}.

The physical picture for R_{II} is a resonance line with a sharp lower level and an upper level, of width a, broadened by natural damping only. Hence, in the atom's frame the scattering is coherent and the profile is a Lorentzian. For absorption frequencies larger than a few Doppler widths, the angle-averaged laboratory frame redistribution function shows a single peak, centered about the absorption frequency, which varies like $\exp|-(x-x')^{2/2}|$, times a slowly varying function of (x+x'). Scattering of large frequency photons is thus almost coherent. This produces a large scale behaviour which is of the diffusive type, the diffusion being not only in space (as for pure coherent scattering), but also in frequency because of the finite width of the peak in the redistribution function (Harrington, 1973). One consequence of this diffusion is that large frequency photons do not usually escape with one very long flight, as Rybicki and Hummer (1969) have shown it to be the case for complete redistribution, but through a random walk in the course of which their frequency is shifted toward the line center.

A systematic asymptotic analysis of R_{II} redistribution can be carried out on the integral equation for the angle averaged source fucntion (Frisch, 1980a), b). Once the optical depth has been rescaled by a factor

$\tau_c(\varepsilon)$ and the frequency by a factor $x_c(\varepsilon)$, a Taylor expansion of the scattering term in the integral equation is carried out to second order in space and frequency. One obtains, in the limit $\varepsilon\to 0$, a space and frequency diffusion equation for the leading term of the source function and the scaling laws

$$\tau_c(\varepsilon) \sim 1/\varepsilon \quad \text{and} \quad x_c(\varepsilon) \sim (a/\varepsilon)^{1/3} \quad . \tag{5.1}$$

The Taylor expansion is made possible by the Gaussian behaviour of R_{II}. The characteristic scales τ_c and x_c are usually referred to as the thermalization length and thermalization frequency. The frequency profile of the source function will be essentially flat up to x_c where it starts decreasing rapidly. The frequency x_c provides also a rough position of the two peaks in the emergent intensity profile.

The asymptotic technique developed by Larsen and Keller (1974) for neutron transport has been generalized to R_{II} by Frisch and Bardos (1981). It has the advantage over the other medhod of (i) proving the isotropisation of the radiation field at large optical depths and (ii) of providing boundary conditions for the large scale diffusion equation. The starting point is the transfer equation for the radiation field, written with the full frequency and angle-dependent redistribution function. The rescaled transfer problem, which turns out to be a singular perturbation problem, is obtained by introducing a rescaled optical depth $\tilde{\tau}=\varepsilon\tau$ and a rescaled frequency $\tilde{x}=\varepsilon^{1/3}x$. Then one assumes a solution of the form

$$I^{\varepsilon}(\tilde{\underline{r}},\tilde{x},\underline{\Omega}) \simeq I_o(\tilde{\underline{r}},\tilde{x},\underline{\Omega}) + \eta I_1(\tilde{\underline{r}},\tilde{x},\underline{\Omega}) + \eta^2 I_2(\tilde{\underline{r}},\tilde{x},\underline{\Omega}) + \ldots \tag{5.2}$$

The expansion parameter η is equal to $\varepsilon^{1/3}$. A hierarchy of equations is obtained by collecting terms with like powers of η in the transfer equation (the scattering term is being Taylor expanded with respect to frequency and the absorption profile and redistribution function are being expanded in powers of η). The three first equations in the hierarchy demonstrate that the leading term I_o is isotropic and that it obeys the diffusion equation,

$$\delta(\tilde{x})\tilde{B}(\tilde{\tau}) - \delta(\tilde{x})I_o + \frac{\pi\tilde{x}^2}{3a}\frac{\partial^2 I_o}{\partial\tilde{\tau}^2} + \frac{a}{2\pi}\frac{\partial}{\partial\tilde{x}}\left(\frac{1}{\tilde{x}^2}\frac{\partial I_o}{\partial\tilde{x}}\right) = 0 \quad , \tag{5.3}$$

first given by Harrington (1973). The two first terms show that creation and destruction of photons occur almost exclusively at line center. The third term describes diffusion in space and the last one diffusion in frequency together with a shift toward the line core due to an asymmetry in the redistribution function. In the case of conservative scattering the destruction term is zero. It is equal to I_o when the destruction of photons is dominated by the continuous absorption due to an overlapping continuum. In the latter case there is no more preferred frequency for the destruction of photons. The same diffusion equation holds for three-dimensional problems provided the operator $\partial^2/\partial\tilde{\tau}^2$ is replaced by $k^{-1}\underline{\nabla}.(k^{-1}\underline{\nabla})$.

The boundary conditions for (5.3) cannot be obtained from the large scale analysis alone. They are given by a boundary layer analysis which is described in Sect.6.

We now give a simple random walk argument which leads to the scaling laws (5.1). It is necessary to introduce the mean number of scatterings $\langle N\rangle$ undergone by the photons and the mean path length $\langle \ell\rangle$, which is proportional to the mean time spent by the photons in the medium (Ivanov, 1970). We introduce also N_W the mean number of scatterings undergone by photons with frequency of order x_c. Using the space and frequency diffusive nature of R_{II} and using the fact that the frequency shift at each scattering is of order one, we can write

$$x_c \sim N_W^{1/2} , \tag{5.4}$$

and

$$\tau_c \sim N_W^{1/2}/\phi(x_c) . \tag{5.5}$$

Assuming now that the dominant contribution to the mean path length comes from large frequency photons, we find

$$\langle \ell\rangle \sim N_W/\phi(x_c) . \tag{5.6}$$

Finally the definitions of $\langle N\rangle$ and $\langle \ell\rangle$ give us the relation (Frisch, 1980b)

$$\langle \ell\rangle \sim x_c\langle N\rangle . \tag{5.7}$$

Combining the relations (5.4) to (5.7) and using $\phi(x_c) \simeq a/(\pi x_c^2)$, which is the asymptotic form of the profile, we obtain

$$\tau_c \sim \frac{\pi}{a} x_c^3 \sim \langle N\rangle . \tag{5.8}$$

When the destruction of photons is controlled by inelastic collisions $\langle N\rangle \simeq 1/\varepsilon$ and the scaling laws (5.1) are recovered. The relations (5.7) and (5.8) provide also the scaling laws for conservative scattering in a finite slab or destruction by continuous absorption. In the first case $\tau_c \sim T$, which leads to $x_c \sim (aT)^{1/3}$, $\langle N\rangle \sim T$ and $\langle \ell\rangle/T \sim (aT)^{1/3}$. In the second case $\langle \ell\rangle \simeq 1/\beta$ (Hummer and Kunasz, 1980) and the scaling laws are $\tau_c \sim \langle N\rangle \sim a^{-1/4}\ \beta^{-3/4}$ and $x_c \sim a^{1/4}\ \beta^{-1/4}$. The relations (5.4) to (5.7) show also that $N_W \sim (a\langle N\rangle)^{2/3}$. The mean number of scatterings undergone by large frequency photons is thus negligible compared to the total mean number of scatterings, although it also goes to infinity as the expansion parameter goes to zero.

Hummer (1969) has shown that R_{II} redistribution behaves like coherent scattering for frequencies which are much larger than $\varepsilon^{-1/2}$. This regime is not included in the diffusive equation (5.3), which holds at frequencies of order $\varepsilon^{-1/3}$ only, but which nevertheless describes the most significant features of R_{II} redistribution (Adams, 1972,1975 ; Harrington, 1973 ; Hummer and Kunasz, 1980).

6. BOUNDARY LAYER SOLUTIONS.

The large scale equations (2.3) and (5.3) describe the radiation field away from boundaries. In the boundary layer the radiation field is solution to another transfer problem which is extracted from the rescaled transfer problem by stretching the slow space variable. As in all boundary layer problems, the purpose of the stretching is to introduce a new variable, say $\hat{\tau}$, in which the boundary layer has a thickness of order unity. The boundary layer coordinate is not necessarily the same as the primitive τ variable. The boundary layer equations for non diffusive (complete redistribution) and diffusive behaviour (R_{II}) differ as drastically as the interior equations and they will be discussed separately.

6.1 Complete Redistribution.

To leading order, the boundary layer source function is solution to the homogeneous integral equation

$$S^{bl}(\hat{\tau}) = \int_0^\infty K_1(|\hat{\tau}-\hat{\tau}'|)\, S^{bl}(\hat{\tau}')\, d\hat{\tau}' \quad . \tag{6.1}$$

For complete redistribution in a one dimensional semi-infinite medium the boundary layer variable $\hat{\tau}$ reduces to τ. Equation (6.1) has first been obtained heuristically (cf. Hummer and Rybicki, 1971) and then by asymptotic arguments (Ivanov, 1973, p. 231 ; Frisch, 1982). The solution of (6.1) is defined within a multiplicative constant which can be determined by applying the general technique of asymptotic matching between interior and boundary layer solutions (Cole, 1968). This amounts to write that the boundary layer solution in the limit of large $\hat{\tau}$, and the interior solution, in the limit of $\tilde{\tau}\to 0$, have the same behaviour. The asymptotic behaviours of $\tilde{S}(\hat{\tau})$ and $S^{bl}(\hat{\tau})$ can be established analytically by methods of the Wiener-Hopf type (Ivavov, 1973, p. 231 ; Frisch and Frisch, 1982). An asymptotic matching of this kind is explicitely worked out in Frisch (1982) for the case of conservative scattering in a slab.

6.2 R_{II} Partial Redistribution.

The boundary layer radiation field, for frequencies which are of the order of the thermalization frequency, can be obtained by a "superposition" method, first introduced by Larsen and Keller. The idea is to write the radiation field in the boundary layer as

$$I^{bl} = I^{i}(\tilde{\tau},\tilde{x},\underline{\Omega}) + I^{c}(\hat{\tau},\tilde{x},\underline{\Omega}) \quad , \tag{6.2}$$

where I^i is the interior field (solution of the large scale diffusion equation) and I^c the boundary layer term which goes to zero in the interior and which is function of a conveniently defined boundary layer coordinate $\hat{\tau}$. The surface boundary condition, of zero incident field for instance, must now be applied to the sum I^i+I^c. For a semi-infinite one dimensional medium one will thus have at the boundary $\tau=0$,

$$I^i(0,\tilde{x},\underline{\Omega}) + I^c(0,\tilde{x},\underline{\Omega}) = 0 \text{ for } \underline{\Omega}.\underline{n}<0 \ , \tag{6.3}$$

where $\underline{n}$ is the outward normal to the surface.

The thickness of the boundary layer in the primitive variable τ is of order $1/\phi(x_c)\sim x_c^2\sim\varepsilon^{-2/3}$. The boundary layer variable is obtained by rescaling τ by a factor $\varepsilon^{2/3}$. This amount to "enlarge" the slow variable $\tilde{\tau}=\varepsilon\tau$ by a factor $\eta^{-1}=\varepsilon^{-1/3}$. In this variable $\hat{\tau}$ the boundary layer has thus a thickness of order one.

The boundary layer solution is obtained by writing I^c as a perturbation expansion in powers of η, viz.

$$I^c = I_o^c(\hat{\tau},x,\underline{\Omega}) + \eta I_1^c(\hat{\tau},x,\underline{\Omega})+\ldots\ldots \quad . \tag{6.4}$$

The two first terms in (6.4) satisfy (Frisch and Bardos, 1981) a coherent scattering equation

$$\mu\frac{\partial I^c}{\partial\hat{\tau}} = -\frac{a}{\pi\tilde{x}^2}\left[I^c - \frac{1}{2}\int_{-1}^{+1} d\mu'\ I^c(\hat{\tau},x,\mu')\right] \ . \tag{6.5}$$

Here, $\mu=\underline{\Omega}.\underline{n}$. Hence, for large frequency photons and close to the surface, the changes in frequency at each scattering become negligible compared to the changes in direction, at least for the two lowest orders in the perturbation expansion. Equation (6.5) reduces actually to a diffuse reflection (or Milne) problem, with the opposite of the interior field playing the part of the "incident" field. Diffuse reflection has been quite extensively studied (Chandrasekhar, 1960, Benssoussan et al. 1979). One knows for instance how to express the emergent field in terms of the incident field. One knows also that (6.5) has a unique bounded solution which goes to constant at infinity, the value of which is related in a well defined way to the incident field. The condition that I^c goes to zero in the interior at all order in η, which is actually the matching condition between the interior and the boundary layer solution, provides thus surface values for all the terms in the expansion of the interior radiation field.

When the matching is carried out at leading order, one finds that the interior solution is zero on the boundary, i.e

$$I_o^i(0,\tilde{x}) = 0 \ , \tag{6.6}$$

and that the leading term in the boundary correction is zero. The condition (6.6) may serve as boundary condition for the diffusion equation (5.3). When the matching is carried out to order η, one finds that the mean (angle averaged) intensity at the boundary is given by

$$J_{em}(0,\tilde{x}) \simeq -\frac{\eta}{\sqrt{3}}\,\frac{\pi\tilde{x}^2}{a}\,\frac{\partial I_o^i}{\partial\tilde{\tau}}\Big|_{\text{surface}} \quad . \tag{6.7}$$

As for the emergent intensity $(I(0,x,\mu),\mu>0)$ it varies like $H(\mu)J_{em}$, where $H(\mu)$ is the Chandrasekhar H-function. The approximation (6.7) compares very favourably with numerical calculations (Harrington, 1973 ; Frisch and Bardos, 1981) and can be used to construct approximate solutions of the R_{II} radiation field (cf. Section 7). One can also obtain an equation and a boundary condition for the interior radiation field accurate to order η. The argument, first developed for coherent scattering (Bardos et al. 1983) and non grey radiative transport (Larsen et al. 1983 ; Frisch and Faurobert, 1984) relies on the fact that the approximation of order η of the angle average intensity, $J=I_0^i+\eta\int I_1^i\, d\omega/4\pi$, satisfy also the diffusion equation (5.3). A mixed type boundary condition of the form,

$$J(0,\tilde{x}) = \eta\,\frac{\pi\tilde{x}^2}{a}\,Q\,\frac{\partial J}{\partial\tilde{\tau}}\Big|_{\text{surface}} \tag{6.8}$$

is then readily obtained for J. The constant Q=0.7104 is the value at infinity of the Hopf function.

For multidimensional media the boundary layer coordinate is obtained by stretching the space variable in the direction perpendicular to the boundary (Larsen and Keller, 1974 ; Bensoussan et al., 1979) and the analysis described above applies identically.

7. NUMERICAL IMPLICATION.

Some implications of the asymptotic analysis for the numerical calculation of subordinate lines have been mentioned in the preceeding Sections. Here, we shall concentrate on resonance lines for which the general form of the redistribution function is (Omont et al., 1972)

$$R(x,\underline{\Omega},x',\underline{\Omega}')=\gamma R_{II}(x,\underline{\Omega},x',\underline{\Omega}')+(1-\gamma)R_{III}(x,\underline{\Omega},x',\underline{\Omega}') . \tag{7.1}$$

The branching ratio γ goes to one in the limit of low densities and to zero in the limit of high densities.

When the redistribution reduces to R_{II}, the diffusion equation (5.3) provides an approximation for the interior radiation field, provided of course the expansion parameter η be much smaller than unity. By adding the expression (6.6) for the mean emergent intensity to the interior solution one can construct an approximation which is good everywhere, except at small frequencies and small optical depths. For instance, emergent fluxes integrated over frequencies can be reproduced with an accuracy of a few percent as soon as the characteristic frequency is around 10 (Faurobert, 1984).

To simplify numerical work with R_{II}, Jefferies and White (1960) have introduced an approximation, subsequently revised by Kneer (1975), which is of the form

$$R_{II}(x,x')\simeq|1-\xi(x,x')|\phi(x)\phi(x')+\xi(x,x')\phi(x')\delta(x-x') \quad . \qquad (7.2)$$

$\xi(x,x')$ is chosen to give complete frequency near the line center and coherent scattering beyond a few Doppler widths. This partial coherent approximation (PCS) predicts a radiation field which is much too low at large frequencies, as soon as the space and frequency diffusion phenomenon becomes important (Basri, 1980, Hubeny, 1984). The reason is that this approximation inhibits the coupling between wings and line core. Large frequency photons can no more be shifted to the line center by diffusion in the frequency space. Also, as pointed out by Milkey (1976), this approximation cannot reproduce the correct asymptotic scaling laws of R_{II}. Hubeny (1984) has suggested to let the cut-off frequency x^* between complete frequency redistribution and coherent scattering increase monotically with optical depth according to the law $x^*=(a\tau)^{1/3}$. This procedure, though it does in some cases improve the emergent profile, makes the redistribution function depth dependent and does not really account for the space and frequency diffuse nature of R_{II} and hence can hardly lead to the R_{II} scaling laws.

In the general case ($\gamma\neq 1$), there is a competition between complete redistribution on the one hand and almost coherent scattering on the other hand. The situation thus resembles closely R_{IV} redistribution. Presumably when γ falls below some critical value γ_c complete redistribution will take over the control of the large scale behaviour. When this happens the PCS approximation may be just fine. This critical value of γ, if it exists, should be easy to obtain, numerically if not analytically.

REFERENCES

Adams, R.F., 1972, *Astrophys. J.* **174**, 439 ; 1975, op. cit. **201**, 350.

Avrett, E.H., Hummer, D.G., 1965, *Mon. Not. Roy. Astr. Soc.* **130**, 295.

Bardos, C., Santos, R., Sentis, R., 1983, 'Diffusion approximation and and computation of the critical size', *Trans. Am. Math. Soc.* (in press).

Basko, H.M., 1978, *Sov. J.E.T.P.* **48**, 644.

Basri, G.S., 1980, *Astrophys. J.* **242**, 1133.

Bensoussan, A., Lions, J.L., Papanicolaou, G., 1979, *Publ. R.I.M.S. Kyoto Univ.* **15**, 53.

Chandrasekhar, S., 1960, *Radiative Transfer*, Dover.

Cole, J.D., 1968, *Perturbation Methods in Applied Mathematics*, Blaisdell Publ. Co. Waltham, Massachusetts.

Faurobert, M., 1984, Thèse de 3ème cycle, Université de Paris VII.

Frisch, H., 1980a, *Astron. Astrophys.* **83**, 166 ; 1980b, op. cit. **87**, 357 ; 1982, *J. Quant. Spectrosc. Radiat. Transfer* **28**, 377.

Frisch, H., Bardos, C., 1981, *J. Quant. Spectrosc. Radiat. Transfer* **26**, 119.

Frisch, H., Faurobert, M., 1984, 'Boundary layer conditions for the transport of radiation in stars', *Astron. Astrophys.* (in press).

Frisch, H., Frisch, U., 1982, *J. Quant. Spectrosc. Radiat. Transfer* **28**, 361.
Frisch, H., Froeschlé, Ch., 1977, *Mon. Not. Roy. Astr. Soc.* **181**, 281.
Frisch, U., Frisch, H., 1975, *Mon. Not. Roy. Astr. Soc.* **173**, 167 ; 1977, op. cit. **181**, 273.
Harrington, J.P., 1973, *Mon. Not. Roy. Astron. Soc.* **162**, 43.
Heinzel, P., Hubeny, I., 1981, *J. Quant. Spectrosc. Radiat. Transfer* **27**, 1 ; 1983, op. cit. **30**, 77.
Hubeny, I., 1984, 'A modified Rybicki method and the partial coherent scatterings approximation', preprint.
Hubeny, I., Heinzel, P., 1984, 'Non-coherent scattering in subordinate lines V. Solutions of the transfer problem', *J. Quant. Spectrosc. Radiat. Transfer* (in press).
Hummer, D.G., 1962, *Mon. Not. Roy. Astron. Soc.* **125**, 21 ; 1969, op. cit. **145**, 95.
Hummer, D.G., Kunasz, P.B., 1980, *Astrophys. J.* **236**, 609.
Hummer, D.G., Rybicki, G., 1971, *Annual Review of Astronomy and Astrophyscs* **9**, 237 ; 1982, *Astrophys. J.* **263**, 925.
Ivanov, V.V., 1970, *Astrophysics* **6**, 355 ; 1973, *Transfer of Radiation in Spectral Lines*, transl. D.C. Hummer, NBS Spec. Publ. 385, US Dept. of Commerce, Washington, D.C.
Jefferies, J.T., White, O.R., 1960, *Astrophys. J.* **132**, 767.
Kneer, F., 1975, *Astrophys.J.* **200**, 367.
Larsen, E.W., Keller, J.B., 1974, *J. Math. Phys.* **15**, 75.
Larsen, E.W., Pomraning, G.C., Badham, V.C., 1983, *J. Quant. Spectrosc. Radiat. Transfer* **29**, 285.
Milkey, R.W. 1976, *Interpretation of Atmospheric Structure in the Presence of Inhomogeneities*, IAU General Assembly 1976, Ed. C.J. Cannon, University of Sydney, p. 55.
Omont, A., Smith, E.W., Cooper, J., 1972, *Astrophys. J.* **175**, 185.
Rybicki, G., 1984, *Methods in Radiative Transfer*, Ed. W. Kalkofen, Cambridge University Press (U.K.), p. 21.
Rybicki, G., Hummer, D.G., 1969, *Mon. Not. Roy. Astron. Soc.* **144**, 313.
Woolley, R., Stibbs, D., 1953, *The Outer Layer of a Star*, Clarendon Press, Oxford.

SOME COMMENTS UPON THE LINE EMISSION PROFILE ψ_ν

R. Freire Ferrero

Observatoire de Strasbourg
11, Rue de l'Université
67.000 - Strasbourg - FRANCE

ABSTRACT.- The mathematical expression of the emission profile ψ_ν is given in the case of a two-level atom plus continuum: the result is that ψ_ν is independent of populations and abundances, depending only on T_e, N_e and J_ν. In this way, the problem to resolve by iteration, the coupled statistical equilibrium and transfer equations (Milkey and Mihalas; Heasley and Kneer), transforms in the resolution by iteration, of the transfer equations for each frequency considered, including the mathematical expression of ψ_ν.

1. INTRODUCTION.-

The problem to numerically solve the coupled system of transfer and statistical equilibrium equations including partial redistribution (PR) was treated in a pioneer paper by Milkey and Mihalas (1973) and quoted later by Mihalas (1978).

Two important astrophysical consequences had such a work, both on profile line computations and on eventual changes in the structure of the stellar atmosphere models (mainly for the case of hydrogen lines).

Another important subproduct of this work (that develops the complete linearization technique for a 2 or 3 level-atom plus continuum) was the computation, by iteration, of the emission line profile ψ_ν by the fact of its inclusion in the abovementionned coupled system of equations (Mihalas, 1978, p. 435). This could be done by dividing the upper level in many sublevels each of them with $n_2(\nu) = n_2 \psi_\nu$ upper state numbers. In this way the ψ_ν values follow immediately.

In this work, we give the mathematical expression of ψ_ν as a function of electronic temperature T_e, frequency ν and mean intensity J_ν (and having atomic transition rates as parameters) and a simpler iteration method, than those proposed early (Milkey and Mihalas, 1973 ; Heasley and Kneer, 1976), to solve the coupled system of statistical equilibrium equations and the transfer equation in the case of partial redistribution.

J. E. Beckman and L. Crivellari (eds.), Progress in Stellar Spectral Line Formation Theory, 101–108.

2. THE MATHEMATICAL EXPRESSION OF THE EMISSION PROFILE ψ_ν .-

Let us consider a two-level atom plus continuum. The rate equation giving the upper state occupation number for atoms that emit at frequency ν in the observer's frame is :(Baschek, Mihalas, Oxenius, 1981):

$$n_2 \psi_\nu \left(A_{21} + B_{21} \int \psi_{\nu'} J_{\nu'} d\nu' + C_{21} + R_{2K} + C_{2K}\right) =$$

$$= n_1 \left(B_{12} \int R(\nu',\nu) J_{\nu'} d\nu' + C_{12} \phi_\nu\right) + n_2^* \phi_\nu \left(R_{K2} + C_{2K}\right) \qquad (1')$$

where we have written $\psi_\nu^* = \phi_\nu$ in the right hand side.Whith a simplified notation, we can write :

$$- a_{12\nu}\, n_1 + a_{22} \psi_\nu\, n_2 = \phi_\nu B_2 \qquad (1)$$

We obtain immediately

$$\psi_\nu = \frac{\frac{n_1}{n_2}\left(B_{12} \int R(\nu',\nu) J_{\nu'} d\nu' + C_{12} \phi_\nu\right) + \frac{n_2^*}{n_2} \phi_\nu \left(R_{K2} + C_{2K}\right)}{A_{21} + B_{21} \int \psi_{\nu'} J_{\nu'} d\nu' + C_{21} + R_{2K} + C_{2K}} \qquad (2')$$

or,

$$\psi_\nu = \frac{n_1 a_{12\nu} + \phi_\nu B_2}{n_2 a_{22}} \qquad (2)$$

This formula gives an implicit definition of the emission profile ψ_ν :

$$\psi_\nu = f\left(\frac{n_1}{n_2}, \frac{n_2^*}{n_2}, \psi_{\nu'}, J_{\nu'}\right)$$

When stimulated emission is negligible, we can ignore this term in the denominator or we can replace $\psi_{\nu'}$ by $\phi_{\nu'}$.

We now proceed to the elimination of the population ratios in (2) . From the statistical equilibrium equations for the two atomic sharp levels,

$$n_1 \left(B_{12} \int \phi_{\nu'} J_{\nu'} d\nu' + C_{12} + R_{1K} + C_{1K}\right) - n_2 \left(A_{21} + B_{21} \int \psi_{\nu'} J_{\nu'} d\nu' + C_{21}\right) =$$

$$= n_1^* \left(R_{K1} + C_{1K}\right) \qquad (3)$$

$$-n_1 \left(B_{12} \int \phi_{\nu'} J_{\nu'} d\nu' + C_{12}\right) + n_2 \left(A_{21} + B_{21} \int \psi_{\nu'} J_{\nu'} d\nu' + C_{21} + R_{2K} + C_{2K}\right) =$$

$$= n_2^* \left(R_{K2} + C_{2K}\right) \qquad (4)$$

and the total number equation : $n_1 + n_2 + n_K = n_{el}$ (5)
we can obtain the desired ratios.

Equations (3) and (4) can be written as :

$$a_{11}\, n_1 - a_{21}\, n_2 = B_1 \qquad (6)$$

$$- a_{12}\, n_1 + a_{22}\, n_2 = B_2 \qquad (7)$$

which have the solutions :

$$n_1 = \frac{\begin{vmatrix} B_1 & -a_{21} \\ B_2 & a_{22} \end{vmatrix}}{\begin{vmatrix} a_{11} & -a_{21} \\ -a_{12} & a_{22} \end{vmatrix}} \qquad n_2 = \frac{\begin{vmatrix} a_{11} & B_1 \\ -a_{12} & B_2 \end{vmatrix}}{\begin{vmatrix} a_{11} & -a_{21} \\ -a_{12} & a_{22} \end{vmatrix}} \tag{8}$$

If we replace n_1 on (2), putting D as the determinant of the coefficients, we have :

$$\psi_\nu = \frac{(B_1 a_{22} + B_2 a_{21})\, a_{12\nu} + B_2 \phi_\nu (a_{11} a_{22} - a_{12} a_{21})}{n_2\, a_{22}\, D}$$

The terms : $B_2 a_{21} (a_{12\nu} - a_{12} \phi_\nu)$ become

$$= B_2 a_{21} B_{12} \left(\int R(\nu',\nu) J_{\nu'} d\nu' - \int \phi_\nu \phi_{\nu'} J_{\nu'} d\nu' \right) = B_2 a_{21} B_{12} \int \left(R(\nu',\nu) - \phi_\nu \phi_{\nu'} \right) J_{\nu'} d\nu'$$

so

$$\psi_\nu = \frac{(B_2 a_{11} + B_1 a_{12\nu} \phi_\nu^{-1})\, a_{22} + B_2 a_{21} G_\nu}{n_2\, D\, a_{22}} \tag{9}$$

with

$$G_\nu = B_{12} \phi_\nu^{-1} \int \left[R(\nu',\nu) - \phi_\nu \phi_{\nu'} \right] J_{\nu'} d\nu' \tag{10}$$

Again, (9) becomes

$$\frac{\psi_\nu}{\phi_\nu} = \frac{\begin{vmatrix} a_{11} & B_1 \\ -a_{12\nu} \phi_\nu^{-1} & B_2 \end{vmatrix} + B_2 \dfrac{a_{21}}{a_{22}} G_\nu}{\begin{vmatrix} a_{11} & B_1 \\ -a_{12} & B_2 \end{vmatrix}} \tag{11}$$

The expressions of B_1 and B_2 are given by :

$$\begin{aligned} B_1 &= n_1^* (R_{K1} + C_{1K}) \\ B_2 &= n_2^* (R_{K2} + C_{2K}) \end{aligned} \tag{12}$$

where the LTE populations n_1^* , n_2^* are :

$$\begin{aligned} n_1^* &= n_e\, n_K\, \Phi_1(T) \\ n_2^* &= n_e\, n_K\, \Phi_2(T) \end{aligned} \tag{13}$$

where $\Phi(T)$ is the appropriate Boltzmann-Saha factor.
In this way,

$$\frac{B_1}{B_2} = \frac{\Phi_1(T)\ (R_{K1} + C_{1K})}{\Phi_2(T)\ (R_{K2} + C_{2K})} = d \tag{14}$$

Finally, dividing by B_2 we have :

$$\frac{\psi_\nu}{\phi_\nu} = \frac{\begin{vmatrix} a_{11} & d \\ -a_{12\nu} \phi_\nu^{-1} & 1 \end{vmatrix} + \dfrac{a_{21}}{a_{22}} G_\nu}{\begin{vmatrix} a_{11} & d \\ -a_{12} & 1 \end{vmatrix}} \tag{15}$$

Equation (15) gives the expression of the emission profile ψ_ν in the case of a two-level atom with a continuum : G_ν and $d(T_e)$ are given respectively by (10) and (14) and the a_{ij} parameters are :

$$a_{11} = B_{12}\int \phi_{\nu'} J_{\nu'} d\nu' + C_{12} + R_{1K} + C_{1K} \qquad (16)$$

$$a_{12} = B_{12}\int \phi_{\nu'} J_{\nu'} d\nu' + C_{12} \qquad (17)$$

$$a_{12\nu} = B_{12}\int R(\nu',\nu)\, J_{\nu'}\, d\nu' + C_{12}\,\phi_\nu \qquad (18)$$

$$a_{21} = A_{21} + B_{21}\int \psi_{\nu'} J_{\nu'} d\nu' + C_{21} \qquad (19)$$

$$a_{22} = A_{21} + B_{21}\int \psi_{\nu'} J_{\nu'} d\nu' + C_{21} + R_{2K} + C_{2K} \qquad (20)$$

If we would to obtain the exact values of the function ψ_ν ,we can compute it by iteration, giving for example an initial value $\psi_\nu^{(o)} = \phi_\nu$ for the induced emission profile, and using the appropriate J_ν values obtained from the equation of transfer (see section 3.).

We remark that :

(1) the expression (15) is independent of the occupation numbers (n_1,n_2, n_k, n_1 ,n_2) or the abundance (n_{el}).

(2) ψ_ν is a function of temperature T_e,electron density N_e and mean intensity of radiation J_ν and also depends on ν and on the transition rates involved (A_{21},B_{21},C_{12},C_{21},etc.) : a general result that ww was obtained early by Oxenius (1965) and Hubeny (1981).

(3) ψ_ν is a normalized function,as can be easily verified: $\int_0^\infty \psi_\nu d\nu = 1.$ (knowing that $\int R(\nu',\nu)\, d\nu' = \phi_\nu$).

(4) In the case of complete redistribution (CR), $R(\nu',\nu) = \phi_{\nu'}\phi_\nu$, $a_{12\nu} = a_{12}\,\phi_\nu$ and $G_\nu = 0$: from (15) $\psi_\nu = \phi_\nu$.

(5) In the general case,given $R(\nu',\nu)$ we can calculate ψ_ν .

(6) In the case of a two-level atom without continuum $R_{1k} = R_{2k} = 0$, $C_{1k} = C_{2k} = 0$ and neglecting collisional processes , $C_{12} = C_{21} = 0$, we have,from (16) to (20) :

$$a_{11} = a_{12} = B_{12}\int \phi_{\nu'} J_{\nu'}\, d\nu'$$

$$a_{21} = a_{22} = A_{21} + B_{21}\int \phi_{\nu'} J_{\nu'}\, d\nu'$$

$$a_{12\nu} = B_{12}\int R(\nu',\nu)\, J_{\nu'}\, d\nu'$$

and

$$\psi_\nu = \frac{\begin{vmatrix} a_{11}\,\phi_\nu & d \\ -a_{12\nu} & 1 \end{vmatrix} + \phi_\nu G_\nu}{a_{11}\begin{vmatrix} 1 & d \\ -1 & 1 \end{vmatrix}} = \frac{a_{11}\,\phi_\nu + a_{12\nu}\, d + \phi_\nu G_\nu}{a_{11}\,(1+d)} =$$

$$= \frac{(1+d)\int R(\nu',\nu)\, J_{\nu'} d\nu'}{(1+d)\int \phi_{\nu'} J_{\nu'} d\nu'} \quad \text{Thus :} \quad \psi_\nu = \frac{\int R(\nu',\nu)\, J_{\nu'} d\nu'}{\int \phi_{\nu'} J_{\nu'} d\nu'} \tag{21}$$

In this particular case, we obtain equation (2-13) quoted by Mihalas (1978) . In conclusion,we must be very cautious in using simplified expressions of ψ_ν that involve particular hypothesis (pure scattering).

3. THE SOLUTION OF THE COUPLED SYSTEM OF STATISTICAL EQUILIBRIUM EQUATIONS AND THE TRANSFER EQUATION IN THE PR CASE.-

To compute ψ_ν we need J_ν which is given by the transfer equation. Given the emission coefficient,

$$\eta_\nu = \frac{2h\nu^3}{c^2} \alpha_{12} \frac{g_1}{g_2} n_2 \psi_\nu + E_\nu^{cont} \tag{22}$$

and the absorption coefficient,

$$\chi_\nu = \alpha_{12}\left(n_1 \phi_\nu - \frac{g_1}{g_2} n_2 \psi_\nu\right) + \chi_\nu^{cont} \tag{23}$$

the line source function becomes,

$$S_{\nu\ell} = \frac{\eta_{\nu\ell}}{\chi_{\nu\ell}} = \frac{2h\nu^3}{c^2} \frac{g_1}{g_2} \frac{n_2 \psi_\nu}{n_1\phi_\nu - \frac{g_1}{g_2} n_2 \psi_\nu} = \frac{2h\nu^3}{c^2}\left(\frac{g_2 n_1 \phi_\nu}{g_1 n_2 \psi_\nu} - 1\right)^{-1} \tag{24}$$

where

$$\frac{n_1}{n_2} = \frac{\begin{vmatrix} B_1 & -a_{21} \\ B_2 & a_{22} \end{vmatrix}}{\begin{vmatrix} a_{11} & B_1 \\ -a_{12} & B_2 \end{vmatrix}} = \frac{\begin{vmatrix} d & -a_{21} \\ 1 & a_{22} \end{vmatrix}}{\begin{vmatrix} a_{11} & d \\ -a_{12} & 1 \end{vmatrix}} \tag{25}$$

and finally :

$$S_{\nu\ell} = \frac{2h\nu^3}{c^2}\left[\frac{g_2}{g_1} \frac{\begin{vmatrix} d & -a_{21} \\ 1 & a_{22} \end{vmatrix}}{\begin{vmatrix} a_{11} & d \\ -a_{12\nu}\phi_\nu^{-1} & 1 \end{vmatrix} + \frac{a_{21}}{a_{22}} G_\nu} - 1\right]^{-1} \tag{26}$$

To obtain J_ν we solve the transfer equation

$$\mu \frac{dI_\nu}{d\tau_\nu} = I_\nu - S_\nu \tag{27}$$

with
$$d\tau_\nu = -\chi_\nu \, dz \tag{28}$$
using the Feautrier's technique.

The iteration should proceed as follows : suppose that we know an estimation of the J_ν .These initial values $J_\nu^{(0)}$ allow us to compute $\psi_\nu^{(0)}$ (equation (15), assuming CR in the a_{ij} only for these initial values) and n_1, n_2 (from (8)) after what we could obtain η_ν , χ_ν and S_ν , τ_ν (equations (26) and (28).

Solving the transfer equation with the Feautrier's technique, gives us $J_\nu^{(1)}$ .Now the iteration proceeds : the $\Psi_\nu^{(0)}$ values enter in the stimulated emission terms and we can compute $\Psi_\nu^{(1,0)}$ .We can iterate over (15) introducing these values into a_{21} and a_{22}, to arrive at a final $\Psi_\nu^{(1)}$ or, we can continue directly, computing n_1, n_2 and solving the transfer equation for another $J_\nu^{(2)}$.

4. SOME COMMENTS ABOUT THE MILKEY AND MIHALAS ITERATION METHOD.-

We can now compare our proposed iteration method with that of Milkey and Mihalas (1973). The M - M method starts with an estimate of the occupation numbers for all depths: $n_1^{(0)}, n_{2i}^{(0)}, n_K^{(0)}$;then $\chi_\nu^{(0)}, \eta_\nu^{(0)}$ and $J_\nu^{(0)}$ may be computed. Now they compute the corrections $\delta\chi_j, \delta\eta_j$ and δJ_j to be applied, from the linearization of both transfer equation (in terms of $\delta J_j, \delta\chi_j, \delta\eta_j$ expressed in fonction of corrections for the populations $\delta n_1, \delta n_{2i}, \delta n_K$) and the statistical equilibrium equations (in terms of $\delta n_1, \delta n_{2i}, \delta n_K$ expressed in fonction of δJ_j).

The hole system is of the Feautrier form :

$$-A_d\, \delta J_{d-1} + B_d\, \delta J_d - C_d\, \delta J_{d+1} = L_d$$

The solution of the Feautrier system of equations give them δJ_j and $J_\nu^{(1)} = J_\nu^{(0)} + \delta J_\nu^{(0)}$. Now, using $J_\nu^{(1)}$ the solution of the statistical equilibrium equations for the lower level 1 and the upper sublevels 2 i jointly with the total number equation , give $n_1^{(1)}, n_{2i}^{(1)}, n_K^{(1)}$.With these values, χ_ν and η_ν can be computed and from a formal solution of the transfer equation, new Eddington factors f_ν can be obtained. The iteration proceeds over the Feautrier system of equations (coupled transfer and statistical equilibrium equations solved for the δJ_j) and the system of statistical equilibrium and total number equations.

Basically, our proposed method (section 3.) exempts the numerical resolution of the statistical equilibrium equations, because :
(a) the solutions of such a system of equations are known a priori ;
(b) we have obtained the mathematical expression of Ψ_ν (equation (15)).

To demonstrate the assertion (a), we write the system of statistical equilibrium equations (lower sharp level and upper level divided into N sublevels) :

$$\begin{array}{lllll}
a_{11}\, n_1 & - \omega_1 a_{21}\, n_{2,1} & - \omega_2 a_{21}\, n_{2,2} - \cdots & - \omega_N a_{21}\, n_{2,N} & = B_1 \\
- a_{12,1}\, n_1 & + a_{22}\, n_{2,1} & & & = B_2 \phi_1 \\
- a_{12,2}\, n_1 & & + a_{22}\, n_{2,2} & \Phi & = B_2 \phi_2 \\
\vdots & \Phi & \ddots & & \vdots \\
- a_{12,N}\, n_1 & & & + a_{22}\, n_{2,N} & = B_2 \phi_N
\end{array}$$

which are linearized (i =1,...,N) from:

$$a_{11}\, n_1 - a_{21}\int n_2(\nu)\, d\nu = B_1$$
$$-a_{12\nu}\, n_1 + a_{22}\, n_2(\nu) = B_2\, \phi_\nu$$

The solution of such a system of linear equations is given by the quotient of two determinants: for n_1 ,

$$\begin{vmatrix} B_1 & -\omega_1 a_{21} & \cdots & -\omega_i a_{21} & \cdots \\ B_2\phi_1 & a_{22} & & \phi & \\ \vdots & & \ddots & & \\ B_2\phi_i & \phi & & a_{22} & \\ \vdots & & & & \ddots \end{vmatrix} = a_{22}^{N-1}\left(a_{22}\, B_1 + \sum_{i=1}^{N} \omega_i a_{21} B_2\, \phi_i\right)$$

over :

$$\begin{vmatrix} a_{11} & -\omega_1 a_{21} & \cdots & -\omega_i a_{21} & \cdots \\ -a_{12,1} & a_{22} & & & \\ \vdots & & \ddots & & \\ -a_{12,i} & & & a_{22} & \\ \vdots & & & & \ddots \end{vmatrix} = a_{22}^{N-1}\left(a_{22} a_{11} + \sum_{i=1}^{N} \omega_i a_{21}(-a_{12,i})\right)$$

Thus :

$$n_1 = \frac{\begin{vmatrix} B_1 & -\sum_{i=1}^{N} \phi_i \omega_i a_{21} \\ B_2 & a_{22} \end{vmatrix}}{\begin{vmatrix} a_{11} & -a_{21} \\ -\sum_{i=1}^{N} a_{12,i}\omega_i & a_{22} \end{vmatrix}} = \frac{\begin{vmatrix} B_1 & -a_{21} \\ B_2 & a_{22} \end{vmatrix}}{\begin{vmatrix} a_{11} & -a_{21} \\ -a_{12} & a_{22} \end{vmatrix}}$$

which is the same (formal and numerical) solution (8) of the system (3)(4) or (6)(7).
The solution for $n_{2,i}$ follows immediately from each rate equation (1)

$$n_{2,i} = n_2 \psi_i = \frac{n_1 a_{12,i} + B_2 \phi_i}{a_{22}}$$

If we multiply by ω_i we obtain

$$\sum_{i=1}^{N} \omega_i n_{2,i} = n_2 \sum_{i=1}^{N} \omega_i \psi_i = n_2 = \frac{n_1 a_{12} + B_2}{a_{22}}$$

and

$$\psi_i = \frac{n_{2,i}}{n_2} = \frac{a_{12,i}\; n_1 + B_2 \phi_i}{a_{12}\; n_1 + B_2} =$$

$$= \frac{a_{12,i}\begin{vmatrix} B_1 & -a_{21} \\ B_2 & a_{22} \end{vmatrix} + B_2\phi_i \begin{vmatrix} a_{11} & -a_{21} \\ -a_{12} & a_{22} \end{vmatrix}}{a_{12}\begin{vmatrix} B_1 & -a_{21} \\ B_2 & a_{22} \end{vmatrix} + B_2 \begin{vmatrix} a_{11} & -a_{21} \\ -a_{12} & a_{22} \end{vmatrix}}$$

and after some algebra, the denominator becomes :

$$a_{22} \begin{vmatrix} a_{11} & B_1 \\ -a_{12} & B_2 \end{vmatrix}$$

and we obtain the same expression (11).

5. SUMMARY AND CONCLUSIONS.-

For the case of a two-level atom plus continuum, we have obtained the mathematical expression of the emission line profile Ψ_ν from the rate equations of the sublevels of the upper atomic level : Ψ_ν is, in reality a function of ν , T_e , N_e , J_ν and varies from transition to transition due to different transition rates values.

A first good approximation, can be obtained if Ψ_ν is replaced by Φ_ν in the stimulated emission terms. But the exact values of Ψ_ν could be obtained by iteration over the mathematical expression of Ψ_ν and by coupling the Ψ_ν formula with the transfer equation. This iteration method resolves the problem of the coupled statistical equilibrium and transfer equations including PR, treated by Milkey and Mihalas (1973), in a more simplest form. There is no need to solve the N + 1 statistical equililibrium equations to obtain $n_{2,i}$ or Ψ_i because :

(a) the Ψ_i solutions could be known a priori as algebric expressions (equation (14)) and,

(b) the solutions (n_1 or n_{2i}, i =1,...,N) of the system of statistical equilibrium equations are also known a priori.

The proposed iteration method should reduce greatly computational time.

6. REFERENCES.-

Baschek B, Mihalas D, Oxenius J, 1981, Astron Astrophys 97,43.

Heasley J N, Kneer F, 1976, Astrophys J 203,660.

Hubeny I, 1981, Bull Astron Inst Czechosl 32,271.

Milkey R W, Mihalas D, 1973, Astrophys J 185,709.

Mihalas D, 1970, Stellar Atmospheres, Freeman, San Francisco, 1st ed.
1978, " " " " 2nd ed.

Oxenius J, 1965, J Quant Spectrosc Radiat Transfer 5,771.
1979, Astron Astrophys 76,312.

A MODIFIED RYBICKI METHOD WITH PARTIAL REDISTRIBUTION

Ivan Hubeny
Astronomical Institute
Czechoslovak Academy of Sciences
251 65 Ondřejov
Czechoslovakia

ABSTRACT. A new approximate numerical method is presented that retains the basic computational advantages of the Rybicki method while still being capable of handling partial redistribution transfer problems. This method avoids calculations of the redistribution functions and, moreover, the computer time scales linearly with the number of frequency points. The crucial point of this method is to consider the frequency which separates the complete redistribution core region from the coherent wing region, to be depth-dependent. The present method yields excellent agreement with exact calculations and gives much better results than any depth-independent version of the partial coherent scattering approximation. Moreover, it may also be applied to moving and/or multidimensional media.

1. INTRODUCTION

The present status of the partial redistribution (PRD) theory, the relevant numerical methods of solving PRD transfer problems, as well as the motivation for developing suitable approximate numerical schemes have been thoroughly discussed by Hubeny (in these Proceedings; hereinafter referred to as H 84), and will not be repeated here. In fact, Section 7.3 of H 84 may be viewed as an introduction to the present topic.

In this paper, we shall briefly outline the basic ideas of the modified Rybicki method, together with the physical significance of the adopted treatment; a more comprehensive discussion and an explicit formulation will be given elsewhere (Hubeny, 1985 - H 85).

2. FORMAL PRESENTATION OF THE METHOD

2.1. Line Source Function in the Partial Coherent Scattering Approximation (PCSA)

Here, we shall outline how one may employ the idea of the PCSA to re-

J. E. Beckman and L. Crivellari (eds.), Progress in Stellar Spectral Line Formation Theory, 109–114.

cast the line source function to a form that enables easy incorporation into the Rybicki-type (Rybicki, 1971) elimination scheme for solving the transfer equation

We assume the line source function to have the form

$$S_\nu = [\bar{J} + \mu(\bar{R}_\nu - \bar{J}) + \eta]/(1 + \varepsilon^*_\nu) \ . \tag{1}$$

In the context of H 84, Sec. 8.1, it is the isotropic, linear source function appropriate to multilevel PRD problems without redistribution interlocking. The parameters η, ε^*_ν represent the sources and sinks of line photons, μ accounts for the destruction of photon correlation. For the purposes of this paper, we shall simply consider these quantities as a priori given parameters. The only radiation-dependent quantities are the frequency-averaged mean intensity, $\bar{J}$, and the redistribution integral, $\bar{R}_\nu$, which are defined as

$$\bar{R}_\nu = \int_0^\infty R(\nu',\nu)\ J_{\nu'}\ d\nu'/\phi_\nu \ ; \ \bar{J} = \int_0^\infty \phi_\nu\ J_\nu\ d\nu \ . \tag{2}$$

J_ν is the mean intensity of radiation, and ϕ_ν is the (natural-excitation) absorption profile; $R(\nu',\nu)$ is the (angle-averaged) redistribution function.

We consider the OSC redistribution function (Omont et al., 1972) written in the partial coherent scattering approximation, viz.

$$R(\nu',\nu) = (1 - \Lambda)\ \phi_{\nu'}[\xi a_{\nu'\nu}\ \delta(\nu' - \nu) + (1 - \xi a_{\nu'\nu})\phi_\nu] + \\ + \Lambda\phi_{\nu'}\phi_\nu \ , \tag{3}$$

where the last term follows from the usual replacement $R_{III}(\nu',\nu) \simeq \phi_{\nu'}\phi_\nu$; Λ is the incoherence fraction and ξ is a parameter that accounts for the differences of redistribution in resonance and subordinate lines ($\xi = 1$ for resonance lines, $\xi < 1$ for subordinate lines - see H 85); δ is the Dirac function.

In order to satisfy the requirements of normalization and symmetry, function $a_{\nu'\nu}$ should be given by (Kneer, 1975)

$$a_{\nu'\nu} = \bar{a}[\max(|x'|,|x|)] \quad \text{for} \quad \nu' \neq \nu, \\ a_{\nu\nu} = \int_0^\infty a_{\nu'\nu}\ \phi_{\nu'}\ d\nu' \ ; \tag{4}$$

where $\bar{a}(x)$ is a somewhat arbitrary function of x , $x = (\nu - \nu_0)/\Delta\nu_D$ being the frequency displacement from the line center measured in units of the Doppler width, $\Delta\nu_D$. In the following, we shall call x simply frequency.

As explained in H 85, it is quite sufficient to take function $\bar{a}(x)$ as a simple step function,

$$\bar{a}(x) = 0 \quad \text{for} \quad x < x_D \ ; \quad \bar{a}(x) = 1 \quad \text{for} \quad x \geq x_D \ , \tag{5}$$

where x_D is the so-called division frequency.

Substituting Eqs (3) - (5) into (1) yields

$$S_\nu = (\bar{J} + \lambda\Phi_0 J_\nu - \lambda\bar{J}_0 + \eta)/(1 + \varepsilon^*_\nu) \quad \text{for} \quad x < x_D \ ;$$
$$S_\nu = [(1 - \lambda)\bar{J} + \lambda J_\nu + \eta]/(1 + \varepsilon^*_\nu) \quad \text{for} \quad x \geq x_D \ ; \tag{6}$$

where λ is a modified incoherence fraction (see H 85), $\lambda = \mu\xi(1 - \Lambda)$, and

$$\bar{J}_0 = 2\int_{x_D}^{\infty} J_x \, \phi_x \, dx \ ; \quad \Phi_0 = 2\int_{x_D}^{\infty} \phi_x \, dx \ . \tag{7}$$

Equation (7) is only valid for the radiation field which is symmetric about the line centre; nevertheless, it may simply be generalized to account for any frequency dependence of the radiation.

2.2. Solution of the Transfer Equation

An inspection of Eq. (6) shows that one may apply the idea of the Rybicki (1971) method. (For a general discussion, refer to Mihalas, 1978). Indeed, a coupling of different frequencies in the source function is contained only in two quantities, $\bar{J}$ and $\bar{J}_0$, similarly to the case of complete redistribution where there is just one function $\bar{J}$. Introducing the discretization of frequencies, x_i $(i = 1, \ldots, NF)$, and depths, τ_d $(d = 1, \ldots, ND)$, and grouping the variables into column vectors $J_i = (J_i^1, J_i^2, \ldots, J_i^{ND})^T$, $\bar{J} = (\bar{J}^1, \bar{J}^2, \ldots, \bar{J}^{ND})^T$, and $\bar{J}_0 = (\bar{J}_0^1, \bar{J}_0^2, \ldots, \bar{J}_0^{ND})^T$, the overall matrix of the transfer problem may be written as

$$\begin{bmatrix} T_1 & & & & U_1 & X_1 \\ & T_2 & & 0 & U_2 & X_2 \\ & & \ddots & & \vdots & \\ & 0 & & \ddots & \vdots & \\ & & & T_{NF} & U_{NF} & X_{NF} \\ V_1 & V_2 & \ldots & V_{NF} & E & 0 \\ Y_1 & Y_2 & \ldots & Y_{NF} & 0 & E \end{bmatrix} \times \begin{bmatrix} J_1 \\ J_2 \\ \vdots \\ \vdots \\ J_{NF} \\ \bar{J} \\ \bar{J}_0 \end{bmatrix} = \begin{bmatrix} L_1 \\ L_2 \\ \vdots \\ \vdots \\ L_{NF} \\ 0 \\ 0 \end{bmatrix} \tag{8}$$

Here, matrices T_i are NF×NF tridiagonal matrices, V_i and Y_i are diagonal matrices, E is the negative unit matrix, and 0 is the zero matrix. Matrices U_i and X_i may be either diagonal (if one uses the usual difference form of the transfer equation, i.e. the original Feautrier scheme), or, preferably, tridiagonal NF×NF matrices (if one applies the Hermitian scheme - Auer 1976). The latter approach yields considerably more accurate results with a negligible increase of computational effort.

System (8) may be solved by the straightforward generalization of the original Rybicki method; the explicit formulas are given in H 85. The asymptotic timing estimate is about twice the time required to solve the original Rybicki method, which is still much more favorable than that for the Feautrier method, particularly if a large number of frequency points is required.

3. Depth-dependent Division Frequency

Let us return to Eq. (5) which represents, from the mathematical point of view, the basis of our approximation. At first sight, it might seem that this form is a poorer approximation than the smoothed form of $\bar{a}(x)$ suggested by Kneer (1975). Indeed, the numerical calculations show (see H 85) that Kneer's form yields better agreement with exact R_{II} than our form (5), if a constant value of x_D (e.g., $x_D = 3$) is used. On the other hand, the idea of the Rybicki elimination can only be employed if the step-function form of $\bar{a}(x)$ is used, but not with the smoothed (Kneer's) form. From this point of view, our approximation might be regarded as a somewhat worse approximation than Kneer's, yet it is much more economical computationally.

However, the true physical basis of our approach lies elsewhere. One should bear in mind two features already pointed out in H 84, namely i) the PCSA is basically incapable of providing an accurate approximation due to the neglect of Doppler diffusion phenomena; and ii) PCSA, although commonly viewed as an approximation of the redistribution function (i.e. R_{II}) is, in fact, an approximation of the redistribution integral.

Particularly the latter point indicates the way to physical improvement of the PCSA. Let us first analyse the phenomenon inherently neglected by a depth-independent PCSA. In cases where the optical depth is sufficiently large, a wing photon may suffer a large number of scatterings; each one shifts its frequency towards the line centre rather than away from it. A wing photon may thus ultimately arrive at the line core where its subsequent scatterings can be well described by complete redistribution. We shall call this process "frequency thermalization". The PCSA thus basically neglects the process of frequency thermalization.

Further, one can define a (frequency-dependent) "frequency-thermalization depth", $\tau_{FT}(x)$ (which should not be confused with the usual thermalization depth), as the lowest optical depth (in units of the mean optical depth in the line) from which a wing photon with frequency x has a substantial chance of becoming frequency-thermalized (H 85).

Consider a given frequency, say x_1, and consider a depth $\tau_1 = \tau_{FT}(x_1)$. At depth τ_1, any photon with frequency less than x_1 will be frequency-thermalized, i.e. such a photon will behave as in the CRD before escaping an atmosphere. This means that it will suffer a certain number of scatterings in the line core, eventually being scattered back to the wing and escaping. In strict CRD, the picture is practically the same. Since the basic physical mechanism that controls the behavior of the source function is an exchange of photons through the line wings, we see that it is quite natural to set the division frequency at depth τ_1 to be x_1. In other words, function $x_D(\tau)$ is simply taken to be the inverse function to $\tau_{FT}(x)$. Since the number of scatterings required for frequency thermalization increases monotonically with increasing x, function $x_D(\tau)$ will also be a monotonically increasing function of τ. Obviously, x_D has to be held constant for $\tau \lessapprox \tau_3$, where τ_3 is the depth at which the monochromatic optical depth at $x \simeq 3$ is unity, i.e. $\tau\phi(x = 3) = 1$.

It remains to calculate the frequency-thermalization depth. This has been done in H 85; the result may be approximated by

$$\tau_{FT}(x) \simeq \pi x^3/4a \ , \text{ i.e. } \ x_D(\tau) \simeq (4a/\pi)^{1/3}\tau^{1/3} \ . \qquad (9)$$

The derivation of Eq. (9) is based on the usual random-walk arguments. Roughly, $\tau_{FT} \sim n^{1/2}\,\bar{\ell}$, where n is the number of scatterings necessary for a shift from frequency x to the line core, $\bar{\ell}$ is the (appropriately averaged) photon mean free path, $\bar{\ell} \sim \pi x^2/a$ (a being the damping parameter), and $n \sim x^2$ (mean shift per scattering $\sim 1/x$; total shift $\sim x$).

So far, we have implicitly assumed that a line photon cannot be destroyed by collisional (or other) processes. However, this Phenomenon cannot affect the basic picture outlined above. It should be stressed that the depth-dependent division frequency describes the frequency diffusion of consecutively scattered photons. Thus, if a photon is destroyed, a new photon with a random frequency (according to the absorption profile) is created, regardless of the mechanism of photon scattering (i.e. CRD or PRD). The only new phenomenon , appearing in the presence of a collisional destruction, is that the actual values of the division frequency become inconsequential for optical depths greater than the (usual) thermalization depth (Frisch, 1980). Similar conclusions are also valid for the destruction by continuum processes, or for finite slabs.

This feature also explains the conclusion drawn empirically, viz. that the depth-independent partial coherent scattering approximation works better for larger values of the inelastic destruction probability, for weaker lines (larger probability of destruction by continuum processes), for subordinate lines (due to an additional inherent incoherence of the scattering), and for finite slabs.

4. CONCLUSIONS

Both methods, the original Rybicki and the present one, employ the redundancy of full frequency-dependent information contained in the line source function to devise the numerical schemes that are much more economical than the basic Feautrier one. While the former method employs the exact frequency-independence of the scattering integral, the latter uses asymptotic properties of the scattering integral, based on a physically realistic model of multiple photon scattering. Our approach also provides a new physical insight into the nature of the partial coherent scattering approximation, considering it to be an approximation of the scattering integral rather than an approximation of the redistribution function.

From the computational point of view, this approach, which may also be used in connection with the ordinary Feautrier method, yields excellent agreement with exact solutions (see Fig. 1; for more examples refer to H 85) and, in particular, gives much better results than any depth-independent version of the partial coherent scattering approximation, including Kneer's form, where the use of the much less economical Feautrier method is obligatory.

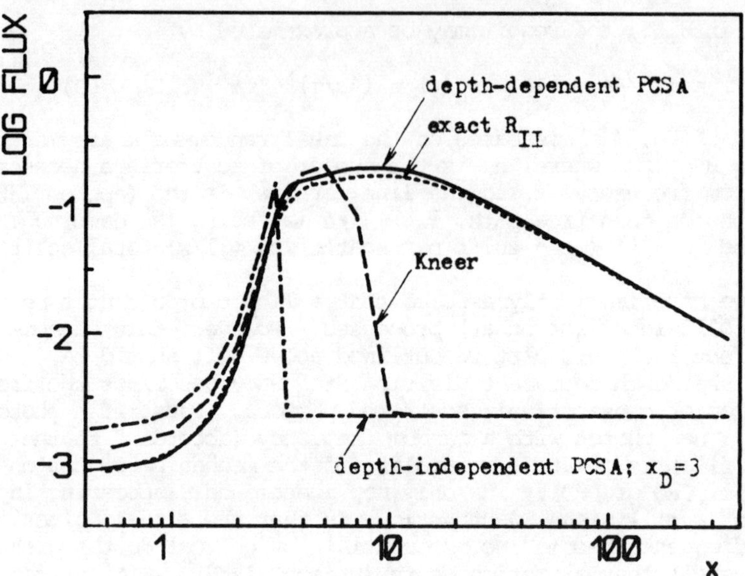

Figure 1. Comparison of the emergent flux from a semi-finite atmosphere with the Voigt parameter $a = 10^{-3}$, destruction probability $\varepsilon = 10^{-6}$, and with no continuum.

REFERENCES

Auer, L.H.: 1976, J. Quant. Spectrosc. Radiat. Transfer 16, 931

Feautrier, P.: 1964, C.R. Acad. Sci. Paris 258, 3189

Frisch, H.: 1980, Astron. Astrophys. 83, 166

Hubeny, I.: 1985, Astron. Astrophys. (submitted) - H 85

Kneer, F.: 1975, Astrophys. J. 200, 367

Mihalas, D.: 1978, Stellar Atmospheres, 2nd Edn., Freeman, San Francisco

Omont, A., Smith, E.W., Cooper, J.: 1972, Astrophys. J. 175, 185

Rybicki, G.B.: 1971, J. Quant. Spectrosc. Radiat. Transfer 11, 589

REDISTRIBUTION FUNCTIONS: A REVIEW OF COMPUTATIONAL METHODS

Petr Heinzel
Astronomical Institute
Czechoslovak Academy of Sciences
251 65 Ondřejov
Czechoslovakia

ABSTRACT. In the present paper, we review and compare the existing computational methods developed to evaluate various types of redistribution functions applicable in astrophysics. We discuss in detail several numerical aspects (codes) of calculating both the angle-dependent and angle-averaged laboratory-frame functions, with particular attention devoted to R_{II}. Finally, we briefly mention the problem of evaluating the scattering integral.

1. INTRODUCTION

The partial redistribution (PRD) problem represents a complicated physical, mathematical and numerical task, and various approaches are now used to solve it (for a review, see Hubený - this volume). One of the most important problems is to obtain suitable forms of astrophysically relevant redistribution functions (RF) which could subsequently be incorporated into the solution of PRD transfer in particular spectral lines. Physically, we have to derive - from quantum electrodynamics - the so-called atomic-frame RF (ARF) for one active atom immersed in a perturber bath. Neglecting possible coupling between the scattering process and the radiator motions, such as velocity-changing collisions, we can transform a given ARF into the observer's frame "simply" by averaging it over the appropriate distribution of atomic velocities. The resulting laboratory-frame RF (LRF) is then used to evaluate the PRD source function.

The present paper is aimed at reviewing and comparing the existing methods developed to evaluate various types of LRF's. We do not intend to discuss here either the applicability of these functions, or their particular properties (shapes).

2. LABORATORY-FRAME FUNCTIONS

Various conceptual aspects of the problem are discussed by Hubený in the present volume. Therefore, we shall primarily concentrate on some math-

J. E. Beckman and L. Crivellari (eds.), Progress in Stellar Spectral Line Formation Theory, 115–124.

ematical and numerical questions.

To obtain the LRF from an a priori known ARF, we first assume that the atomic velocity remains unchanged during the scattering process. The velocity-averaging of ARF $r(\xi',\xi)$ then leads to the corresponding LRF $R(\nu',\vec{n}';\nu,\vec{n})$, formally expressed as

$$R(\nu',\vec{n}';\nu,\vec{n}) = g(\vec{n}',\vec{n}) \int f_1(\vec{v})\, r(\xi',\xi)\, d^3v \, , \tag{1}$$

where $g(\vec{n}',\vec{n})$ is the frequency-independent phase function, ν' and ν denote the absorption and emission frequencies in the laboratory frame, respectively (the atomic-frame frequencies are ξ' and ξ), and $\vec{n}'$ and $\vec{n}$ are the unit vectors in the directions of incoming and outgoing photons, as seen in the laboratory frame (abberation is neglected). The LRF (1) is normalized so that

$$\iiiint R(\nu',\vec{n}';\nu,\vec{n})\, d\nu'\, d\nu\, d\vec{n}'\, d\vec{n} = 1 \, . \tag{2}$$

Velocity-distribution function $f_1(\vec{v})$ for the atoms in an initial (lower) level can generally differ from the Maxwellian distribution $f_M(v)$, particularly in subordinate lines (due to the selective absorption which populates the lower level of a subordinate line). However, as discussed by Heinzel and Hubený (1982) (Paper II), an additional assumption $f_1 \equiv f_M$ seems to be appropriate to many astrophysical situations. Moreover, no method has yet been developed to deal with the non-Maxwellian distributions in Eq. (1). Since we are basically interested in the coherence properties of different RF's (i.e. in transfer effects due to coherent scattering in the line wings), we shall concentrate only on the existing LRF's based on the Maxwellian velocity distribution. Details of the velocity-averaging procedures are discussed by Hummer (1962) and Heinzel (1981) (Paper I). As to microturbulent velocities, one has to modify the Doppler width in the usual manner (retaining LRF's unchanged), if the so-called microturbulent limit is to be considered [for this point, see also the discussion following the paper of Magnan (1976)].

Hummer (1962) has discussed the ordinary LRF's $R_{I,II,III,IV}$, while Heinzel (1981) introduced an additional function R_V which describes the resonance scattering between two radiatively broadened atomic levels (a subordinate line). It has been proved that R_{I-IV} are the limiting cases of R_V. Subsequently, Hubený (1983) (Paper III) generalized these RF's to the case of an arbitrary two-photon process taking place between three atomic levels (e.g., Raman scattering). These generalized LRF's have been denoted as $P_I - P_V$. If the initial and final levels coincide, $P_i \to R_i$. We note that LRF P_{IV} (and $\bar{P}_{IV}$) is physically meaningful (Paper III), whereas its counterpart R_{IV} is not. In Paper I, we retained R_{IV} in the system of mathematical LRF's to avoid any confusion with the correct R_V. On the other hand, we can speak of physical LRF's which appear as various linear combinations of these mathematical functions (see Omont et al., 1972, and Papers II and III).

Hitherto we have discussed the so-called angle-dependent LRF's for which the directional dependence is caused by (i) the angular dependence of the phase function, and (ii) by the correlation between the Doppler shifts of the incoming and outgoing photons, as seen in the laboratory frame. However, as demonstrated in several studies, the frequency dependence of these LRF's dominates over the angular and, therefore, many transfer problems can be simplified considerably by using the so-called angle-averaged LRF's defined as

$$R(\nu',\nu) = \iint R(\nu',\vec{n}';\nu,\vec{n})\, d\vec{n}'\, d\vec{n} = 8\pi^2 \int_0^\pi R(\nu',\nu;\Theta)\sin\Theta\, d\Theta \quad (3)$$

with Θ being the scattering angle.

Finally, let us return to the first assumption made in this section. The effects of velocity-chaging collisions on the astrophysical source function have very recently been studied by Cooper et al. (1982) in terms of the "strong collision model". In this model, the collisions either redistribute the atomic velocity to a Maxwellian, or else leave the velocity unchanged, depending on the relevant branching ratios. The source function is then composed of the LRF's discussed above.

3. COMPUTATIONAL METHODS

3.1. Angle-dependent LRF's

In this section, we shall discuss and compare the most important methods developed to evaluate various types of R_i and P_i. We shall start with the "Fourier transform method" (FTM) which covers all these LRF's, while other methods only deal with one or more of their specific types. The basic principles of FTM are described in Paper I and further details together with applications can be found in the subsequent papers of this series (Papers II - V).

Denoting the two-dimensional Fourier transform of ARF $r(\xi',\xi)$ as $r(t',t)$, we can rewrite Eq. (1) to read (Paper I)

$$R(\nu',\vec{n}';\nu,\vec{n}) = \tfrac{1}{4}\pi^{-2}\, g(\vec{n}',\vec{n}) \iint r(t',t)\, dt'\, dt \,\times$$
$$\times \int f_1(\vec{v}) \exp[i(t'\xi' + t\xi)]\, d^3v \,, \quad (4)$$

where $\xi' = f(\nu',w,\Theta,\vec{v})$ and $\xi = f(\nu,w,\Theta,\vec{v})$ are the transformation relations between the atomic and laboratory-frame frequencies, w is the usual Doppler width. Since $r(t',t)$ does not depend on the velocity [in contrast to $r(\xi',\xi)$!], the averaging procedure takes the simple form

$$\langle \exp[i(t'\xi' + t\xi)]\rangle_{Av} = \int f_1(\vec{v}) \exp[i(t'\xi' + t\xi)]\, d^3v \,. \quad (5)$$

For $f_1 \equiv f_M$, this integral can readily be evaluated analytically (see Paper I). Moreover, even relatively complicated ARF's (from the point of view of averaging) can be transformed into very simple functions in the Fourier domain, as demonstrated in Paper I, and subsequently in Paper III. Therefore, both the velocity-independence and the substantial simplicity of $r(t',t)$ prefer the FTM to the direct application of formula (1). For non-Maxwellian velocity distributions, one can try to expand $f_1(\vec{v})$ into a series of functions which can be transformed analytically into the Fourier domain. By using such an expansion, the problem of averaging the ARF over a non-Maxwellian distribution can, in principle, be resolved in the frame of the FTM formalism.

Starting with ARF $r_V(\xi',\xi)$ and applying FTM, we arrive at LRF R_V (see Paper I)

$$R_V(x',\vec{n}';x,\vec{n}) = \tfrac{1}{4}\pi^{-2}[g(\vec{n}',\vec{n})/\sin\Theta][H(a_j\sec\tfrac{1}{2}\Theta,\tfrac{1}{2}(x+x')\sec\tfrac{1}{2}\Theta)\times$$
$$\times H(a_i\csc\tfrac{1}{2}\Theta,\tfrac{1}{2}(x-x')\csc\tfrac{1}{2}\Theta) + E_V(x',x,\Theta)] \quad (6)$$

where a_i and a_j are the damping parameters (see below), and x and x' are the frequencies expressed in units of the Doppler width. To obtain $r_V(t',t)$, a special form of $r_V(\xi',\xi)$, derived in Paper I, has been applied.

Depending on the actual values of a_l and a_u for the lower and the upper levels, respectively, R_V reduces to other LRF's, $R_{I,II,III}$. The relations between a_i, a_j and a_l, a_u for particular cases are summarized in Tab. 1, where the actual parameters a_l and a_u express the radiative damping, or, in the case of collisional redistribution, the radiative + collisional broadening (see Papers II and III).

TABLE I

I	$a_i = a_l = 0$	$a_j = a_u = 0$
II	$a_i = a_l = 0$	$a_j = a_u \neq 0$
III	$a_i = a_u \neq 0$	$a_j = a_l = 0$
V	$a_i = a_l \neq 0$	$a_j = a_u \neq 0$

Relations between damping parameters

For $R_{I,II}$, we have $E_{I,II} = 0$ and the Voigt functions in the first term of Eq. (6) reduce to Gaussian profiles according to the values of a_l and a_u in Tab. 1. The absorption profiles, corresponding to these RLF's, are generally given by the Voigt function

$$\phi(x') = H(a_l + a_u, x')/\sqrt{\pi} \ . \quad (7)$$

Whereas the first term of Eq. (6) (called the main term) can simply be evaluated by using any suitable numerical method for function $H(a,x)$ [e.g., the algorithm of Matta and Reichel (1971), or Hui et al. (1978)], the second "E-term" has to be evaluated in a special manner as described in detail in Appendix C of Paper I (further refinements are described in Paper IV, Section 2). Although our E-term method seems to be sufficiently accurate, it remains computationally prohibitive, even if a special algorithm is used, which consists in evaluating R_V and R_{III} along paths where $x' + x = \text{const.}$ In its precise version, the time requirements and accuracy have been found comparable with those of the "error-correction" method of Reichel and Vardavas (1975), devised for $R_{III,IV}$

(see below). However, in the transfer calculations with R_V or R_{III}, one can choose the quadrature parameters for the E-term (see Paper I) so that LRF's $R_{III,V}$ will be less precise (the time requirements are then greatly improved), retaining the overall accuracy of the emergent line profiles. In other words, the line transfer with $R_{III,V}$ is only weakly dependent on the choice of the quadrature parameters for the E-term and these can be optimized accordingly by making some transfer tests (also see Paper V).

Due to the relative complexity of the E-term method (i.e. its numerical evaluation), we shall not present further details here; a potential user may obtain a copy of our FORTRAN IV code "UNIRED", together with a brief manual in which some application remarks can be found in detail. We shall only note that UNIRED was developed to evaluate the matrix of RF's $R_{I,II,III,V}(i,j)$, where the indices i and j represent equidistant points with equal spacings in both x' and x. An arbitrary phase function can be incorporated to evaluate the angle-dependent or angle-averaged matrix $R(i,j)$. The symmetry relations valid for these LRF's [see Hummer (1962) and Paper I] have been used in the code.

In Paper III, FTM was applied to derive LRF's P_{I-V}. Here, the function P_V can no longer be expressed in terms of a two-dimensional Fourier transform of an even function (as for R_V), so that the resulting LRF P_V contains the Voigt K-functions (i.e. the imaginary part of the complex probability function). Moreover, the presence of two different Doppler widths also complicates the matter to some extent.

An alternative method for the numerical evaluation of the angle-dependent LRF's $R_{III,IV}$ was developed earlier by Reichel and Vardavas (1975) (their "error-correction" method - ECM) who started with Hummer's forms (1962) of $R_{III,IV}$ formally expressed as

$$R_{III,IV} \approx \int f(u)\, H(a,u)\, du\ . \tag{8}$$

Reichel and Vardavas proved that this integral could be computed, to any degree of accuracy, by using a simple quarature supplemented by a correction term which depends on the mesh size in the quadrature formula. These authors have presented the appropriate formulas for this correction, together with some practical computation hints. To make a comparison with the FTM, we have tested the ECM numerically with the following results: the time requirements and overall accuracy of both methods are comparable; however, the ECM is much more sensitive to a particular choice of quadrature parameters - an inappropriate choice may lead to significant instabilities of the method. The original method is not able to generate R_V or P_V, but as shown in Paper III (and recently tested numerically by Heinzel and Hubený), the ECM is also suitable, after simple improvements, for evaluating LRF's P_{III}, P_{IV} and $\bar{P}_{IV}$. In principle, the ECM could also be generalized to calculate R_V or P_V, but the resulting expressions would be extremely complicated and danger of numerical instabilities would exist. Nevertheless, the ECM can be used to an advantage for other LRF's as discussed above.

Functions R_{III} and R_V were also studied by McKenna (1980) who inserted the original form of ARF $r_V(\xi',\xi)$ (Omont et al., 1972) directly into Eq. (1) and performed a numerical quadrature to obtain the LRF's (for R_{III} he used Hummer's formula; function R_V was denoted as R_{IV}). McKenna's appraoch does not seem to be very suitable for fast transfer calculations and, moreover, it is not sufficiently precise; we have found discrepancies in important peaks of his R_V reaching a factor of 1.5. In fact, we developed our FTM just to avoid tedious calculations resulting from the direct application of Eq. (1) to the original r_V of Omont et al. (some aspects of McKenna's treatment are also discussed in Paper II).

For completeness, we shall also mention some other attempts aimed at evaluating LRF's, in particular R_{III}. Firstly, Lee (1977) has made Monte Carlo simulations of angle-dependent LRF's $R_{I,II,III}$ and presented their plots. The Monte Carlo technique was subsequently applied to the transfer calculations with PRD in resonance lines (Lee and Meier, 1980; Meier, 1981). Since the Monte Carlo treatment is quite different from those commonly employed in stellar atmospheric calculations, we shall not discuss it in more detail, although it can prove to be a powerful tool in some specific problems. Secondly, some authors obtained various series expansions for the angle-dependent LRF's (Magnan, 1975; Haruthyunian, 1979), namely for R_{III} (also see the next subsection).

All the methods discussed so far are based either on Hummer's (1962) formulas, or on Eq. (6) and its limiting cases (for P_V see Paper III). However, none of these relations can be applied directly if the scattering angles $\Theta = 0$ and $\Theta = \pi$. For these "critical" angles, one has to derive other formulas which, as shown by Heinzel (unpublished), can be obtained in an analytical form. As the respective expressions indicate, R_{III} and R_V tend to be sharply peaked at $\Theta = 0$, or $\Theta = \pi$, but they do not converge to a delta function as stated by Reichel and Vardavas (1975). In astrophysical situations, these angles may be of some importance when dealing with angle-dependent PRD transfer.

3.2. Angle-averaged LRF's

As pointed out in the previous section, the PRD transfer problem is much more easily tractable if the angle-averaged LRF's, defined by Eq. (3), are used. However, these LRF's are usually more complicated than their angle-averaged counterparts; as a typical example, we can probably mention the most important function R_{II}. Again, various methods devised to evaluate the angle-averaged RF's have appeared in the literature; we shall review some of them.

The simplest way of evaluating $R_i(\nu',\nu)$ is to apply the second line of definition equation (3) directly. Nevertheless, this possibility was not fully explored until recently, although some calculations were made by Reichel and Vardavas (1975) who used their ECM to generate angle-dependent LRF's R_{III} and R_{IV}. In our recent Paper IV, we have tested in detail the possibility of applying the common Gaussian quadrature to the integration over the scattering angles in Eq. (3). It has been found that, although LRF's R_i have sharply peaked maxima at $\nu' = \nu$ when $\Theta \to 0$, or $\Theta \to \pi$, their product with the term $\sin\Theta$ yields a sufficiently smooth integrand in the interval $<0,\pi>$. Therefore, the Gaussian quadrature method is stable and accurate, leading to a simple summation of angle-dependent LRF's evaluated at the appropriate abscissas and multiplied by the corresponding weights. Even the 8-point quadrature yields satisfactory results applicable in transfer calculations. The accuracy of this approach was discussed in Paper IV, and its application to the PRD transfer is the subject of Paper V. In dealing with generalized RF's P_i, one has to divide the interval $<0,\pi>$ into two sub-intervals $<0,\varepsilon>$ and $<\varepsilon,\pi>$, and apply the Gaussian quadrature to each of them separately. The parameter ε is either equal to $\arccos(w_1/w_2)$ for $w_1/w_2 < 1$, or to $\arccos(w_2/w_1)$ for $w_1/w_2 > 1$ (w_1 and w_2 being the respective Doppler widths of the two transitions involved).

Another possibility of evaluating $R(\nu',\nu)$ is to start with the special formulas derived using Hummer's (1962) treatment. Hummer (1962) himself presented the angle-averaged forms of $R_{I,II,III,IV}$ and derived a series expansion for R_{III}. Similar expansions were also obtained by Rees and Reichel (1968) for all $R_{I,II,III}$. However, as stated by the above authors, their series expansions converge only slowly in the line core region and, therefore, should rather be used as <u>approximate asymptotic forms</u> in the line wings. For example, the first term of the R_{II}-series of Rees and Reichel (1968) (which is exactly Unno's (1952) asymptotic expression) is very suitable as an approximation of R_{II}, as we have found by direct comparison with the exact R_{II} in the region $|x'|,|x| \gtrsim 10$. Using Hummer's procedure, Hubený (Paper III) derived exact expressions for the angle-averaged LRF's $P_I - P_{IV}$ which represent a simple generalization of Hummer's functions mentioned above. His function P_{II} is identical with the so-called "cross-redistribution" function of Milkey et al. (1975).

3.3. Angle-averaged Function R_{II}

We shall now discuss two approaches dealing with the very important function R_{II} in deatil. The first method was developed by Adams et al. (1971) starting with the formula first derived by Unno (1952). The authors have expressed $R_{II}(x',x)$ with an isotropic phase function as

$$R_{II}(x',x) = (a/2\pi) \int_{\frac{1}{2}|x-x'|}^{\infty} \operatorname{erfc}(t)\, dt \,\{[a^2 + (t + \underline{x})^2]^{-1} + [a^2 + (t - \overline{x})^2]^{-1}\}, \quad (9)$$

where $\overline{x}$ ($\underline{x}$) is the larger (smaller) of x and x', and a is the damping parameter. To evaluate this integral numerically, a piecewise polynomial approximation to the erfc, obtained by the Chebyshew fitting procedure, is used. However, the resulting formulas are rather complicated to code. Although Adams has written a convenient code which is available to some users, for others it would generally be useful to have the possibility of writing their own code. However, here one meets a serious difficulty: the coefficients of the Chebyshew fit to the erfc, summarized in Table 1 of Adams et al., have an insufficient number of significant figures (only four) so that even the erfc itself cannot be calculated using them. On the other hand, Adams' code uses

14-digit coefficients which are quite sufficient to approximate the erfc and, subsequently, also R_{II}.

As an alternative method to evaluating $R_{II}(x',x)$ with an arbitrary phase function, one can use the Gaussian quadrature method, discussed above, which is very simple to code. Moreover, to improve the computer-time requirements significantly, one can employ, to an advantage, the fact that the angle-dependent R_{II} depends on $x + x'$ and $x - x'$ [see also Eq. (6)]. Therefore, for each scattering angle in the Gaussian quadrature, we firstly calculate and store the Voigt functions for all possible combinations of $x + x'$ and the Gaussian functions for all $x - x'$, using equidistant steps in both x and x'. Subsequently, $R_{II}(\theta)$ is evaluated as a simple product of these pretabulated values. This procedure avoids unnecessary multiple evaluation of the Voigt and Gaussian functions along the path where $x + x'$ and $x - x'$ are constant. As a result, we obtain an equdistant matrix R_{II} in which we interpolate (the steps in x and x' may be different, but their ratio must be an integer). The symmetry relations valid for R_{II} can also be used to an advantage. An important feature of this Gaussian quadrature technique is that the function R_{II}, calculated in this way, is not too sensitive to the choice of the number of quadrature points in the coherence peaks. Therefore, for transfer calculations, even the 8-point quadrature is sufficient and the 16-point one is as accurate as Adams' code, but has a better time efficiency. In the far wings, one can use Unno's approximate form mentioned above.

4. CONCLUSIONS

The evaluation of LRF's, discussed in this review, represents one basic step in solving the PRD line formation problem. Having calculated the appropriate LRF's, we proceed further by evaluating the corresponding scattering (or redistribution) integral which appears as a crucial term in the PRD source function. The angle-dependent scattering integral has been treated, for example, by Milkey et al. (1975), while the angle-averaged case has been studied by Adams et al. (1971) who used the cubic-spline representation of the radiation field. In fact, Adams has written the appropriate code used by several authors (in Paper V, we have applied his code to R_{II} and, after small improvements, also to R_V). In contrast to these more or less exact treatments, there exist some approximate forms of the scattering integral. Their development has been stimulated by the fact that the direct numerical evaluation of all important LRF's is still rather costly, especially for depth-dependent parameters. The well-known approximation is that originally propsed by Jefferies and White (1960) for R_{II} [also see Kneer (1975)]. A generalization of this approach to R_V was the subject of Papers IV and V [for a review of these methods, refer to Hubený (1984)]. The angle-averaged R_{III} is then commonly approximated by simple complete redistribution, which seems to be numerically (Finn, 1967), as well as physically (see Cooper et al., 1982) well founded.

The redistribution functions discussed here have relatively wide applicability not only to classical PRD problems, but also in treating polarized light (see Ballagh and Cooper, 1977; Paper II; Rees and Saliba, 1982), and they can also be used to describe collisional redistribution in a non-impact line-wing region (see, e.g., Roussel-Dupré, 1983). Therefore, further refinements of the numerical methods described in this paper are highly desirable, as well as studying the more sophisticated LRF's such as those based on non-Maxwellian velocity distributions. Moreover, we are interested in the further development of various approximate methods which can simplify considerably the solution of more complex PRD problems involving velocity fields, multidimensional geometry, etc.

REFERENCES

Adams, T.F., Hummer, D.G., Rybicki, G.B.: 1971, J. Quant. Spectrosc. Radiat. Transfer 11, 1365
Ballagh, R.J., Cooper, J.: 1977, Astrophys. J. 213, 479
Cooper, J., Ballagh, R.J., Burnett, K., Hummer, D.G.: 1982, Astrophys. J. 260, 299
Finn, G.: 1967, Astrophys. J. 147, 1085
Haruthyunian, H.A.: 1979, Byurakan Obs. Rep. LII, 137
Heinzel, P.: 1981, J. Quant. Spectrosc. Radiat. Transfer 25, 483 (Paper I)
Heinzel, P., Hubený, I.: 1982, J. Quant. Spectrosc. Radiat. Transfer 27, 1 (Paper II)
Heinzel, P., Hubený, I.: 1983, J. Quant. Spectrosc. Radiat. Transfer 30, 77 (Paper IV)
Hubený, I.: 1982, J. Quant. Spectrosc. Radiat. Transfer 27, 593 (Paper III)
Hubený, I., Heinzel, P.: 1984, J. Quant. Spectrosc. Radiat. Transfer 32, (Paper V)
Hubený, I.: 1984, Astron. Astrophys. (submitted)
Hui, A.K., Armstrong, B.H., Wray, A.A.: 1978, J. Quant. Spectrosc. Radiat. Transfer 19, 509
Hummer, D.G.: 1962, Mon. Not. R. A. S. 125, 21
Jefferies, J., White, O.: 1960, Astrophys. J. 132, 767
Kneer, F.: 1975, Astrophys. J. 200, 367
Lee, J.-S.: 1977, Astrophys. J. 218, 857
Lee, J.-S., Meier, R.R.: 1980, Astrophys. J. 240, 185
Magnan, Ch.: 1975, J. Quant. Spectrosc. Radiat. Transfer 15, 979
Magnan, Ch.: 1976, in Physique des mouvements dans les atmosphères stellaires (Eds Cayrel, R. and Steinberg, M.), CNRS, 179
Matta, F., Reichel, A.: 1971, Math. Comput. 25, 339
McKenna, S.J.: 1980, Astrophys. J. 242, 283
Meier, R.R.: 1981, J. Quant. Spectrosc. Radiat. Transfer 25, 137
Milkey, R.W., Shine, R.A., Mihalas, D.: 1975, Astrophys. J. 199, 718
Milkey, R.W., Shine, R.A., Mihalas, D.: 1975, Astrophys. J. 202, 250
Omont, A., Smith, E.W., Cooper, J.: 1972, Astrophys. J. 175, 185
Rees, D. Reichel, A.: 1968, J. Quant. Spectrosc. Radiat. Transfer 8, 1795
Rees, D., Saliba, G.J.: 1982, Astron. Astrophys. 115, 1
Reichel, A., Vardavas, I.M.: 1975, J. Quant. Spectrosc. Radiat. Transfer 15, 929

Roussel-Dupré, D.: 1983, Astrophys. J. 272, 723
Unno, W.: 1952, Publ. Astron. Soc. Japan 4, 100

THE EFFECT OF ABUNDANCE VALUES ON PARTIAL REDISTRIBUTION LINE COMPUTATIONS

R.Freire Ferrero
Observatoire de Strasbourg
11, Rue de l'Université
67.000 - Strasbourg
F R A N C E

P. Gouttebroze
Laboratoire de Physique
Stellaire et Planétaire
BP N° 10 - F - 91.370
Verrières-le-Buisson- FRANCE

ABSTRACT.- Chromospheric diagnostic needs the analysis of the line cores of resonance or of strong lines, belonging to atomic species that have high abundance values, and that are collision-dominated: the so-called chromospheric indicators. But in the external stellar atmospheric layers the density may be sufficiently low so that natural broadening dominates collisional broadening. In this way, PR could be necessary to interpret properly high resolution observations of line cores and near wings. Nevertheless, the line formation will be conditionned by the abundance value of the corresponding atomic specie (via the optical depth) i.e., the abundance can determine whether the CR assumption is valid or not. In this work, we discuss this influence as well as the practical validity of CR profiles,from some examples corresponding to A dwarf stars.

1. INTRODUCTION.-

The interpretation of stellar spectra is made by comparison of observed and theoretical computed fluxes for both continuum and line profiles, using stellar atmospheric models and appropriate functions for the continuum and line source functions.

When the interest is centered on the outer stellar layers,i.e.,if we want to check the existence of chromospheres or transition regions, the most convenient method consists in analyzing the line cores of the so-called chromospheric indicators: the line cores of resonance or strong lines that are collision-dominated and that belong to atomic species with high abundance values.

We treat here, the formation of these lines, in particular the cores of Ca II, Mg II, C II and Si II resonance lines in the case of A dwarf stars. Some of these lines must be interpreted as computed partial redistribution (P R) line profiles and others as complete redistribution (C R) ones. We analyze how abundances play a role in this theoretical interpretation.

J. E. Beckman and L. Crivellari (eds.), Progress in Stellar Spectral Line Formation Theory, 125–136.

2. A ROUGH CRITERION TO ESTIMATE THE VALIDITY OF CR COMPUTATIONS.-

Let us undertake the interpretation of an observed line profile. We assume that we have at our disposal a suitable atmospheric model that gives a good theoretical continuum flux. Now, we can compute theoretical CR line profiles (assuming an upper and a lower expected abundance values of the element concerned) and compare the position of the optical depth of line formation relative to the broadening values in the stellar atmosphere.

We consider the collisional (γ_C) and radiative (γ_R) damping parameters where :

$$\gamma_C = \gamma_e + (\gamma_{H+} + \gamma_{He+}) + (\gamma_H + \gamma_{He}) = f(\tau_\ell)$$

$$\gamma_R = A_{21} \quad \text{(for resonance transitions)} = \text{constant}$$

where τ_ℓ is the line optical depth at the central wavelength λ_o ($\Delta\lambda = \lambda - \lambda_o = 0$) of the transition. In general, τ_ℓ ($\Delta\lambda = 0$) = 1. occurs for $\gamma_C << \gamma_R$.

Now, we need the τ_ℓ ($\Delta\lambda = 0$) value corresponding to $\tau_\ell(\Delta\lambda = 3\Delta\lambda_D) = 1$. and an estimate of this τ_ℓ can be obtained from the approximation $\varphi_\nu \propto e^{-v^2}$; $\left(v = \frac{\nu - \nu_o}{\Delta\nu_D} = \frac{\lambda - \lambda_o}{\Delta\lambda_D} \right)$, as follows.

Let us consider two τ-scales : τ_{λ_o} and τ_{λ_1}, being λ_1 other λ on the line: τ is given by

$$\tau(\lambda, h) = \int_0^h \alpha_\lambda(x)\, \rho(x)\, dx \simeq \alpha\rho h = C e^{-v^2} \rho h$$

where $\alpha\rho$ are mean values of α_λ (x) ρ(x) for $x \cong h$. We choice the unities of both τ-scales (or the zeros of both log τ-scales) at h and H respectively :

$$\tau(\lambda_o, h; \Delta\lambda = 0) \simeq C\rho(h).h \qquad \text{that we take equal to 1.}$$

$$\tau(\lambda_o, H; \Delta\lambda = 0) \simeq C\rho(H).H$$

$$\tau(\lambda_1, h; \Delta\lambda = 3\Delta\lambda_D) \simeq C e^{-9} \rho(h)\, h \qquad \text{becomes } e^{-9}.$$

$$\tau(\lambda_1, H; \Delta\lambda = 3\Delta\lambda_D) \simeq C e^{-9} \rho(H)\, H \qquad \text{that we take equal to 1.}$$

Thus :

$$\frac{\tau(\lambda_o, H; \Delta\lambda = 0)}{\tau(\lambda_o, h; \Delta\lambda = 0)} = \frac{\tau(\lambda_1, H; \Delta\lambda = 3\Delta\lambda_D)}{\tau(\lambda_1, h; \Delta\lambda = 3\Delta\lambda_D)} = \frac{\rho(H).H}{\rho(h).h} \simeq e^9$$

and $\tau(\lambda_o, H; \Delta\lambda = 0) \simeq \quad e^9 \simeq 10^{3.91} \simeq 8100.$

Thus, when the optical depth at $\Delta\lambda = 3\Delta\lambda_D$ (taken at $\lambda_1 = \lambda_o + 3\Delta\lambda_D$) has the unity value, the optical depth at $\Delta\lambda = 0$ (taken at $\lambda = \lambda_o$) has a value of the order of 10^4 .

We can now establish a rough criterion to estimate the validity of complete (C R) or partial (P R) redistribution as follows :

IF THE COMPARISON OF γ_C AND γ_R AT $\tau(\lambda_0;\Delta\lambda=0) = e^9 = 8100$ OR AT $\tau(\lambda_1;\Delta\lambda = 3\Delta\lambda_D) = 1$ GIVES :	
$\gamma_C \ll \gamma_R$	THEN, PR INFLUENCES THE NEAR WINGS.
$\gamma_C \gtrsim \gamma_R$	THEN, PR DO NOT INFLUENCES GREATLY THE NEAR WINGS AND IN CONSEQUENCE, CR IS A GOOD APPROXIMATION.
$\gamma_C > \gamma_R$	THEN, CR PREVAILS ANYWHERE.
We recall that the Doppler core ($\Delta\lambda \lesssim 3\Delta\lambda_D$) is formed pratically in C R , as well as far wings.	

This criterion arises from the approximate formula for the redistribution function :

$$R(\nu',\nu) = \gamma R_{II}(\nu',\nu) + (1-\gamma)\,\phi_{\nu'}\phi_\nu$$

(Omont et al, 1972 ; Milkey,Mihalas, 1973)

and :
$$R_{II}(\nu',\nu) \cong \begin{cases} \phi_{\nu'}\phi_\nu & \text{for } \Delta\lambda \lesssim 3\Delta\lambda_D \\ \phi_\nu \delta(\nu'-\nu) & \text{for } \Delta\lambda > 3\Delta\lambda_D \end{cases}$$

(Thomas, 1957)

with :
$$\gamma = \frac{\gamma_R}{\gamma_R + \gamma_C}$$

The approximate behaviour of $R(\nu',\nu)$ is shown in Table 1.

TABLE 1

	γ	$R(\nu',\nu)$		
			$\Delta\lambda < 3\Delta\nu_D$	$\Delta\nu > 3\Delta\nu_D$
$\gamma_C \ll \gamma_R$	$\simeq 1.$	$\simeq R_{II}$	$\phi_{\nu'}\phi_\nu$ CR	$\phi_\nu\delta(\nu'-\nu)$ coherence
$\gamma_C \simeq \gamma_R$	$\simeq .5$	$\simeq \frac{R_{II}+\phi_{\nu'}\phi_\nu}{2}$	$\phi_{\nu'}\phi_\nu$ CR	$\frac{\phi_\nu}{2}\left[\delta(\nu'-\nu)+\phi_{\nu'}\right]$ PR
$\gamma_C \gg \gamma_R$	$\ll 1.$	$\simeq \phi_{\nu'}\phi_\nu$	$\phi_{\nu'}\phi_\nu$ CR	$\phi_{\nu'}\phi_\nu$ CR

The most frequent situation is given by (see Table 1) :

S_ν of line core in C R ,because the line core is formed in the τ-region where $\gamma_C < \gamma_R$,but dominated by Doppler redistribution ;

S_ν of near wings in P R,because the near wings are formed in the τ-region where $\gamma_C < \gamma_R$ or $\gamma_C \simeq \gamma_R$;

S_ν of far wings in C R ,because they are formed in the τ-region where $\gamma_C > \gamma_R$.

Thus, redistribution plays a role pratically on near wings,depending if $\gamma_C \ll \gamma_R$ or $\gamma_C > \gamma_R$ at $\tau(\lambda_o;\Delta\lambda = 3\Delta\lambda_D) \simeq 1$ where they begin to form.

3. APPLICATION OF THE CRITERION TO RESONANCE LINES IN "A" DWARF STARS .-

Let us now consider some examples concerning resonance lines in A type stars. The formation of the Mg II resonance lines, has been studied in the case of Vega (α Lyr, A 0 V) using several atmospheric models (Freire Ferrero,Gouttebroze,Kondo 1983). The atmospheric model, allow us to establish a relation between temperature and density with optical depths. Considering the Schild et al (1971) model, we have plotted on Figure 1 the variations of γ_C and γ_R for the Mg II lines at 2800 A, as a function of the optical depth (τ_c , τ_ℓ) at the center of the k line.

Three region may be distinguished in the stellar atmosphere :

(a) in the upper region ($\tau_\ell < 10^3$) where the line cores are formed, P R effects are negligible since Doppler redistribution dominates ;

(b) in the lower layers ($\tau_\ell > 10^5$) the increase of N_e results in a γ_C/γ_R ratio > 1, so that, the far wings ($|\Delta\lambda| > 1$ A) may be also expected to be free from PR effects ;

(c) between these two regions, there remains a large gap where the near wings are formed (in Vega, between 0.1 to 1 A from the line center) and in which, scattering is mainly coherent .

As a result, the corresponding parts of the profiles are significantly different in P R than in C R computations (Fig. 2). As a general feature, the fluxes corresponding to P R calculations are lower in these near wings than in the C R ones, and a double reversal appears with peaks near $|\Delta\lambda| \simeq 0.1$ A. When stellar rotation is taken into account, these peaks disappear but P R fluxes remain substantially lower than C R ones in the whole range $|\Delta\lambda| < 1$ A (Fig. 2). (v.sin i = 17 Km/s).

For the Mg II lines in Sirius (Freire Ferrero,Gouttebroze,Talavera, 1984a), the situation is similar than in Vega (Fig. 3). The coherent parameter γ is nearly unity in external layers where the Doppler core is formed. But Doppler redistribution dominates, and the line core is formed in C R. At depth of formation of the near wings ($4.<\log\tau_{\ell k}<6.$) γ varies from unity to .4 and P R becomes important.At great depth γ tends to 0. and C R profiles are satisfactory. Figure 4 shows the case for Mg II lines in Altair (Freire Ferrero et al, 1984 b).

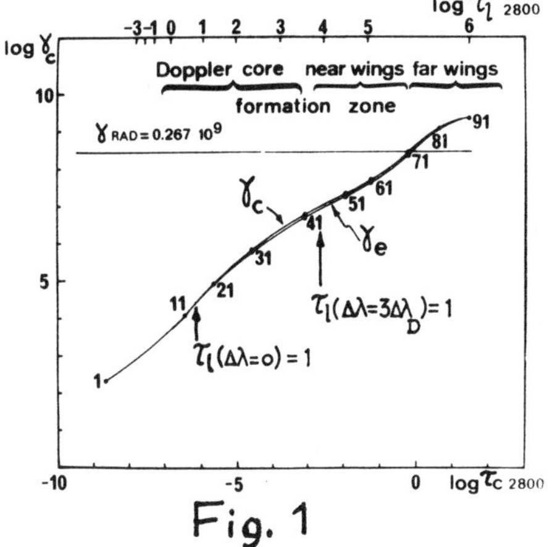

Fig. 1

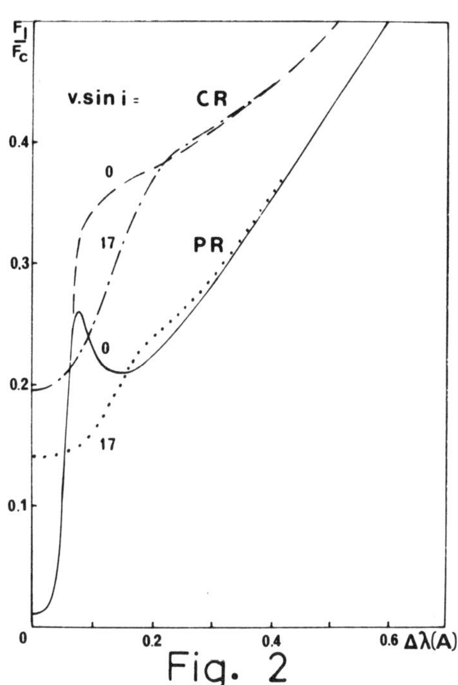

Fig. 2

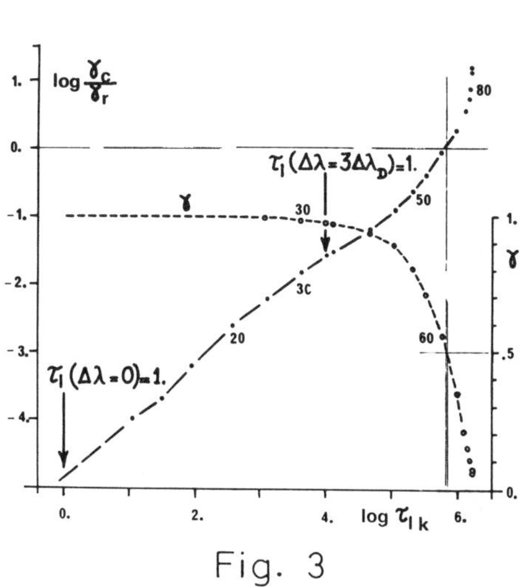

Fig. 3

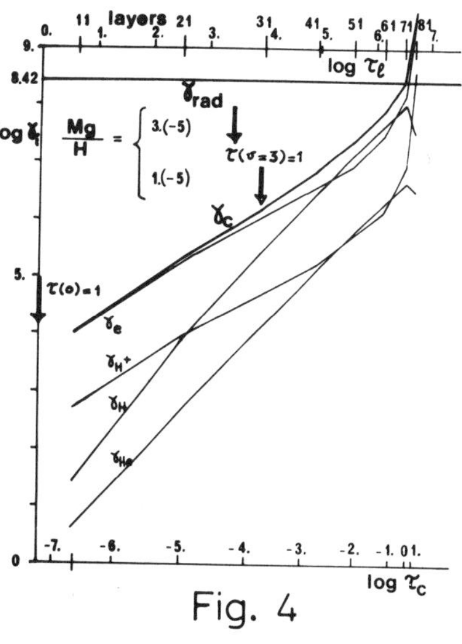

Fig. 4

For the resonance lines of Ca II, Fig. 5 to 7 show γ_C as function of τ_c in Vega, Sirius and Altair respectively (Freire Ferrero 1977 ; Freire Ferrero et al 1977,1978). For normal values of the Ca abundance, we see that C R is valid anywhere.

Furthermore, we consider the resonance C II and Si II lines,which have large A_{ji} values and in consequence,this lead us to suspect that departures from C R exists. In fact γ_R is larger than γ_C over much of the atmosphere where these lines are formed (Freire Ferrero 1977,1979).

The equality $\gamma_C = \gamma_R$ for C II lines (Fig. 8) occurs at a layer where $\tau_\ell(\Delta\lambda = \beta\Delta\lambda_D) = 1$ corresponds to $\beta = 22$ and thus $\Delta\lambda = 0.36$ A ($\Delta\lambda_D = 0.0165$ A): in the case of Si II lines, $\beta = 35$ and $\Delta\lambda = 0.37$ A ($\Delta\lambda_D = 0.0106$ A). As a result of great abundances of C and Si in relation to Ca in the stellar atmosphere, the τ_ℓ-scale is both expanded and displaced relative to fixed geometrical scale.Then the inner wings are formed in regions where $\gamma_R > \gamma_C$. Also, the fact that resonance lines of C II and Si II,here studied,have a great A_{ji}, leads the atmospheric region where $\gamma_R = \gamma_C$ to be displaced down in the stellar atmosphere.

But in practice, the computed P R and C R line profiles,expressed in relative fluxes to the continuum,are roughly similar. The explanation is that C II resonance profiles are very deep because of the carbon great abundance (Table 2). Thus,near wings have a very low intensity and,even if P R near wings are lower than C R ones, their difference is pratically not detected. Comparing C II and Mg II source line functions, we have $S_{C\ II} \ll S_{Mg\ II}$ for near wings.

As a particular conclusion of this case,we can say that for very great abundances,and when A_{ji} has also a great value,even if P R is the correct formation theory,in practice,C R gives a satisfactory or the same profile (in relative fluxes)as P R (for the same abundance value).

TABLE 2

element	lines	NLTE abundance	τ_{5000} for $\tau_{line} = 1.$ (Vega)
C II	1334.53 - 1335.66 - 1335.71	2. (-4)	1. (-7)
Si II	1304.37 - 1309.27	3. (-5)	4. (-6)
Mg II	2795.53 - 2802.7	1.to 3.(-6)	1. (-6)
Ca II	3933.7 - 2968.5	1.to 3.(-6)	3. (-5)

4. THE INDIRECT ABUNDANCE EFFECT UPON P R COMPUTED LINE PROFILES IN A STELLAR ATMOSPHERE .-

As we have shown in Table 2, the higher the abundance,the lower the optical depth of line formation $\tau_{\nu f}$. In other words, the $\tau_{\nu f}$ depends on abundance values and $\tau_{\nu f}$ move to the upper atmospheric layers when we rise the abundance values. This assertion is valid for any frequency ν of the line profile and in particular, the $\tau_{\nu f}$ of the edge of Doppler core move up in the atmosphere when the abundance rises. These considerations lead us to propose the computation of a line profile,with two abundance

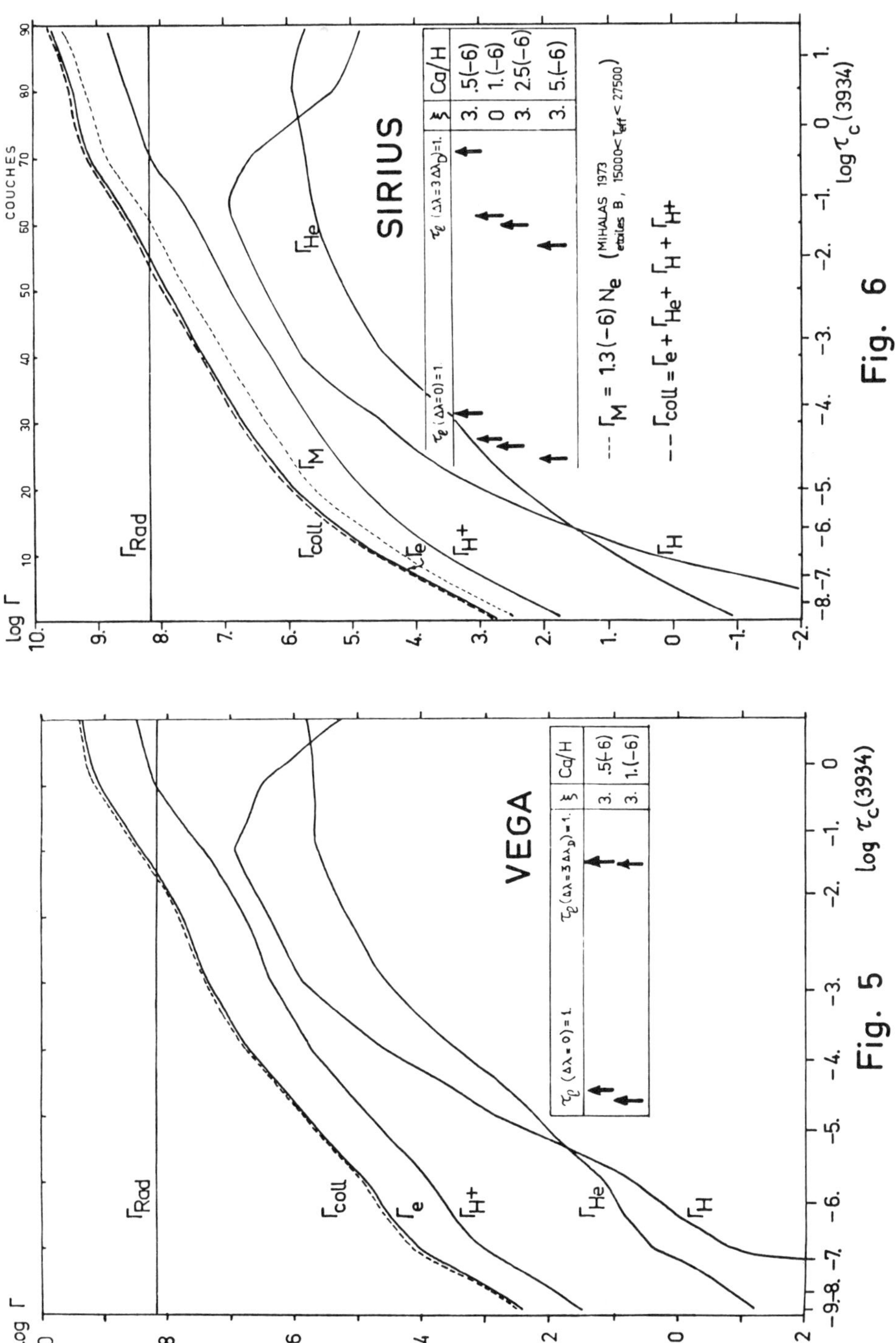
SIRIUS
log Γ
COUCHES
Γ_Rad
Γ_coll
Γ_e
Γ_H+
Γ_M
Γ_He
Γ_H
τ_ℓ (Δλ=0)=1
τ_ℓ (Δλ=3Δλ_D)=1.
ξ Ca/H
3. .5(−6)
0 1.(−6)
3. 2.5(−6)
3. 5.(−6)
--- Γ_M = 1.3(−6) N_e (MIHALAS 1973 étoiles B, 15000 < T_eff < 27500)
-- Γ_coll = Γ_e + Γ_He+ Γ_H + Γ_H+
log τ_c(3934)
VEGA
log Γ
Γ_Rad
Γ_coll
Γ_e
Γ_H+
Γ_He
Γ_H
τ_ℓ (Δλ=0)=1.
τ_ℓ (Δλ=3Δλ_D)=1.
ξ Ca/H
3. .5(−6)
3. 1.(−6)
log τ_c(3934)

Fig. 6

Fig. 5

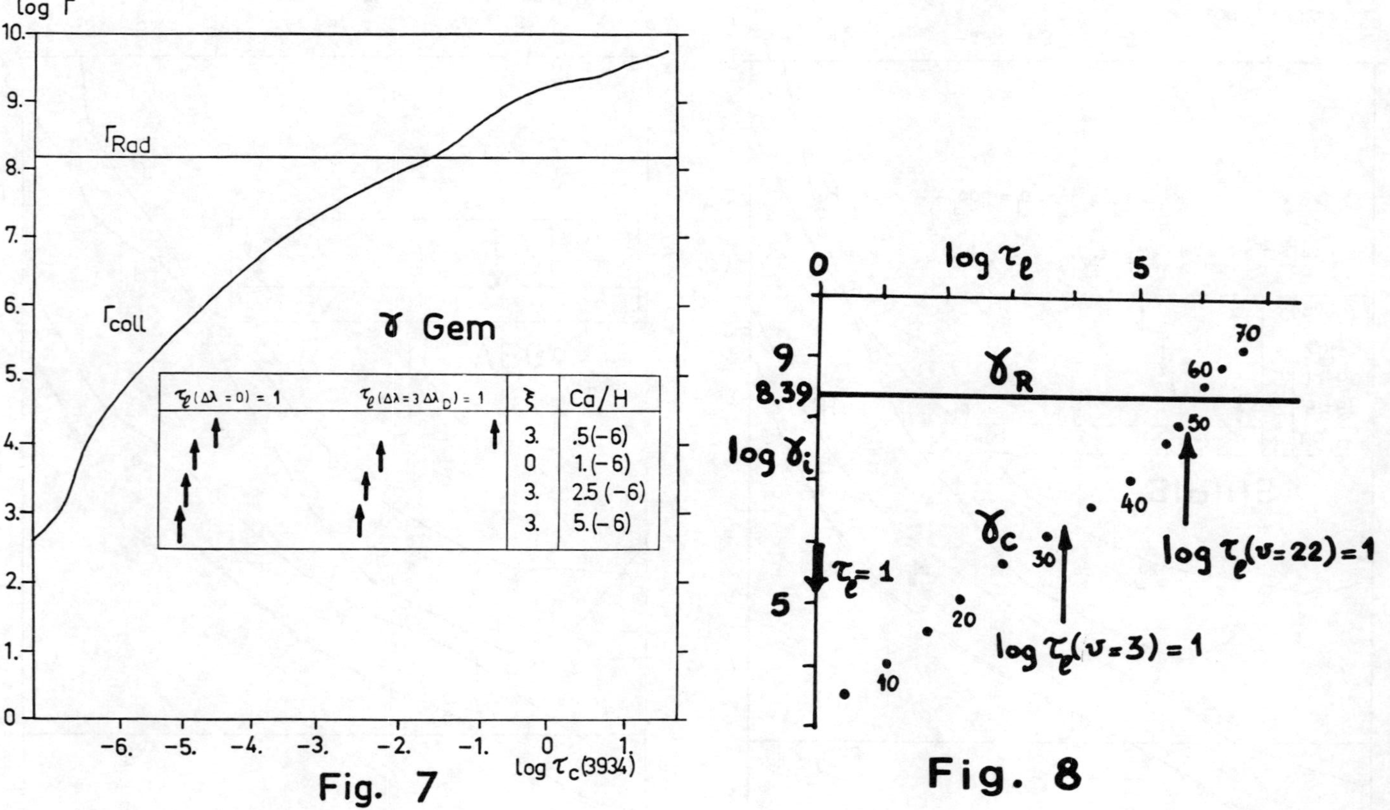
log Γ
Γ_{Rad}
Γ_{coll}
γ Gem
$\tau_\ell(\Delta\lambda = 0) = 1$
$\tau_\ell(\Delta\lambda = 3\,\Delta\lambda_D) = 1$
ξ
Ca/H
3. .5(−6)
0 1.(−6)
3. 2.5(−6)
3. 5.(−6)
log τ_c(3934)
log τ_ℓ
γ_R
8.39
log γ_i
γ_c
$\tau_\ell = 1$
log $\tau_\ell(v=22) = 1$
log $\tau_\ell(v=3) = 1$

Fig. 7

Fig. 8

values in the expected range of abundances of the element concerned,because contradictory results could be derived from the two values: for the greater, the near wings could be formed in P R and for the lower,in C R. But we know that the absorption Φ_ν and the emission Ψ_ν (that depends on redistribution) line profiles, are independent of level populations and on abundance values : they depend only on ν or ν, N_e, T_e and J_ν respectively (Freire Ferrero 1985).

The explanation is that stellar atmospheres are not composed by an unique isothermal layer.Classical photospheric models are a set of plane parallel layers,where T and ρ (or τ_ρ) decreases towards the exterior.

Thus,if we assume a given stellar atmosphere where we have little (in relation to solar values) abundance of element X (for example Mg) and we begin to rise its abundance,we could observe how the spectral line develops in a classical way,from a little core to a core plus wings,C R remaining sufficient to represent the observed profile. At some limiting abundance value, C R is no longer sufficient to explain the observed profile and this one, moves away more and more from the C R one,when the abundance rises.

P R or C R become the right line formation theory depending on where the optical depth of formation $\tau_{\nu f}$ of the near wings takes place in the stellar atmosphere (in relation to τ_c or to the γ_C / γ_R ratio),as was shown in section 2.: $\tau_{\nu f}$ depending on abundance value, we see that great values of the abundance could give P R line profiles and that low abundance values could give C R line profiles for the same transition. As an example,let us consider the Mg II lines in Altair:to obtain Mg II C R line profiles we should reduce by 400 the Mg abundance.

5. THE NON VALIDITY OF C R WHEN $\gamma_R > \gamma_C$ FOR $\tau(\lambda_0 + 3\Delta\lambda_D) = 1$.

Let us consider again the Mg II lines in Vega.The computed C R profiles do not fit neither the cores nor the near wings of the observed ones. Must be conclude that P R profiles are the only possible interpretation ? From Fig. 1,we have deduced (section 2.) that P R acts on near wings, and computed P R profiles,with convenient atmospheric models, give a satisfactory (but not complete)fit (Fig. 9 a,b ;Fig. 10) (Freire Ferrero Gouttebroze,Kondo 1983).

To illustrate the major differences between the computed C R and P R line profiles,we have compared on Fig. 2 both kind of profiles assuming stellar rotational velocities v.sin i = 0. and 17. Km/s.The resultant P R line profiles are more deeper and enlarged (near wings)than the C R ones.

Now the question is: can be construct a C R profile that approaches the P R one ? The only possibility is to modify the stellar atmospheric model,changing the microturbulence dependence with depth $(\xi(\tau))$. The idea is to enlarge the Doppler core,because,after rotational convolution it can become lower than the normal C R core (with ξ = 1 or 3 Km/s):if we have a higher ξ we can expect to enlarge the Doppler core obtaining a similar effect as if we enlarge the near wings in the normal C R profile. The rising of ξ must be done in the external layers because they contri-

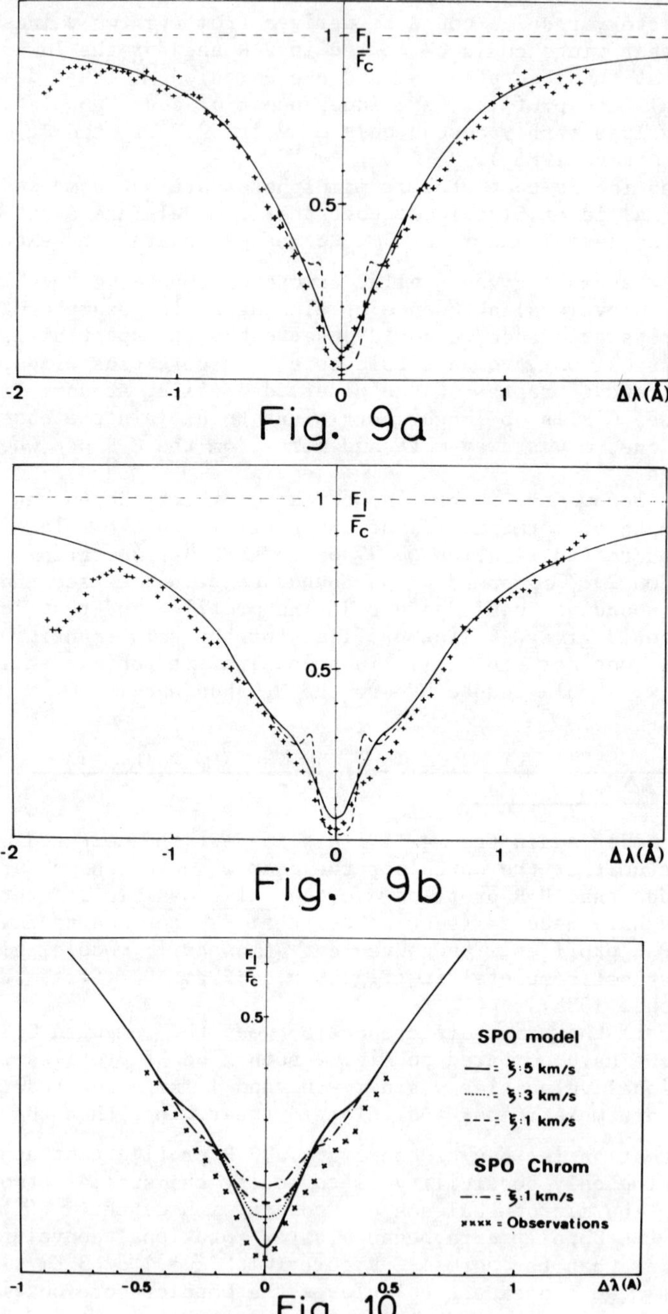

Fig. 9a

Fig. 9b

Fig. 10

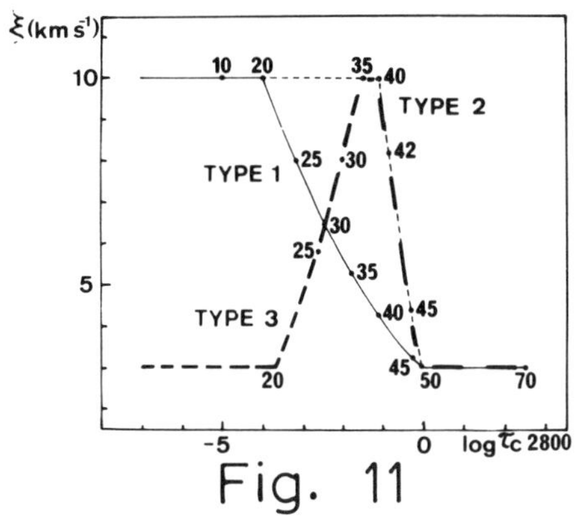

Fig. 11

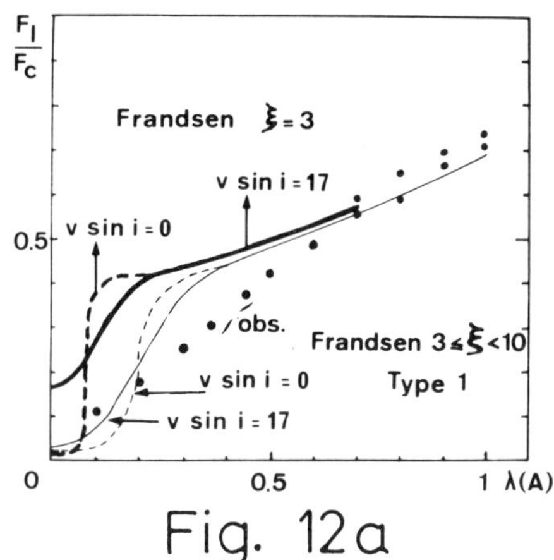

Fig. 12a

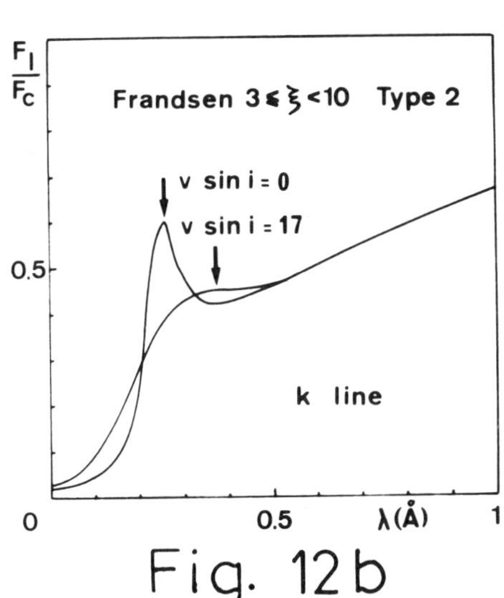

Fig. 12b

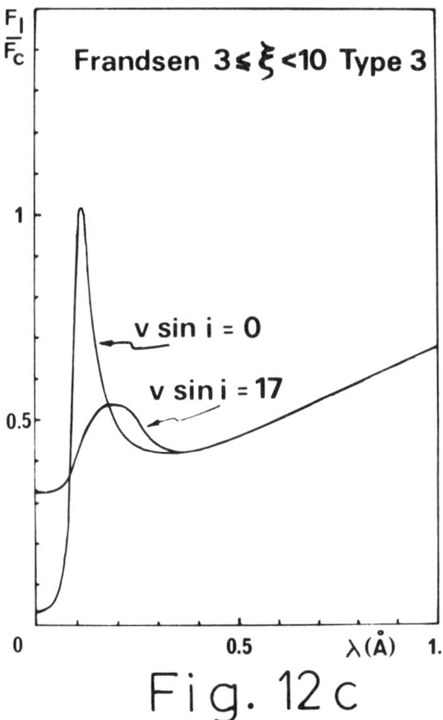

Fig. 12c

bute mainly to the formation of the line core.

Three types of functions $\xi(\tau)$ were considered (Types 1,2 and 3:Fig. 11). The results are shown in Fig.12 (Freire Ferrero,Gouttebroze,Kondo 1983): Type 1 give us a line profile with a core width greater than that observed and with higher fluxes between $0.2\ A \leqslant \Delta\lambda \leqslant 0.6\ A$ (Fig.12 a). Types 2 and 3 give emissions around the central core,which either disappears (Type 2) or does not disappear (Type 3)after rotational convolution (Fig. 12 b and c).But still more important is the result that central fluxes are lowered via the increase of ξ in the external layers (Types 1 and 2). This is because the profiles with ξ = 3 Km/s gives a width of 0.15 A in the inner core,while that for ξ = 10 Km/s (Types 1 and 2) gives a width of 0.4 A. As a comparison,rotational width is $\Delta\lambda$= 0.16 A.

In brief,$\xi(\tau)$of Type 1 accounts for a better fit on line core but nevertheless this kind of function cannot explain the gap between observed and calculated profiles for $0.2\ A < \Delta\lambda < 0.6\ A$. Thus,no C R calculation of this kind gives a satisfactory fit of the observations.

It should be noted that,since ξ is an ajustable parameter that is difficult to link with atmospheric physical conditions,the explanation of line profiles via the variation of ξ with depth can hide the real physical conditions in a stellar atmosphere.

In addition,the modification of ξ ,that could be suitable for the Mg II line cores,it could not be for other ones,showing that this kind of solution is only ad hoc.

6. SUMMARY AND CONCLUSIONS.-

The interpretation of an observed line profile by a C R or P R computed one,depends on the abundance value needed to obtain a reasonable fit. From the computed C R profile,we can prove the validity of such a calculus or to establish the need to perform a P R computation to give a better account of the observations.This can be done using a rough criterion based on the simplified form of the redistribution function $R(\nu', \nu)$. Moreover,this kind of analysis lead us to reject other possible explanations like a microturbulence $\xi(\tau)$ variable with optical depth.

7. REFERENCES.-

Freire Ferrero R 1977,Thèse Doctorat d'Etat,Université Paris VII.
" " 1979,Astron Astrophys 78,148.
" " 1985,"Comments on the emission profile ψ_ν",these Proceed
Freire Ferrero R,Czarny J,Felenbok P,Praderie F 1977,Astron Astro 61,785.
" " " " " 1978, " " 68,89.
" " ,Gouttebroze P,Kondo Y 1983,Astron Astrophys 121,59.
" " " ,Talavera A 1984a,submitted Astron Astroph.
" " " ,Catalano S,Kondo Y 1984b,in preparation.
Hummer D 1962, M.N.R.A.S. 125,21.
Milkey R W,Mihalas D 1973,Astrophys J 185,709
Omont A,Smith E,Cooper J 1972,Astrophys J 175,185.
Schild R,Peterson D M,Oke J B 1971,Astrophys J 166,95.
Thomas R N 1957, Astrophys J 125,260.

PARTIAL REDISTRIBUTION INTERLOCKING IN THE SOLAR CHROMOSPHERE

Petr Heinzel and Ivan Hubený
Astronomical Institute
Czechoslovak Academy of Sciences
251 65 Ondřejov
Czechoslovakia

ABSTRACT. Starting with the model of a quiet solar chromosphere, we have calculated the relative probabilities of radiative and natural population of the second and third hydrogen levels, pertinent to various population processes. Our analysis indicates that, while the Lα line is formed by resonance scattering between the first two levels, the third hydrogen level, from which Lβ and Hα are generated, is populated partly by direct photoexcitation 1 → 3 (about 55%), and partly by two-photon absorption 1 → 2 → 3 (about 45%). These results suggest the correct form of the general emission profile for the hydrogen Lα-Lβ-Hα-system which contains various types of redistribution functions, recently derived in the literature. The partial redistribution line transfer, with the emission profile obtained in this way, should lead to an improvement of synthetic line profiles, namely for the solar Lβ line.

1. INTRODUCTION

In interpreting space observations of the solar hydrogen Lyman α (Lα) and Lyman β (Lβ) lines [OSO-8, NRL High Resolution Telescope and Spectrograph - for references see Bonnet (1981), Basri et al. (1979)], various problems are encountered which, as it seems, will require a more sophisticated approach to stellar atmospheric spectrum synthesis than those commonly used. Similarly, the recent polarimetric observations of the solar Hα line (Stenflo et al., 1983) have given rise to some controversial views of the formation of this important line.

It is intuitively clear that the above-mentioned interpretational uncertainties may be due to the behaviour of the photons absorbed or emitted in a particular line of the Lα-Lβ-Hα-system not being independent of the photons involved in the other lines of this system. In other words, the existence of correlated chains of radiative processes, like those of 1 → 2 → 3, 1 → 3 → 2, 2 → 3 → 1, 1 → 2 → 3 → 1, 1 → 2 → 3 → 2 (the numbers denote the principal quantum numbers of the corresponding levels; we neglect the level degeneracy for convenience), leads to complicated coupling between these lines which is called <u>redistribution interlocking</u>.

J. E. Beckman and L. Crivellari (eds.), Progress in Stellar Spectral Line Formation Theory, 137–142.

In this paper, we present a preliminary study of this problem. Using the recently published semiempirical model of the quiet solar chromosphere (model C of Vernazza et al., 1981), we try to estimate the importance of the redistribution interlocking effect under the actual physical conditions in the solar chromosphere.

2. PHYSICAL CONSIDERATIONS

A semiclassical formulation of the spectral line formation in multilevel atoms, allowing for multiphoton processes, has recently been given by Hubený et al. (1983-HOS). By applying this formalism, one can, in principle, obtain the complete self-consistent solution of the problem. However, this is an extremely difficult numerical task and, moreover, the relevant atomic parameters are not yet sufficiently well known. Nevertheless, as a first step, one may simply estimate the relative contributions of the individual processes that populate the particular atomic energy levels (in our case, the second and third levels of hydrogen, from which the emission in lines $L\alpha$, $L\beta$ and $H\alpha$ arises). Having specified the contributions of the individual processes, one may decide whether the consecutive radiative chains of correlated transitions (and thus the corresponding generalized redistribution functions) are to be taken into account in actual applications.

Strictly speaking, in order to be able to determine these relative contributions, one should know the radiation field in the corresponding lines. However, this radiation field is not known precisely until the emission (absorption) coefficients in the lines under study are specified. These coefficients depend, in turn, on the relative contributions of the individual processes. Nevertheless, the relative probabilities of occurrence of the individual processes only depend on the radiation field through the radiative rates. The dominant contribution to the radiative rates comes from the line cores where the possible departures from complete redistribution are rather small (due to the predominance of the Doppler redistribution; see, e.g., Mihalas, 1978).

Therefore, one may proceed using the following iteration pattern: i) determine the relative contributions of the individual processes, assuming complete redistribution (i.e. the absorption and emission profiles in a given transition are assumed equal to the natural population profile, and all the velocity distributions are assumed Maxwellian); ii) determine the form of the emission (absorption) coefficients; and iii) solve the radiative transfer equation in order to obtain a new estimate for the radiation field.

3. NUMERICAL RESULTS

In this paper, we have carried out an approximate numerical study of item i). Adopting model C of Vernazza et al. (1981), we have calculated the radiation field in the hydrogen lines and continua for an 8-level model atom, using the suitably modified non-LTE stellar atmosphere computer code TLUSTY (see Hubený, 1983). Subsequently, we determined the depth-dependent transition rates which appear in the formulas for the relative probabilities of the individual processes. In calculating the bound-bound radiative rates, we have considered the Doppler cores only.

The relevant probabilites are denoted as p_2^* and p_{12} (corresponding to the second level), and p_3^*, p_{13}, p_{23}, and p_{123} (corresponding to the third level); for explicit formulas, refer to HOS. The physical meaning of these quantities is the following:

p_2^* - the probability of natural population of the second level, i.e. the rate of collisional excitation $1 \rightarrow 2$ plus the rate of collisional and radiative de-excitations $n \rightarrow 2$ ($n > 2$; including continuum), divided by the total rate of transitions populating level 2. Analogously, the other p's represent the relative probabilities for the following processes:

p_{12} - radiative excitation $1 \rightarrow 2$,
p_3^* - natural population of the third level,
p_{13} - radiative excitation $1 \rightarrow 3$,
p_{23} - radiative excitation $2 \rightarrow 3$ that starts from the naturally populated level 2,
$\overline{p}_{123}$ - consecutive radiative transitions $1 \rightarrow 2 \rightarrow 3$ (i.e. the resonance two-photon absorption).

By definition, these probabilities satisfy the normalization conditions $p_2^* + p_{12} = 1$, and $p_3^* + p_{13} + p_{23} + p_{123} = 1$. As explained in HOS, the corresponding quantities for the ground state are not needed as level 1 is assumed to be populated naturally, $p_1^* = 1$.

The probabilities p_2^* - p_{123} as functions of depth are displayed in Fig. 1. The depth points 1 - 52 correspond to the actual geometrical heights in the atmosphere, according to Table 12 of Vernazza et al.

4. DISCUSSION

The principal objective of this preliminary study was to determine the probabilities p_2^* - p_{123} in those regions of the quiet solar chromosphere where the wings of Lα and Lβ, and simultaneously the core of the Hα line are formed. According to model C, these regions are located at heights between 800 and 1800 km, which correspond to depth points in the range from 37 to 27 (approximately). Figure 1 immediately indicates that p_{12} is practically equal to unity in these regions, hence the dominant mechanism for Lα line formation is the resonance scattering (resonance fluorescence), characterized by the process $1 \rightarrow 2 \rightarrow 1$ (an absorption of an Lα-photon in the transition $1 \rightarrow 2$ and subsequent spontaneous re-emission of a new Lα-photon in the de-excitation $2 \rightarrow 1$). Other processes, populating the second level, contribute with the probability $p_2^* \ll p_{12}$ and may thus be neglected.

For the third level, the picture is more complicated. As follows from Fig. 1, probabilities p_{13} and p_{123} are nearly equal in these regions. This implies that Lβ is generated partly by scattering $1 \rightarrow 3 \rightarrow 1$ (about 55%) and partly as a consequence of the three-photon process $1 \rightarrow 2 \rightarrow 3 \rightarrow 1$ (about 45%), i.e. a two-photon absorption $1 \rightarrow 2 \rightarrow 3$ of photons in Lα and Hα, followed by a subsequent spontaneous re-emission of an Lβ-photon. Hα is formed quite analogously, partly due to Raman

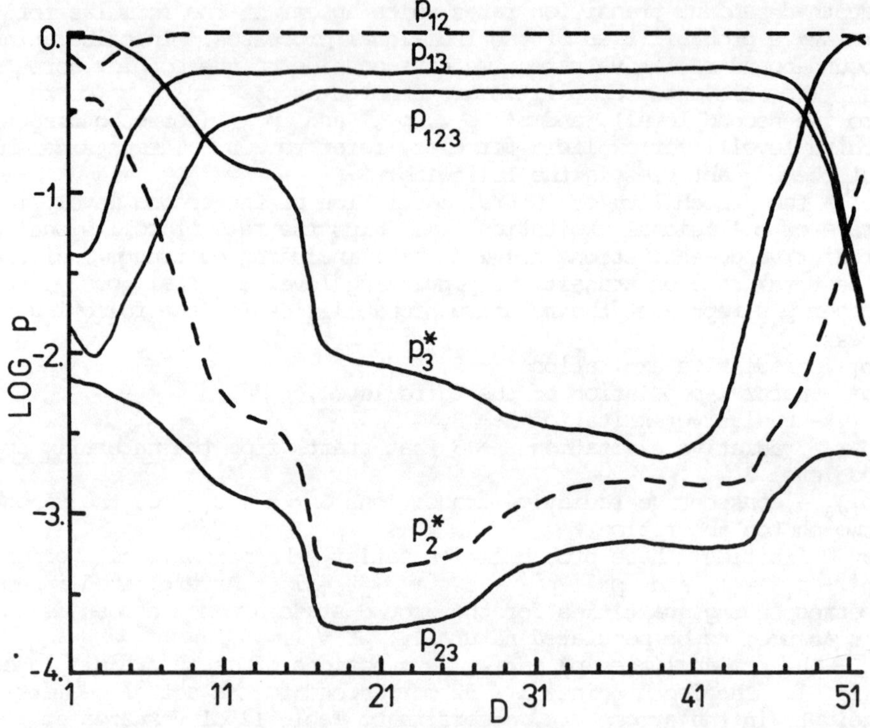

Figure 1. Probabilities p_2^*-p_{123} as functions of depth in the solar atmosphere. Ordinate: probabilities on logarithmic scale; abscissa: depth points corresponding to the actual geometrical height in the atmosphere (see text).

scattering 1 → 3 → 2 (about 55%) and partly due to the similar three-photon process 1 → 2 → 3 → 2 (about 45%). Note also that the dominant contribution to the absorption coefficient for Hα is given by the two-photon absorption 1 → 2 → 3.

Under the assumption of complete redistribution, probabilities p_2^* - p_{123} are of no consequence, because the absorption and emission coefficients for all the considered lines are given by the corresponding natural population profiles. However, if one takes the photon correlation processes into account (i.e. if one considers the non-Markovian behavior of the consecutive radiative transitions), the emission and absorption coefficients should be given by the above probabilities multiplied by the corresponding redistribution integrals (for details refer to HOS). In our case, these redistribution integrals will contain various types of ordinary (R_i) as well as generalized (P_i) redistribution functions, toegether with more complicated three-photon redistributions (for a discussion of the redistribution functions, refer to Hubený - this volume). In the first approximation, one can use the well-known function R_{II} (process 1 → 2 → 1) for Lα-formation, while for Lβ the general emission coefficient will contain redistributions R_{II} and P_{IV}, and for Hα one can apply $\bar{P}_{IV}$ and R_V (collisional redistribution can be accounted for by means of functions R_{III} and P_{III} according to the respective branching ratios).

5. CONCLUSIONS

In the present paper, we have estimated the probabilities of the individual processes that populate the second and third levels of hydrogen under solar chromospheric conditions. With a view to the above results we may conclude that, whereas Lα is formed as a typical resonance line, the Lβ and Hα lines are generated partly by resonance and partly by Raman scattering. Therefore, for these lines we can no longer use a two-level approach to study partial redistribution effects. The emergent profiles of all these lines will generally be affected by the coherence properties of the redistribution functions involved (namely R_{II}, P_{IV} and $\bar{P}_{IV}$) which can lead to important effects in the line wings. We may thus speculate that a transfusion of Hα photons into the wings of Lβ (i.e. the process 2 → 3 → 1 with wing coherence described by the redistribution function P_{IV}) could lead to an enhancement of the Lβ-line wing intensity, which is actually observed without being satisfactorily explained (see Vernazza et al., 1981, or Lemaire et al., 1981).

At present, we are trying to tackle the second part of the problem, i.e. to determine the emission and absorption coefficients of these lines explicitly, and to solve the radiative transfer problem in order to obtain the emergent line profiles to be compared with the observations. This can be done since suitable numerical methods of evaluating the above-mentioned redistribution functions are now available (for a review, see Heinzel - this volume). Moreover, we also intend to apply this procedure to other stellar chromospheric models.

Acknowledgements. The EC 1040 computer of the Ondřejov Observatory Computing Center was used to perform the numerical calculations presented in this paper.

REFERENCES

Basri, G.S., Linsky, J.F., Bartoe, J.-D.F., Brueckner, G., Van Hoosier, M.E.: 1979, Astrophys. J. 230, 924

Bonnet, R.M.: 1981, Space Sci. Rev. 29, 131

Hubený, I., Oxenius, J., Simonneau, E.: 1983, J. Quant. Spectrosc. Radiat. Transfer 29, 477 and 495

Hubený, I.: 1983, in Stellar Spectra and Their Interpretation, Publ. Astron. Inst. Czechosl. Acad. Sci. No. 57 (Eds Hubený, I. and Onderlička, B.), 32

Lemaire, P., Gouttebroze, P., Vial, J.C., Artzner, G.E.: 1981, Astron. Astrophys. 103, 160

Mihalas, D.: 1978, Stellar Atmospheres, W.H. Freeman, San Francisco (2nd edition)

Stenflo, J.O., Twerenbold, D., Harvey, J.W., Brault, J.W.: 1983, Astron. Astrophys. Suppl. Ser. 54, 505

Vernazza, J.E., Avrett, E.H., Loeser , R.: 1981, Astrophys. J. Suppl. 45, 635

PANNEL DISCUSSION ON PARTIAL REDISTRIBUTION.

Pannel members: R. Freire-Ferrero, H. Frisch,
J. Linsky, J. Oxenius, E. Simonneau.

What follows is a summary and a review by the editors of the contents of the September 4th pannel discussion, which followed the first two sessions of the Workshop, devoted to general aspects of partial redistribution and its astrophysical implications. In presenting here our version of the main outlines of a long and articulate discussion, we hope we have done justice to the original ideas expressed by thc participants.

When the Workshop was conceived, it seemed to us to be critically important to include within the programme a set of open discussions which would raise key topics in the application of radiative transfer to stellar atmospheres, of which PRD is a topic of prime importance. Before the open discussion, the invited papers and a series of contributed papers had set out the PRD problem from the theoretical and observational viewpoint. The discussion itself then fall naturally under four headings:

A) the impact of PRD on the physics of stellar atmospheres;

B) atomic and kinematic aspects of PRD;

C) the interplay between laboratory physics and astrophysical results;

D) the use of PRD in interpreting the physics of moving media, and in astrophysical contexts which are non-stellar.

The discussion was rounded off with a series of concluding remarks.

A) The impact of PRD on the physics of stellar atmospheres.

J. E. Beckman and L. Crivellari (eds.), Progress in Stellar Spectral Line Formation Theory, 143–151.

Four major questions were raised.

1) To which extent can the effects of PRD on the profiles of strong resonance lines alter the overall atmospheric structure?

It is not surprising that the discussion started with a debate about this question.

One clear-cut answer was offered: the role of the Ca II and Mg II resonance lines in the temperature distribution and the energy balance of the outer layers of solar-type stars is of dominant importance, and in this context the core of very strong resonance lines is Doppler redistributed, therefore CRD and PRD results are identical. If most of the flux in the line is located in the core, even under the assumption of CRD the temperature distribution derived, at least in the upper part of the chromosphere, will be reasonably close to the correct model. On the other hand, when deriving the temperature distribution of the lower part of the chromosphere, where the inner wings form and where PRD or coherent scattering effects are extremely important, the results can be quite wrong.

To get a better fit between observed and computed profiles of solar Mg II h and k and Ca II H and K lines, Ayers and Linsky (1976, Ap.J., 205, 874) had to raise the minimum of the temperature distribution from 4200°K to about 4500°K. Although this difference seems quite small, it implies dramatic changes in the atmosphere enery balance, and therefore in terms of mechanical (or magneto-mechanical) energy generation and deposition. With the amount of mechanical heating necessary to raise the temperature minimum from 4200°K to 4500°K it is possible to heat the chromosphere, the transition zone and the whole corona, to drive the solar wind and still have a lot of energy left over.

For similar reasons an accurate treatment of redistribution effects in the profiles of the C IV resonance lines is required. These lines strongly affect the structure of the outer layers of Be stars. The optical depth of the C IV resonance lines is very large and they are already thermalized very high in the outer atmosphere. The thermalization lenght is correlated with the temperature gradient and the global thermodynamic properties of the atmosphere. The interplay between the latter and the effects due to redistribution is very dependent on detailed initial assumptions.

A key question, whether it is satisfactory to take a simplified approach, assuming CRD in the line core and coherent scattering in the wings without taking into account a depth dependence of the dividing frequency led to an exchange of views about the physical regimes under which such a simplified treatment is still valid.

It was concluded that the "rule of thumb" of a three $\Delta\nu_{Doppler}$-wide core can be safely used when turbulence or systematic motions are not important. On the other hand when flows (expecially non-monotonic) are present no such rule of thumb can be valid, and each single case must be carefully investigated. Similarly it was affirmed that partial coherent scattering in the wings can be completely wrong if the line is very thick.

2) To what extent can theoretical profiles reproduce observed profile?

The question was raised explicitly in the following way: when, as a result of calculations, you obtain a synthetized profile which agrees quite well with the observed one, can you be sure that your model is right?

The response was to use great caution, since there is in general no proof of uniqueness in this kind of modelling. In these circumstances it is of the greatest importance to formulate the problem in terms of the underlying physics. Simply performing increasingly complicated radiative transfer calculations does not give a better insight into the problem.

This discussion raised the tart comment that a perfect agreement between observations and theoretical predictions is the worst thing that can happen for the progress of any field of science.

A case was quoted of a perfect theoretical match to a very complex profile. Someone had succeeded in this at first sight impossible task by playing with a two-stream atmospheric model and a set of free parameters. To emphasize the point of non-uniqueness the audience was reminded that with two free parameters you can make an elephant, and with three you can make its left eye wink.

The parameters needed to interpret line profile observations can be divided into two categories:

i) the atmospheric parameters, namely the basic physical quantities which characterize the model, especially temperature, electron density and pressure

ii) the parameters which affect the line transfer solution: the intrinsic profile, broadening factors, frequency redistribution, etc.

Non-linear effects can occur, when a more realistic (and complex) physical situation is described. In such cases an inadequate initial solution can prevent convergence to the correct final solution.

3) What is the best strategy to investigate the role of PRD in stellar atmosphere?

Astrophysical situations are intrinsically complex. Is it advantageous to perform a systematical analysis by adopting, instead of a realistic but complex model, simplified ones which may allow a better understanding of basic physical effects?

Two somewhat conflicting answers were given by a theorist and an observationalist.

According to the former the adoption of simplified models is a useful idea. People who try to interpret particular observations are of course concerned with many different physical problems at the same time. On the other hand a simplified treatment is of the highest value and a great deal can be learnt this way about basic physical effects. In this sense the early work by Hummer was fundamental.

The latter rejoined that the simplified approach is certainly of great value, but concentrating esclusively on very general solutions one can by-pass important parts of the astrophysics. To give an example he quoted the case of ionization equilibrium in a stellar atmosphere. In many cases of interest this depends on the ionization of hydrogen, which is strongly dependent on the population of the HI second level. The latter is in turn controlled by the radiation field in the Lyα line. The result is a very interesting coupling between Lyα, the hydrogen ionization equilibrium, the electron density and the PRD of some other lines of possible interest. For this problem, and for many others, there is little point in considering general solutions, for which the atmosphere is assumed to have either constant temperature or constant density.

So both general and specific solutions have to be investigated, to get a complete insight into any specific physical situation.

4) What are the limiting circumstances in which PRD reduces to CRD?

A parallel was drawn, here, with the reconduction of non-LTE to LTE under certain physical regimes. In certain cases it is indeed possible to obtain accurate solutions with CRD instead of PRD.

The analysis of the two-level atom suggests the rule of thumb that for all redistribution functions but R_{II} (including any mixed expression containing R_{II}) PRD reduces to CRD, in astrophysical situations.

When dealing with the more general case of a multi-level atom, involving a lot of radiative transitions, an overall sum rule for the whole chain of transitions should apply. It was nevertheless the feeling of one theorist that under certain circumstances the CRD approach might still be a very good approximation.

A comparison between PRD and CRD formulations, based on work in the early 1970's, was then quoted. Using perturbation theory, the PRD operator had been expanded about the much simpler CRD operator. A point of information about this work was then raised. It seems that the original author had not used the full form of the R_{II} distribution function but an approximation involving coherent scattering plus CRD. On the contrary in 1969 an analysis of the type referred to, based on simplified models, had in fact been published.

B) Atomic and kinetic aspects of PRD.

1) The semi-classical approach.

The discussion was started by a remark about the semi-classical formulation of the kinetic equations, in which it was pointed out how line narrowing during velocity changing collisions is not considered, and one of the theorist involved answered that this quantum mechanical non-coherent effect was indeed not treated in their semi-classical approach.

Attention was then focused on the introduction of the concept of natural population of an atomic level. It was pointed out that this concept is not necessarily linked with the concept of sublevels. However, the sublevel picture allows the suggestive interpretation of natural population as corresponding to a uniform population of sublevels. Specialists in quantum mechanics do not like the formalism of sublevels, even though it gives correct results for the two-level atom. It might be attractive, however, to see how far one can go with this picture. That would allow the use of this formalism in the general case of a multilevel atom, with all the consequent implications. It would then be possible to consider the system as Markovian.

It was made clear that the formalism of sublevels does give a correct results for the two-level atom, not only in the case of sharp lower level, but also for a broadened lower level. It has in fact been demonstrated that even for a broadened lower level the results obtained using a model with sublevels agree with those using the concept of natural populations.

The origin of the levels (and sub-levels) can be understood by the structure of the Hamiltonian of the system, as was pointed out by an atomic physicist from the audience, who noted that three or four years ago a very interesting quantum mechanical approach to the problem of frequency redistribution had been published. In this, the Hamiltonian of the emitter, including the interaction with the field (not in the high, but in the low field limit), was written down and a Master equation drawn. From that the elements of the density matrix were inferred. The frequency redistribution terms were simply the diagonal elements of the density matrix. The problem had been solved in the case of pure radiative processes, with only a hint that a similar approach might work for collisional processes.

In this quantum mechanical description of the emitter/radiation field interaction the apparence of sublevels is quite natural and correct, as the density matrix takes into account the sub-states of the ensemble: atom plus field. This method offers a unified way to introduce all the physical processes (radiative, collisional and any other contributing to frequency redistribution).

C) Interplay between laboratory and astrophysical results.

The point of view of the astrophysicist, who introduced the subject was that radiative transfer methods are tools to interpret the structure of stellar atmospheres via the diagnostics of stellar spectral lines. Astrophysicists have many questions to put to theoreticians, but they also need the collaboration of laboratory physicists because laboratory plasmas can be made much simpler than astrophysical plasmas, and this simplicity helps to elucidate the basic physics.

A caustic comment came from one of the audience, from personal experience with laboratory physicists. The latter work with physical system under completely different regimes (e.g. high pressure plasmas with density in the range of 10^{17} cm-3). It is not easy to fix their attention on plasmas of hydrogen or helium instead of zirconium or titanium, to say nothing about communication difficulties due to specialized jargon.

The differences between conditions in laboratory and astrophysical plasmas were stressed by one of the atomic physicist. Essentially in astrophysics one is concerned with long distances and diffuse media, in the laboratory with very short distances and high density plasmas.

To the explicit question, of whether PRD effects have ever been observed in laboratory experiments, he answered that the first results about PRD in a dense plasma were presented in 1984 at the Aussois Conference on Spectral Line Shapes. Even if these results are very preliminary, one can predict that these effects are important mainly in the line wings. In some cases (e.g. Lyα) they could effect the line core, too.

It was reported that PRD effects were observed in laboratory also for neutral gas.

In a concluding remark to this section, it was noted that while in other fields of physics laboratory experiments can guarantee the validity of theory, in the case of frequency redistribution the laboratory results are still in their infancy. On the other hand astrophysical results cannot be used to prove theories on PRD, as many interacting physical processes are occuring in stellar atmospheres.

D) PRD in non-stellar environments and moving media.

The ball was set rolling by a reference to work done in the fifties explaining the 21 cm hydrogen line hyperfine structure, and to calculations made in order to interpret the galactic Lyα field, and more recently approaches to understand Lyα emission in QSO's.

Mention was made of the importance that PRD effects may assume in moving media, and the need for a systematic investigation. It was agreed that the validity of the Sobolev approximation in the context of PRD/CRD has not yet been properly checked for cool stars. Expecially in red giants, where the wind velocity is of the order of few times the Doppler velocity such effects may be important, but only a few studies have been so far performed. It was noted that PRD effects may be important in fast-moving winds of hot stars also. In particular angle-frequency coupled electron scattering could produce a red-tail in a line profile.

Conclusions.

In a pannel "round-robin" at the end of the discussion, a few consensus points clearly emerged. It was agreed that to some extent there is a dichotomy between more and more detailed work on line profiles, including the effects of PRD, and the real situation in stellar atmospheres where inhomogeneities and velocity fields may affect the observed profiles even more than the difference between CRD and PRD. Every one felt that it was essential that the atomic physics input should be as nearly correct as pratically possible, and that in this context PRD methodology and its developement are of the highest importance.

Where divergences did appear was in the priority to be assigned to each part of the physical problem: a realistic atmospheric model with simplified line properties, or a simplified model with highly developed intrinsic line properties. No one denied that in the end developed model atmospheres with the full range of atomic physics will be needed, and differences lie only in the question of where theorists should devote their immediate effort most predominantly.

Acknowledgement.

The editors wish to express their thanks to Dr. Drake, whose accurate notes proved very helpful when writing this summary of the discussion.

NUMERICAL METHODS IN RADIATIVE TRANSFER

Wolfgang Kalkofen
Harvard-Smithsonian Center for Astrophysics
60 Garden Street
Cambridge, Massachusetts 02138
USA

ABSTRACT. We discuss the operator perturbation method for the solution of radiative transfer problems in the integral equation formulation. The example given is that of line transfer in complete redistribution for a two-level atom in statistical equilibrium.

The essence of the method is the separation of the calculation into two parts: the calculation of corrections to a solution with the aid of an approximate integral operator; and the calculation of the error with which the solution satisfies the conservation equation. The accuracy of the converged solution depends only on the accuracy of the error calculation.

The error is computed by solving the (scalar) transfer equation in differential form, either as the Feautrier equation with second-order accuracy, or in Auer's formulation with fourth-order accuracy. Thus the approximate integral operator is the only matrix that needs to be constructed.

A convergence criterion for the series expansion of the solution vector is discussed, and the solution is shown to be independent of the initial estimate. The approximate integral operator must satisfy a normalization condition and, for a line transfer problem for example, it must correctly describe the radiative transfer in the line wings. These conditions are satisfied by approximate matrix operators that can be constructed very rapidly. General guidelines are given for procedures that increase the speed of the method for linear and non-linear problems in static and time-dependent media.

1. INTRODUCTION

Of the numerical methods for solving transfer problems that have recently been developed, the operator perturbation method (Cannon 1973, 1984; Scharmer 1981, 1984) achieves the greatest advance in speed without sacrificing accuracy. It can therefore be used in a wide class of physical problems. It is the method that will be discussed in this review.

The solution of transfer problems with the aid of the so-called probabilistic transfer equation is even faster, but the speed is purchased at the expense of accuracy. The method describes the transfer in a spectral line by means of a single first-order differential equation for the line source function or for the mean integrated intensity. But the accuracy is only modest, as follows from the fact that a first-order differential equation transports information only in one direction whereas radiation and hence information in an atmosphere flow both inward and outward. Thus, in a typical problem, events in the outer layers of a medium cannot influence conditions in deeper layers. In addition, the medium must satisfy certain stringent conditions: it must be isothermal so that the profile function is depth-independent, and there must not be any macroscopic motions. Although the method appears to yield fair accuracy when these conditions are satisfied, the problems that call

J. E. Beckman and L. Crivellari (eds.), Progress in Stellar Spectral Line Formation Theory, 153–168.

for such fast methods tend to be those for which these conditions are not met. A further drawback is that the method does not provide for a way to improve the numerical result — it is not the zeroth-order of an expansion — so that another method must be used if the solution is to be improved. Nevertheless, the probabilistic transfer equation has its use in certain applications, but I will not describe it further. It has recently been treated by Athay (1984) and Canfield *et al.* (1984) (*cf.* also related discussions by Frisch 1984; and Rybicki 1984, 1985).

In the operator perturbation method to be discussed below I will stress the application to integral equations although the method can be used with both differential and integral equations. But the greatest strides in recent years have been made with integral equations, which have been applied to linear as well as non-linear problems.

Methods similar to the operator perturbation method have been known for some time in applied mathematics as deferred correction or relaxation methods (*cf.* Fox 1962, chapter 23), and they are related to the Born approximation in quantum mechanics (*cf.* Newton 1966, p233ff). In radiative transfer they were first used by Cannon (1973; for later references, *cf.* Cannon 1984) who applied them initially to differential equations.

Cannon suggested this method as a means of solving the transfer equation for complicated problems by means of a perturbation series. The method promised significant savings in computer time without loss of accuracy. For a line transfer problem with partial frequency redistribution of the photons over the line, for example, the perturbation about the isotropic case required perhaps only a single angle point for the approximate operator. Thus, the part of the problem in which the intensities are coupled explicitly has only low order, resulting in a major economy. One may also expect that in some circumstances the solution of the problem by operator perturbation can be more accurate than the direct solution. For problems in media with high differential velocity, or in spherical geometry, the perturbation method turns the troubling derivative of the intensity into a known source in a series expansion. Thus, the basic strategy of Cannon's approach is to simplify the operator acting on the unknown function and to treat the correction operator as a known source term in a hierarchy of equations.

The operator perturbation method has been applied to the integral equation formulation of transfer problems by Scharmer (1981). It is described here in a form similar to one suggested by Scharmer (1984) and Nordlund (1984), where the correction to a solution is based on the error made by the solution in the conservation equation. This reference of the correction to the error in a conserved quantity makes the approach particularly useful for non-linear problems (*cf.* Fox, 1962, on this point).

The basic strategy of the operator perturbation method in the integral equation formulation is to divide the transfer problem into the calculation of corrections to the solution vector, and the computation of the error made by the current estimate in the conservation equation. There are three places were economy in the computation must be achieved for the iterative solution of the perturbation equations to present an advantage in speed over the direct solution method: (i) in the construction of the approximate operator, (ii) in the solution of the set of linear equations for the correction terms, and (iii), in the calculation of the driving term of the correction equation, which is the error in the conservation equation.

We describe the operator perturbation method for the case of a linear line transfer problem, which would ordinarily be solved in a single pass by means of differential or integral equations. Instead, we use the iterative approach of the perturbation method for integral equations. In section 2 we recapitulate the formulation of a line transfer problem in terms of differential equations; in section 3 we formulate the problem in terms of integral equations; in section 4 we treat the operator perturbation method, show under what conditions the iterations will converge, and demonstrate that the solution is independent of the initial estimate; in section 5 we describe the construction of approximate integral operators; in section 6 we discuss the solution of the set of coupled equations and the calculation of the error in the conserved quantity and show where the short-range and long-range properties of the transfer problem are treated; in section 7 we derive a convergence criterion for the source function expansion; and in section 8 we summarize the method, give a list of procedures for speeding up the calculation for both linear and non-linear problems, and present conclusions.

2. DIRECT SOLUTION WITH DIFFERENTIAL EQUATIONS

A linear line transfer problem can be solved directly by means of differential equations. The formulation is usually based on the second-order differential equation for the intensity mean, $J_{\nu\mu}$, along a ray,

$$\mu^2 \frac{d^2}{d\tau_{\nu\mu}^2} J_{\nu\mu} = J_{\nu\mu} - S_{\nu\mu} \quad , \tag{2.1}$$

which describes the radiative transfer in the interior of a medium, and the first-order equations,

$$\mu \frac{d}{d\tau_{\nu\mu}} J_{\nu\mu} = J_{\nu\mu} - I^-_{\nu\mu,bc} \quad , \tag{2.2}$$

and

$$\mu \frac{d}{d\tau_{\nu\mu}} J_{\nu\mu} = I^+_{\nu\mu,bc} - J_{\nu\mu} \quad , \tag{2.3}$$

which describe the transfer at the upper and lower boundaries, respectively, with the incident intensities $I^-_{\nu\mu,bc}$ and $I^+_{\nu\mu,bc}$. These equations are for the transfer along straight lines defined by frequency ν and angle μ; they are applied here to an atmosphere with plane-parallel stratification, but they may also be used for media with spherical symmetry (*cf.* Auer 1984, Nordlund 1984) or with any other kind of structure.

The derivation of the second-order differential equation from the first-order equation (*cf.* equation 3.1) assumes symmetry of the source function with respect to the angle variable μ, which in a moving medium takes the form

$$S_{\nu\mu} = S_{-\nu,-\mu} \quad . \tag{2.4}$$

In the absence of this symmetry, the transfer equation may be formulated in terms of two coupled first-order equations for the symmetric and antisymmetric parts of the radiation field, with the same accuracy of the solution for a particular choice of the differencing of the two parts on interleaved spatial grids (*cf.* Kalkofen & Wehrse 1982).

The monochromatic source function, $S_{\nu\mu}$, is composed of the line source function, S, and the source function of the background continuum at the frequency of the line, S_c,

$$S_{\nu\mu} = (1 - \rho_{\nu\mu})S + \rho_{\nu\mu}S_c \quad , \tag{2.5}$$

with weighting by the fractional opacity $\rho_{\nu\mu}$ of the background continuum,

$$\rho_{\nu\mu} = \kappa_c/(\kappa_c + \kappa_{\nu\mu}) \quad , \tag{2.6}$$

where κ_c is the opacity of the continuum and $\kappa_{\nu\mu}$ that of the line.

In a typical line transfer problem the source function describes the statistical equilibrium of the populations in the two combining states; if the frequencies of the absorbed and re-emitted photons are uncorrelated (this case is referred to as complete redistribution, or CRD), the equation for the source function is given by

$$S = (1 - \epsilon)\bar{J} + \epsilon B \quad , \tag{2.7}$$

where $\bar{J}$ is the mean integrated intensity,

$$\bar{J}(\tau) = \frac{1}{2}\int_0^\infty d\nu \int_0^1 d\mu\, \varphi_{\nu\mu}(\tau) J_{\nu\mu}(\tau) \quad , \tag{2.8}$$

and $\varphi_{\nu\mu}$ is the profile function of the transition. The part of the excitation that is due to collisions is described in terms of the Planck function at the local electron temperature, B, and the relative frequency of collisions, ϵ, defined by

$$\epsilon = \frac{\epsilon'}{1+\epsilon'} \quad , \tag{2.9}$$

and

$$\epsilon' = \frac{C_{21}}{A_{21}}\left(1 - e^{h\nu/kT}\right) \tag{2.10}$$

(*cf.* Athay 1976), where C_{21} and A_{21} are the collisional and radiative transition coefficients, respectively.

When the absorbed and emitted photons are correlated, *i.e.*, in partial redistribution (referred to as PRD), the profile function $\varphi_{\nu\mu}$ in equation (2.8) must be replaced by a redistribution function, which depends on the frequencies of both the absorbed and the emitted photons. The intensity in the source function equation (2.7) is then frequency-dependent, making the line source function also frequency-dependent. This does not in any way affect the differential equation formulation of the line transfer problem, but it does make the integral equation solution more complicated (*cf.* Scharmer 1983). We will treat here only the CRD problem.

The transfer problem defined by the equations of radiative transfer (2.1) and statistical equilibrium (2.7) is solved by combining these equations, using the definition (2.5), into a single differential equation, which is discretized with respect to depth, frequency, and angle. The second-order differential equation thus becomes a second-order difference equation,

$$-A\bar{J}_{i-1} + B\bar{J}_i - C\bar{J}_{i+1} = (1-\rho)\,\epsilon_i B_i + \rho S_{ci} \tag{2.11}$$

(*cf.* Mihalas 1978, eq. 6-37), for the column vector of the intensity $\bar{J}_i$, which contains all angle and frequency components of $J_{\nu\mu}$ at depth i.

The equation of the transfer problem is written here in the Feautrier form with central differences; another possibility, suggested by Auer (1976; 1984, eq. 42), contains on the right-hand side the thermal line and continuum source term at all three spatial grid points, i and $i \pm 1$; instead of having the second-order accuracy of the Feautrier equation, Auer's equation has fourth-order accuracy. This improvement is obtained at practically no extra cost, provided the depth grid is well chosen and the source terms are well-behaved. (In the iterative solution of the equations, where the transfer equation is solved in scalar form, the approximate source function need not be well-behaved. It may then be safer to begin the calculation with the Feautrier formulation and to use Auer's formulation only in the final iterations; *cf.* section 6.)

Note that the coefficient B of the central term in the finite difference equation (2.11) is a *full* matrix; it contains all the coupling from the scattering term in the source function equation; it's off-diagonal terms are non-zero because of this coupling (CRD and PRD differ only in the values of these non-zero elements). Without it,the equations are uncoupled and can be solved separately — an enormous simplification, of which we will avail ourselves later. It is also worth noting that if the source function had been left in the transfer equation as S without expressing it in terms of $\bar{J}$, i. e., in implicit form, and the problem were solved by iterating back and forth between the transfer equation (2.1) and the source function equation (2.7) — a procedure known as Λ-iteration — convergence would be very slow; for a very small scattering parameter ϵ, the solution might fail altogether to converge. However, the scalar transfer equation will be useful later — and without the risk of the slow convergence of a Λ-iteration — when it will be used integrate the equation for a given source function, which will be left in implicit form.

Note that the order of the system of difference equations, when the coupling of the frequency components via the source function is taken into account, is given by the product of the number of angle points and the number of frequency points. In the numerical solution of the block tri-diagonal equation, a matrix of this order must be inverted at every depth point. The solution time of equation (2.11) scales essentially as the cube of the order of this system, whereas the scaling is linear in the number of depth points.

3. DIRECT SOLUTION WITH INTEGRAL EQUATIONS

Another formulation of the transfer equation uses the first-order differential equation of transfer,

$$\frac{dI^+_{\nu\mu}}{d\tau_{\nu\mu}} = I^+_{\nu\mu} - S_{\nu\mu} \tag{3.1}$$

(the angle cosine has been absorbed into the definition of the optical depth), or, more typically, its integral,

$$I^+_{\nu\mu}(\tau_{\nu\mu}) = \int_{\tau_{\nu\mu}}^{T_{\nu\mu}} dt \exp[-(t-\tau_{\nu\mu})]\, S_{\nu\mu}(t) \; + \; I^+_{\nu\mu,bc} \exp[-(T_{\nu\mu}-\tau_{\nu\mu})] \quad , \tag{3.2}$$

which is usually referred to as the formal integral.

In a discrete optical depth space, the expression for the specific monochromatic intensity at the (reference-)depth τ_i gives the sum over the source function values ranging from the field point i to the boundary. Thus the vector consisting of all depth components of the specific monochromatic intensity in the outward direction can be written as

$$I^+_{\nu\mu,i} = \Lambda^+_{\nu\mu,ij} S_{\nu\mu,j} + g^+_{\nu\mu,i} \quad , \tag{3.3}$$

in which the half-range matrix operator $\Lambda^+_{\nu\mu}$ acts on on the source function vector (note that this formulation does not require the symmetry 2.4 of the source function; *cf.* Kalkofen 1974). The matrix is triangular, indicating that the specific intensity is affected by information from one side only. The quantity $g^+_{\nu\mu}$ is the boundary term; it is zero in semi-infinite atmospheres. In that case the source function is assumed extended from the last depth interval to infinity; the matrix Λ^+ is then not purely triangular, but has an extra element below the principal diagonal in the last row of the matrix.

The expression (3.3) can also be obtained from the differential equation (3.1) of transfer (*cf.* eq.'s 5.1 and 5.2). The elements of the Λ-matrix then have different values from those obtained from the formal integral, but the triangular structure is exactly the same. However, greater care must be taken in this approach since the numerical solution of the first-order equation can be unstable if the monochromatic optical step size is large and if the differencing is half-implicit. Half-implicit differencing is recommended only for small interval length, where it gives second-order accuracy; but for intervals with optical depth steps much larger than unity, this differencing can give highly unstable solutions, which may oscillate in sign and grow in amplitude; for very large intervals the differencing must be fully implicit (*cf.* Kalkofen & Wehrse 1982).

This approach of deriving the Λ-operator from the first-order differential equation can be used to advantage in constructing an approximate Λ-operator when the monochromatic interval length is small, thereby avoiding the time-consuming exponential functions which are required for the discretized integral expression (*cf.* section 5).

By combining the equation for the outward intensity $I^+_{\nu\mu}$ with an analogous equation for the inward intensity $I^-_{\nu\mu}$, which also involves a triangular matrix but now with the complementary structure, we obtain an equation for the mean intensity along a ray,

$$J_{\nu\mu} = \Lambda_{\nu\mu} S_{\nu\mu} + g_{\nu\mu} \quad , \tag{3.4}$$

and by integrating over angle and frequency, using the definition (2.5) of the source function, we find the mean integrated intensity,

$$\bar{J} = \Lambda S + K S_c + g \quad , \tag{3.5}$$

in terms of the integral operators Λ and K of the line and continuum source functions, respectively.

When we insert the transfer equation (3.5) into the equation (2.7) of statistical equilibrium we obtain an integral equation for the line source function,

$$[1-(1-\epsilon)\Lambda]S = \epsilon B + (1-\epsilon)g + (1-\epsilon)KS_c \quad , \tag{3.6}$$

which may be written as

$$\mathcal{L}S = \varphi \quad , \tag{3.7}$$

with the integral operator $\mathcal{L}$,

$$\mathcal{L} = 1-(1-\epsilon)\Lambda \quad , \tag{3.8}$$

and the inhomogeneous term φ,

$$\varphi = \epsilon B + (1-\epsilon)KS_c + (1-\epsilon)g \quad . \tag{3.9}$$

The problem is linear if the driving term φ of the integral equation (3.7) is known on the optical depth scale. The equation can then be solved directly, with the solution time scaling approximately as the cube of the number of depth points. If the sources S_c and B are known only as functions of the geometrical depth, the problem is non-linear. The best approach then is to linearize the problem completely (*cf.* Kalkofen 1984a,b; Scharmer 1984). This leads to the same form (3.7) of the equations, which must now be solved iteratively.

4. ITERATIVE SOLUTION BY OPERATOR PERTURBATION

A variety of transfer problems can be expressed in the form of linear integral equations. This is true not only of the line transfer problem chosen here as the example but also of model atmosphere problems in radiative equilibrium, and both of the linear and non-linear cases (*cf.* Kalkofen 1984b). As discussed in the preceding sections, the linear problems can be solved directly in a single pass. This is often not the most efficient procedure, however. It is usually more economical to perturb the exact operator $\mathcal{L}$ of the problem (3.7) about a simpler operator, to solve the simpler problem, and to base the calculation of corrections to the solution on the error incurred. This is achieved by the operator perturbation method. It will be described here not in the form originally proposed by Cannon (1973, *cf.* 1984, eq.'s 10-13), but more in the spirit of the form proposed by Scharmer & Nordlund (1982) and Scharmer (1984, eq.'s IV6-8) in which the corrections to the solution are based on the error in the conservation equation, a form particularly useful for non-linear problems.

There are two phases in the calculation that make it attractive to seek alternatives to the direct solution: (1) the construction of the accurate integral operator, which can be very time-consuming; and (2), the solution of the system (3.7) of equations, which can be time-consuming as well, especially in multi-level problems. The perturbation method is faster than the direct solution if three calculations are performed very efficiently: the construction of the approximate matrix operator, the solution of the system of linear equations, and the calculation of the error in the conservation equation.

The basic approach of the operator perturbation method is to divide the problem into two parts: in the first, corrections to the source function are calculated with the aid of an approximate integral operator, and in the second, the error made by a solution vector in satisfying a conservation equation is determined.

We describe the operator perturbation method for the case of a line transfer problem, where the integral equation is given by

$$\mathcal{L}S = \varphi \tag{4.1}$$

(*cf.* Cannon 1984, eq. 45; Scharmer 1984, eq. IV-1).

We assume that the operator $\mathcal{L}$ can be approximated by the simpler operator L; later (below and in section 7) we consider what conditions the simpler operator must meet in order to be considered an approximation to the exact operator. We write the exact operator as

$$\mathcal{L} = L + (\mathcal{L} - L) \quad , \tag{4.2}$$

assuming that the difference of the two operators is small compared to the operators themselves; we may imagine that we have an ordering parameter, λ, in front of the expression in parentheses, which we consider the first-order term. And we write the source function as an expansion,

$$S^{(n)} = \sum_{i=0}^{n} s^{(i)} \quad , \quad S = S^{\infty} \quad . \tag{4.3}$$

It is easy to see that the terms in the source function expansion are ordered by the same parameter λ; thus, writing the integral equation (in Cannon's formulation) as

$$[L + (\mathcal{L} - L)]\,(s^{(0)} + s^{(1)} + \ldots) \;=\; \varphi \quad , \tag{4.4}$$

the two lowest orders of the solution are given be

$$\begin{aligned} Ls^{(0)} &= \varphi \\ Ls^{(1)} &= -(\mathcal{L} - L)s^{(0)} \end{aligned} \quad , \tag{4.5}$$

which show that if the first-order matrix $\mathcal{L} - L$ is of order $\lambda \times L$, the first-order source term $s^{(1)}$ is of order $\lambda \times s^{(0)}$. Thus the terms in the source function expansion are ordered by the same parameter.

We introduce the expressions containing the approximate integral operator and the partial sums of the source function expansion into the integral equation, taking all terms to order $(n+1)$. In the resulting equation,

$$LS^{(n+1)} + (\mathcal{L} - L)S^{(n)} = \varphi \quad , \tag{4.6}$$

we carry the term involving the exact operator to the right-hand side. Then

$$L(S^{(n+1)} - S^{(n)}) \;=\; \mathcal{E}^{(n)} \quad , \tag{4.7}$$

where

$$\mathcal{E}^{(n)} = \varphi - \mathcal{L}S^{(n)} \quad , \tag{4.8}$$

which is the error made by the nth approximation of the source function in the original integral equation. Thus, the error $\mathcal{E}^{(n)}$ made by the nth approximate solution in the conservation equation (4.1) becomes the driving term in the integral equation for the $(n+1)^{th}$ partial sum,

$$S^{(n+1)} = S^{(n)} + L^{-1}\mathcal{E}^{(n)} \quad . \tag{4.9}$$

By executing formally the program defined by equations (4.7) and (4.8) we gain insight into the nature of the perturbation and obtain a criterion for the convergence of the equations. For the zeroth-order solution, assume that the source function is zero, $S^{(0)} = 0$. The zeroth-order error is then $\mathcal{E}^{(0)} = \varphi$, and the first-order solution, $S^{(1)} = L^{-1}\varphi$. (Note that if $\mathcal{L}$ were the exact operator, $S^{(1)}$ would be the solution of the problem.) By successive substitution of the solution back into the error expression the nth approximation becomes

$$S^{(n)} = \sum_{i=0}^{n-1}(1 - L^{-1}\mathcal{L})^i L^{-1}\varphi \quad , \tag{4.10}$$

and after an infinite number of steps,

$$S = \sum_{i=0}^{\infty}(1 - L^{-1}\mathcal{L})^i L^{-1}\varphi \quad . \tag{4.11}$$

In the sum we recognize the expansion of an inverse operator. Thus,

$$S = \left[1 - (1 - L^{-1}\mathcal{L})\right]^{-1} L^{-1}\varphi \quad , \tag{4.12}$$

which becomes

$$S = \mathcal{L}^{-1}\varphi \quad ; \tag{4.13}$$

this is again the exact solution. Of course, we expect that it will not take an infinite number of iterations to reach this result.

Equation (4.11) shows that the operator perturbation is essentially an expansion of the inverse of the exact integral operator about a simpler operator.This expansion will converge provided the powers of the operator $(1-L^{-1}\mathcal{L})$ converge to the null matrix. A necessary and sufficient condition for this to be true is that the maximal (absolute) eigenvalue of this operator be less than unity. (This is not as straightforward to check as it may seem since we don't intend to construct the exact operator, except when we are forced to in order to check out a computer program; *cf.* section 7). This condition for convergence to the null matrix is trivially satisfied when the approximate operator is equal to the exact operator.

Before leaving the topic of the operator perturbation proper we consider the application of the method to a time-dependent problem. In that case the initial estimate we have used above, where we supposed that the zeroth-order source function is zero, does not do justice to our knowledge of the solution at any given time step, except the first. In fact, the converged solution at the preceding time step may be a vastly better starting solution than the one we obtain when we begin with the solution $S^{(0)} = 0$, especially when we have been ingenious in cutting the construction time for the approximate operator so that L is not particularly close to $\mathcal{L}$.

Suppose the transfer problem at the preceding time, t_{p-1}, is described by the equation

$$\tilde{\mathcal{L}}\tilde{S} = \tilde{\varphi} \quad , \tag{4.14}$$

with the different integral operator $\tilde{\mathcal{L}}$ and driving term $\tilde{\varphi}$. The best procedure is then to expand the series for the source function at the current time step, t_p, about this solution; in the nth approximation the source function expansion is then

$$S^{(n)} = \tilde{S} + \sum_{i=1}^{n} s^{(i)} \quad , \tag{4.15}$$

with the solution

$$S^{(n)} = \sum_{i=0}^{n-1}(1 - L^{-1}\mathcal{L})^i L^{-1}\varphi + (1 - L^{-1}\mathcal{L})^n \tilde{S} \quad . \tag{4.16}$$

It differs from the previous solution by the additional term arising from the initial solution $\tilde{S}$.

The result shows that the property of the approximate operator that guarantees the convergence of the series (4.11) also guarantees that the converged solution is independent of the initial estimate. Of course, it means also that convergence is assured no matter what that initial estimate is. This is in striking contrast to Newton-Raphson iteration, where a poor starting solution dooms the calculation. The reason for this different behavior is that in the operator perturbation, the approximate operator L and the starting solution $S^{(0)}$ are chosen separately, whereas in the Newton-Raphson method the choice of the starting solution implies a definite choice of the matrix operator. There is still another difference: whereas Newton-Raphson iterations converge, when they do converge, quadratically, the equations in the operator perturbation converge linearly. We must therefor rely on the quality of the approximate operator L to obtain rapid convergence.

5. CONSTRUCTION OF AN APPROXIMATE INTEGRAL OPERATOR

The speed of the operator perturbation method depends on the speed with which we construct the approximate integral operator, solve the system of coupled equations, and calculate the error. We consider here the approximate operator.

The time-consuming part in the construction of the operator $\mathcal{L}$ is the angle and frequency-integrated operator Λ. There are three places where time can be saved, namely, by limiting the number of angle and frequency points and by reducing the number of quadrature points in depth for determining the intensity from the source function.

We know what condition the approximate operator must satisfy in order to ensure convergence of the series expansion for the source function. But that does not tell us how to construct such an operator efficiently. To begin with, recall that the operator $\mathcal{L}$, whether we are concerned with a line transfer problem or a model atmosphere, has its origin in the half-range integral operators $\Lambda^{\pm}$ of the transfer equation. These must correctly reproduce two properties of the transfer: First, no extra photons must be either created or destroyed, at least not at a rate competing with the physical rate of energy emission, which is set by the term ϵB; this property is equivalent to a normalization condition of the approximate Λ-operator, which implies, basically, that the intensity emerging from a semi-infinite isothermal medium in which the source function is constant is equal to that source function. Second, the frequency-integrated operator must correctly describe the photon transport in the line wings — nothing much is happening in the line core, at least far from the outer surface and far from inner surfaces. This suggests that we incorporate the core saturation approximation into the construction of the approximate transfer operator.

Consider a typical line transfer problem for a strong line (*i.e.*, assuming $\kappa_c/\kappa_{\nu\mu} \ll 1$ at line center) in a semi-infinite medium: the depth grid, which might be prescribed in terms of the optical depth at line center (times $\sqrt{\pi}$), has step sizes increasing monotonically with depth, often with a fixed number of grid points per decade in τ. Now, the distance a photon flies between emission and absorption is of the order of unit monochromatic optical distance; and the relation between the reference optical distance and the monochromatic optical distance along a ray is $\delta\tau_{\nu\mu} = \varphi_{\nu\mu}/\mu\,\delta\tau$ (for a strong line). Thus, a photon of frequency ν arriving at a grid point has been emitted a distance $\delta\tau \approx \mu/\varphi_{\nu\mu}$ away. If the distance to the nearest grid point is much larger than $\delta\tau$, the monochromatic intensity in the outward direction is given to a good approximation by the source function plus its gradient; the inward intensity along the same ray is given by the source function minus its gradient; and the mean monochromatic intensity is equal to the source function (plus usually negligible corrections that depend on the curvature of the source function and higher, even derivatives).

Thus, we may assume that the monochromatic mean (along the ray) intensity is equal to the source function, and the contribution of the intensity to the net transition rate is negligible whenever the optical distance (along a ray) to the nearest grid points in both the upstream and downstream directions exceeds a certain value. This condition defines the core of the line; and the approximation that the intensity is equal to the source function is essentially the core saturation approximation (Rybicki 1972), with the modification that it is based not on the optical depth of a given point but on the optical distance to the nearest neighbors (*cf.* Kalkofen & Ulmschneider 1984). This definition facilitates the treatment of surfaces inside the atmosphere, such as those arising from shocks, for example.

If the operator $\mathcal{L}$ is constructed with the aid of the net rate operator $\mathcal{N} \equiv \Lambda - 1$, the contribution to the matrix $\mathcal{N}$ is zero in the line core (except for frequencies for which the contribution of the background continuum cannot be neglected). For typical media (without shocks, for example) this is also true for all depth points that lie deeper than the one just considered.

Outside the line core, *i.e.*, in the line wings, the intensity at the depth point τ_i depends on conditions far from τ_i. At such frequency points the intensity is different from the source function; the radiative transfer contributes to the net transition rate and, hence, to the net rate operator $\mathcal{N}$. But these contributions are weighted by the profile function $\varphi_{\nu\mu}$, which is very small in the line wings. Therefore, sufficiently far in the line wing the contribution of the intensity to the net rate is also small. Thus, only a small number of frequency points near the boundary between line core and line wings contributes significantly to the net transition rate.

Consider a grid point i inside the medium and suppose that the next grid point is at an optical distance of $\delta\tau = 50$ (on the reference τ scale). For a strong Doppler broadened line the monochromatic optical separation $\delta\tau_\nu$ of the two depth points is unity at a frequency of $\nu = 2.0$ (taking $\mu = 1$). Now consider the contribution at other frequencies to the integral for the mean integrated intensity: For $\nu = 3.0$, the contribution of the monochromatic intensity is reduced relative to that for $\nu = 2.0$ by the corresponding ratio of the profile functions, *i.e.*, to less than 1%. Clearly, unless the source function increases very steeply on both sides of the grid point i, monochromatic intensities beyond $\nu = 3.0$ can safely be ignored for an approximate calculation of the mean integrated intensity, or for the construction of an approximate integrated Λ-operator. On the other hand, at a frequency of $\nu = 1.5$, the interval size is about $\delta\tau_\nu = 6$ (*cf.* Figure 1). Unless the source function near τ_i has very strong curvature, the core saturation approximation should be justified, in which the monochromatic intensity is set equal to the source function in the integral for the mean integrated intensity (*cf.* Rybicki 1972), and the contribution to the net rate operator is set equal to zero. Thus, only a small number of frequency points near the boundary between line core and line wings must be considered when the grid spacing is very wide. Of course, near the surface of the medium, all frequencies belong to the wings; but even here not all frequencies are important; because of the rapid decline with frequency of the profile function, frequencies in the far wings make negligible contributions to the integrals. Thus, at $\nu = 2.15$ the profile function is smaller than at line center by a factor of 100. Hence, even at the surface not all frequency points need to be taken into account.

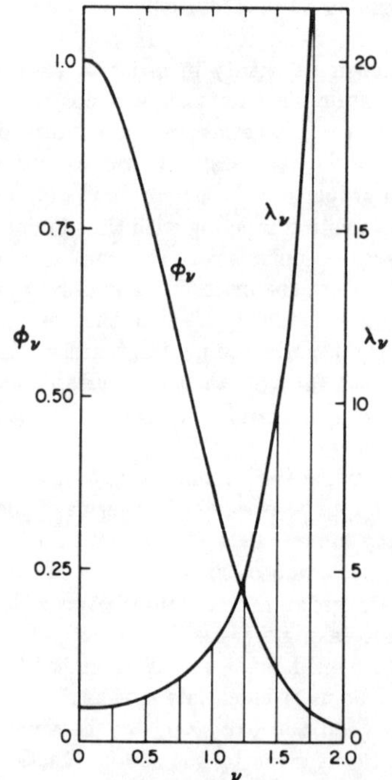

Figure 1. The profile function φ_ν for Doppler broadening and the corresponding mean-free-path λ_ν relative to the line center mean-free-path as functions of the frequency ν measured from line center in units of the Doppler width.

While the exact solution must be calculated with several angle points, a single angle point should be adequate for the approximate operator, especially in a static atmosphere. (In a moving medium somewhat more care may be required if the description is in the rest frame of the observer since then the profile function is anisotropic.)

Now, for the calculation of the matrix elements of the half-range Λ –operators on the depth grid we might discretize the formal solution, but instead of using an expansion of the source function in terms of piecewise quadratic segments with weighted leading and trailing parabolas (*cf.* Kalkofen 1974) we might use the simpler linear expansion (*cf.* Avrett & Loeser 1984).

A much faster construction of Λ and L that achieves essentially the same end is obtained by using not the formal integral but the original differential equation (3.1) on which the formal integral is based. The integration formulae are given by

$$I_i^+ = a I_{i+1}^+ + b S_i + c S_{i+1} \quad , \tag{5.1}$$

and

$$I_i^- = a I_{i-1}^- + b S_i + c S_{i-1} \tag{5.2}$$

(*cf.* Kalkofen & Ulmschneider 1984), where the integration weights are defined in terms of the monochromatic optical distance, δ, to the nearest grid point in either the upstream or the downstream direction (along the ray inclined with the angle cosine μ against the outward normal) by

$$a = \frac{2-\delta}{2+\delta}, \qquad b = c = \frac{\delta}{2+\delta}, \qquad \delta < \delta_0 \, . \tag{5.3}$$

The integration weights correspond to half-implicit differencing. For larger interval length we use either the weights from the differential equation,

$$a = c = \frac{1}{2\delta+1}, \qquad b = \frac{2\delta-1}{2\delta+1}, \qquad \delta > 1 \, , \tag{5.4}$$

which agree with the weights (5.3) for $\delta = 1$, or the weights from the formal integral (3.2) for piecewise linear source function,

$$a = e^{-\delta}, \qquad b = 1 - \frac{1-e^{-\delta}}{\delta}, \qquad c = e^{-\delta}\left(\frac{e^{\delta}-1}{\delta} - 1\right) \quad , \tag{5.5}$$

which may be used for any value of δ. The latter weights are more accurate than the weights (5.4); for $\delta = .1$ they agree with the weights (5.3) to 2% or better.

The formulae (5.1) and (5.2) lead to recursion relations of the form

$$I^+ = \sum_{j=i}^{k} \Lambda_{ij}^+ S_j \quad , \tag{5.6}$$

where the upper limit k is set by the condition that the optical distance along the ray from i to k be suitably large or, that the last contributing interval, $(k-1, k)$ be sufficiently long ($\delta > 5$, perhaps); the latter condition is preferable since it takes into account that the integration weights for the intensities between the points k and i can be determined in the same calculation.

The boundary condition in this integration is that the intensity in the last contributing interval is equal to the corresponding source function. It may also be desirable to truncate the series when a weight Λ_{ij} has become small enough compared to the largest element, typically $\Lambda_{i,i\pm1}$. This is particularly useful for the incoming radiation since it leads to an approximate operator L with nearly triangular structure.

An even faster way of constructing the approximate integral operator is to use Scharmer's (1981; 1984, eq.'s III-2 to 6) prescription for the specific monochromatic intensity,

$$I^{\pm}_{\nu\mu}(\tau_{\nu\mu}) = w^{\pm}(\tau_{\nu\mu})\, S(\tau^{\pm}_{\nu\mu}) \quad , \tag{5.7}$$

which is essentially a generalization of the well-known Eddington-Barbier relation to the interior of a medium. For the outward direction the integration weight and the source point are given by

$$w^{+} = 1 \quad , \quad \tau^{+}_{\nu\mu} = \tau_{\nu\mu} + 1 \quad ; \tag{5.8}$$

for the inward beam the weight and the source point are

$$w^{-} = 1 - \mathrm{e}^{-\tau_{\nu\mu}} \quad , \quad \tau^{-}_{\nu\mu} = \tau_{\nu\mu}/w^{-} - 1 \quad . \tag{5.9}$$

The search for the source points $\tau^{\pm}_{\nu\mu}$ requires a search for the interval in which it is located. This must be carried out fast for the calculation to benefit from the simple prescription. Each row of the monochromatic net rate operator then contains only four or five non-zero elements. The exponential function in the expression (5.9) for the inward radiation takes into account the finite distance to the outer surface; for a finite medium, the expression for the outward beam must be similarly modified.

Thus, the construction time for the approximate integral operator L will be much shorter than that for the exact operator $\mathcal{L}$ since the approximate Λ−operator is determined for only one angle point typically, the frequency points are restricted to the vicinity of the boundary between line core and line wings, and the number of depth points is small either because of the use of the Eddington-Barbier relation or because of the neglect of radiation from optically remote parts of the medium.

6. SOLUTION OF THE SYSTEM OF EQUATIONS AND ERROR CALCULATION

For the solution of equations of the type $LS = \mathcal{E}$ we make two observations: (i) the matrix operator L has band structure and (ii), equations of this type must be solved repeatedly, with the same matrix L but different driving terms $\mathcal{E}$.

The band structure of the matrix is a consequence of the approximations that were made in the construction, which limit the depth range over which the properties of the medium can influence the intensity. We can also modify the structure of the matrix after we have determined it by zeroing out in each row all elements that are sufficiently small compared to the largest element. In this we must pay attention to the physics of the transfer, *i.e.*, we must retain terms that describe the long-range interaction due to scattering, which transfers information over a long distance. For a line in a static medium broadened by the Doppler effect, this distance is of the order $1/\epsilon$. (That the operator perturbation method treats this long-range interaction in the approximate operator L, and not in $\mathcal{L}$, becomes clear below.)

Since the matrix equation with the same matrix operator L must be solved more than once, it is advantageous not to start the solution process from the beginning every time. Since L has band structure, or is nearly triangular (*cf.* Scharmer 1984, Figure 2), it is efficient to write the matrix as the product of two triangular matrices, $L = lu$ (*cf.* Fox 1962, chapter 22; Scharmer & Nordlund 1982), where l is a lower ($l_{ij} = 0,\, i < j$) and u an upper ($u_{ij} = 0,\, i > j$) triangular matrix. This triangular decomposition, or $l \times u$ factorization, needs to be done only once. The solution then proceeds via the equations $l\psi = \mathcal{E}$ and $uS = \psi$, with the auxiliary vector ψ, which must be determined every time the matrix equation is solved with a new right-hand side $\mathcal{E}$. The equations are solved forwards for ψ and backwards for S.

Now, for the calculation of the error (4.8) made by the approximate solution $S^{(n)}$ in the conservation equation, $\mathcal{E}^{(n)} = \varphi - \mathcal{L}S^{(n)}$, we recall the definition of the matrix $\mathcal{L} = 1-(1-\epsilon)\Lambda$, or in terms of the net rate operator $\mathcal{N} \equiv \Lambda - 1$, when it becomes $\mathcal{L} = \epsilon - (1-\epsilon)\mathcal{N}$. The operation with $\mathcal{L}$ on the source function,

$$\mathcal{L}S^{(n)} = S^{(n)} - (1-\epsilon)\Lambda S^{(n)} \quad , \tag{6.1}$$

produces essentially the mean integrated intensity $\bar{J}$,

$$\bar{J}^{(n+\frac{1}{2})} = \Lambda S^{(n)} + KS_c + g \quad . \tag{6.2}$$

The purpose of using the half-integral index on the intensity is to indicate that we intend to solve the transfer equation for the known source function $S^{(n)}$ in a Λ-iteration, which improves the solution, but not by as much as by a full iteration step. This is a very efficient calculation since it involves only the scalar Feautrier equation, *i.e.*, one difference equation for each frequency and angle point. The accuracy of this calculation can be made very high without compromising the efficiency of the operator perturbation method by taking many angle and frequency points and by solving the equation using Auer's Hermite scheme.

The source function that corresponds to this mean integrated intensity may be written as

$$S^{(n+\frac{1}{2})} = (1-\epsilon)\bar{J}^{(n+\frac{1}{2})} + \epsilon B \quad . \tag{6.3}$$

We introduce the transfer equation (6.2) and the source function equation (6.3) into equation (6.1), which may then be written as

$$\mathcal{L}S^{(n)} = S^{(n)} - S^{(n+\frac{1}{2})} + \varphi \quad , \tag{6.4}$$

and the error $\mathcal{E}^{(n)}$ becomes

$$\mathcal{E}^{(n)} = S^{(n+\frac{1}{2})} - S^{(n)} \quad . \tag{6.5}$$

The correction equation for the source function is then

$$S^{(n+1)} - S^{(n)} = L^{-1}(S^{(n+\frac{1}{2})} - S^{(n)}) \quad . \tag{6.6}$$

Thus the difference between the source function estimate $S^{(n)}$ and the improved source function $S^{(n+\frac{1}{2})}$, obtained by a Λ –iteration, is the driving term in the integral equation for the next source function estimate. Because of the use of the Λ-iteration, the difference $S^{(n+\frac{1}{2})} - S^{(n)}$ contains only the short-range interaction of the radiation field. The long-range interaction due to scattering is contained in the approximate operator. Therefore, if the operator L is truncated in order to facilitate the $l \times u$ decomposition of the matrix L into two triangular matrices, care must be taken that this long-range behavior not be destroyed.

To summarize the error calculation: For the known source function $S^{(n)}$ we solve the scalar transfer equation (2.1), one equation for each frequency and angle point. Since the transfer equations are now decoupled, this calculation is fast. It uses either the Feautrier scheme (2.11) or the corresponding Auer scheme, the latter giving higher accuracy but also risking unphysical solutions if the source function estimate varies strongly with depth. Since the Feautrier solution will give only well-behaved intensities when the source function is well-behaved, it may be safer to use the Feautrier equation at the beginning of a calculation. From the specific intensities we then compute the mean integrated intensity (2.8) and with the aid of the source function (6.3), the error term $\mathcal{E}^{(n)}$.

7. CONVERGENCE CRITERION

In section 4 we have seen that the convergence of the perturbation equations depends only on the properties of the approximate integral operator and is independent of the starting solution. Convergence is assured when the eigenvalues of the matrix $\mathcal{M}$,

$$\mathcal{M} = 1 - L^{-1}\mathcal{L} \quad , \tag{7.1}$$

are all smaller than unity in absolute value. They are all zero when L is equal to the exact operator — the equations (4.10) then converge in one iteration; and the maximal eigenvalue is larger than unity when L is a poor approximation to the exact operator $\mathcal{L}$ — then the equations diverge.

If we want to check whether the matrix $\mathcal{M}$ satisfies the convergence criterion of equation (4.10) we must construct the exact operator $\mathcal{L}$, which must be consistent with the differential operator used to solve the transfer equation (6.2). It is straightforward although laborious to determine that operator.

The transfer equation that is solved in the error calculation in order to determine the mean integrated intensity (6.2) is given by

$$T_{\nu\mu}\, J_{\nu\mu}^{(n+\frac{1}{2})} = S_{\nu\mu}^{(n)} \quad , \tag{7.2}$$

where $T_{\nu\mu}$ is the tri-diagonal difference operator and $S_{\nu\mu}^{(n)}$ is the nth approximation of the total source function of the transfer problem. This equation is the monochromatic analogue of the Feautrier equation (2.11). The formal solution of equation (7.2) is

$$J_{\nu\mu}^{(n+\frac{1}{2})} = T_{\nu\mu}^{-1}\, S_{\nu\mu}^{(n)} \quad . \tag{7.3}$$

We obtain the monochromatic full-range Λ –operators by inverting the tri-diagonal operator $T_{\nu\mu}$ of the Feautrier equation. By integrating the equation over frequency and angle we find the integral operators Λ and K. The expression for the mean integrated intensity $\bar{J}$ then differs from the expression (3.5) by the term g. Thus, apart from this boundary term we have constructed the exact integral operator $\mathcal{L}$ by exploiting the relation between the Λ –operator and the inverse differential operators. Finally, we calculate the matrix L using this Λ –operator and hence $\mathcal{M}$, for which we determine the eigenvalues. We can then judge the quality of the approximations made in the construction of the approximate integral operator L.

8. SUMMARY AND CONCLUSIONS

The operator perturbation method in the integral equation formulation leads to fast and accurate solutions of the radiative transfer equation subject to integral constraints. Its basic strategy is to separate the problem into two parts: the calculation of corrections, using an ***approximate integral operator***, and the calculation of an error, using the ***exact differential operator***. It is essential that the approximate operator incorporate the basic physics of the exact operator. In a line transfer problem this concerns the creation and destruction of photons (requiring the normalization of an operator) and the transfer in the line wings (expressed by the core saturation approximation). The speed of the method is achieved: (i) by economies in the construction of the approximate integral operator, which can be obtained with Scharmer's weights for the depth quadrature or with weights from the first-order transfer equation neglecting the influence of optically remote parts of the medium, and by using only one angle point and treating the transfer only near the boundary between line core and line wings; and (ii) by exploiting the nearly triangular structure of the correction matrix. The accuracy is due to the use of the second-order difference equation of transfer, either in the Feautrier or in the Auer version. Another aspect, not considered in this review but important for very small scattering parameters, is the near cancellation of some intensity-dependent terms in the equations which must be carried out analytically if the numerical calculation is performed on a computer with insufficient word length (*cf.* Scharmer 1984, Scharmer & Carlsson 1984).

For the use of the operator perturbation method in time-dependent problems of radiation-hydrodynamics, it may be necessary to squeeze every time advantage out of the method. Strategies to that end might be to proceed as follows: (i) compute the approximate operators Λ and L for only one angle point and few frequency points; (ii) truncate the operator L for the $l \times u$ decomposition; (iii) exploit the band structure in solving the equations utilizing the matrices l and u; (iv) in the error calculation, start with few angle-frequency points and proceed to the full set; (v) determine the error initially with the Feautrier formulation, later with the Auer formulation; (vi) in multi-level problems, freeze early the part of the matrix L referring to resonance lines (or some analogous part of the transfer problem) and freeze the part referring to subordinate lines later; (vii) update the operator not after every time-step but only when needed.

Not all these steps deserve the same weight since they do not affect the computation time to the same extent (for timing estimates, *cf.* Scharmer 1984, Table 1). For problems with many-level atoms, the most time-consuming part of the calculation is the solution of the set of coupled matrix equations; it is the only part that grows with the square of the number of depth points and of the number of levels of the atom. For such problems, the band structure of L and the triangular decomposition deserve the greatest attention.

The operator perturbation method has been described here for a line transfer problem in the integral equation formulation. It can, of course, also be used in other problems that have been completely linearized, where the error made by a solution in a conserved quantity is the driving term in the correction equation.This is generally the case in the complete linearization of the integral equation (*cf.* Kalkofen 1974, Kalkofen & Wehrse 1984, Scharmer & Carlsson 1984). The procedure is then to decouple the calculation of the correction matrix from the calculation of the solution vector. The error of the solution must be computed to the same accuracy as that demanded in the final solution, but for the correction matrix an approximate operator is adequate. The operator perturbation method then introduces an inconsistency between the correction matrix and the error of the solution into the problem. Convergence will therefore not be quadratic as in the Newton-Raphson case, but since the quadratic convergence is found only when the error is already small, the effect on the number of iterations should be slight.

The operator perturbation method should lead to significant savings for many kinds of physical problems. Its basic procedure can be stated concisely with reference to the complete linearization of an integral equation in a Newton-Raphson iteration: Instead of making the ***correction*** of an assumed solution consistent with the ***error*** of the solution, the correction is computed with an operator that contains the essential physics of the problem but is otherwise approximate. Only the error is determined accurately.

ACKNOWLEDGMENTS

This paper is dedicated to the memory of Chris Cannon. I am grateful to George Rybicki for a critical reading of the manuscript.

REFERENCES

Athay, R. G., 1976. ***The Solar Chromosphere and Corona: Quiet Sun***, Reidel, Dordrecht.

———. 1984. ***Methods in Radiative Transfer (MRT)***, Cambridge University Press, Cambridge, 79.

Auer, L. H., 1976. ***J. Quant. Spectr. Rad. Transf.***, **16**, 931.

———. 1984. ***(MRT)***, 237.

Avrett, E. H. & Loeser, R. 1984. ***(MRT)***, 341.

Canfield, R. C., McClymont, A. N., & Puetter, R. C. 1984. ***(MRT)***, 101.

Cannon, C. J., 1973. ***J. Quant. Spectrosc. Rad. Transfer***, **13**, 627.

———. 1984. ***(MRT)***, 157.

Fox, L., 1962. ***Numerical Solution of Ordinary and Partial Differential Equations***, Pergamon Press, Oxford.

Frisch, H. 1984. ***(MRT)***, 79.

Kalkofen, W., 1974. ***Astrophys. J.***, **188**, 105.

———. 1984a. ***(MRT)***, 427.

———. 1984b. This volume.

Kalkofen, W. & Ulmschneider, P., 1984. ***(MRT)***, 131.

Kalkofen, W. & Wehrse, R., 1984. ***(MRT)***, 307.

Mihalas, D. 1978. ***Stellar Atmospheres***, Freeman, San Francisco.

Newton, R. G. 1966. ***Scattering Theory of Waves and Paricles***, McGraw Hill, New York.

Nordlund, Å. 1984. ***(MRT)***, 211.

Rybicki, G. B., 1972. in ***Line Formation in the Presence of Magnetic Fields***, ed. R. G. Athay, L. L. House, & G. Newkirk, Jr. (Boulder: High Altitude Observatory), p 145.

———. 1984. ***(MRT)***, 21.

———. 1985. This volume.

Scharmer, G. B., 1981. ***Astrophys. J.***, **249**, 720.

———. 1983 ***Astron. Astrophys.***, **117**, 83.

———. 1984. ***(MRT)***, 173.

Scharmer, G. B. & Carlsson, 1984. ***J. Comp. Phys.***, in press.

Scharmer, G. B. & Nordlund, Å., 1982. ***Stockholm Obs. Rep.*** **19**.

PARTIAL VERSUS COMPLETE LINEARIZATION

Wolfgang Kalkofen
Harvard-Smithsonian Center for Astrophysics
60 Garden Street
Cambridge, Massachusetts 02138
USA

ABSTRACT. The convergence properties of the partially or completely linearized equations for a grey model atmosphere in radiative equilibrium are compared. The completely linearized equations show the quadratic convergence properties of Newton-Raphson equations. When the opacity depends strongly on temperature, the convergence of the partially linearized equations is very slow initially but improves once the maximal error has moved to the lower boundary.

1. INTRODUCTION

Typical transfer problems in stellar atmospheres involve non-linear equations. Such equations are frequently solved by complete or by partial linearization. In choosing one or the other method one considers that the former achieves the fast convergence of Newton-Raphson iterations whereas the latter requires simpler matrices in the integral equation formulation but may fail to converge if a variable that depends strongly on the solution is excluded from the linearization and, instead, is updated only between iterations. Such may be the case in a model atmosphere where the temperature and density structure of the medium are to be determined if the dependence of the opacity on the kinetic temperature of the gas is not weak compared to that of the Planck function. It is the intent of this paper to show for the case of the construction of a simple model atmosphere with known structure that complete linearization is preferable if the opacity depends strongly on temperature.

The early approach to the problem of constructing a model atmosphere was to assume a temperature structure, to determine the errors in the flux and the flux derivative implied by the assumed solution, and to base the temperature correction on these errors (Unsöld 1955, Avrett & Krook 1963). A more efficient way to proceed is to linearize the equations partially with respect to temperature by perturbing the Planck function only and to solve the equations for the temperature correction self-consistently, *i.e.*, to the extent to which the opacity can be assumed to be temperature-independent. In the differential equation method, this procedure couples the monochromatic transfer equations via the energy equation (Auer & Mihalas 1969); the problem is then to solve this system of coupled differential equations. In the integral equation method, the linearization of the Planck function with respect to temperature leads to a matrix equation for the temperature correction (Böhm-Vitense 1964). If the opacity is in fact temperature-independent (and also pressure-independent), these procedures constitute complete linearizations of the problem and therefore lead to the quadratic convergence of the Newton-Raphson iterations.

The partial linearization is still in use. It is an efficient approach when the opacity depends only weakly on temperature (and pressure) since then the updating of the relative optical

J. E. Beckman and L. Crivellari (eds.), Progress in Stellar Spectral Line Formation Theory, 169–174.

depths in the problem does not disturb the convergence. Numerical methods that proceed in this manner are those of Wehrse (1977) and Schmid-Burgk (1975), for example. But when the temperature dependence of the opacity is stronger than that of the Planck function, the linearization of the equations with respect to temperature only in the Planck function neglects the main effect of the temperature, and the difficulties in the convergence are aggravated by the long "reach" of a temperature perturbation in the opacities, which occurs via the optical depth. The advantage of partial linearization, that the operators for the perturbed equations are nearly the same as those for the unperturbed equations, is then vitiated by the neglect of the most important part of the temperature dependence of the problem. In extreme cases the equations fail to converge. This could arise in late-type stars, for example, if the temperature in the outer layers becomes low enough for dust grains or molecules to form. Because of the steep temperature dependence of the opacity, the partially linearized equations may be unsuitable for the solution of such problems (*cf.* Kalkofen 1984a).

The purpose of this contribution is to investigate the convergence properties of partially and completely linearized equations for a model atmosphere in a case where the temperature dependence of the opacity is not weak compared to that of the Planck function. In order to keep the problem simple we solve the equations for an atmosphere for which the exact solution is known: We take a grey atmosphere in the two-stream approximation in which the opacity depends only on temperature, in the form T^n; the integrated Planck function depends on T as T^4. We solve the problem first in the optical depth space, where the solution for the integrated Planck function is known; in the two-stream approximation it is a linear function of the optical depth. We then solve it in a geometrical space, perturbing either only the Planck function — the partial linearization case — or both the Planck function and the opacity — the complete linearization case — and taking as zeroth-order a solution that differs from the exact one by a known amount. In section 2 we describe the equations of the problem, in section 3 we discuss the numerical solution, and in section 4 we present conclusions.

2. THE BASIC EQUATIONS

Consider a one-dimensional model atmosphere with grey opacity in radiative equilibrium. In the two-stream approximation with the angle cosine $\mu = \pm 1$, the transfer equation for the frequency-integrated specific intensity may be written as

$$\pm \frac{dI^{\pm}}{dz} = \kappa \left(I^{\pm} - B \right) \quad , \tag{1}$$

where I^+ and I^- are the inward and outward directed intensities, respectively, and where κ is the opacity and B the integrated Planck function. The net flux of the atmosphere,

$$H = \frac{1}{2}(I^+ - I^-) \quad , \tag{2}$$

is a prescribed constant, $\mathcal{H}$.

Note that the geometrical depth is measured inward, in the same sense as the optical depth τ, which is defined by the equation

$$d\tau = \kappa(z)\, dz \quad . \tag{3}$$

It is assumed that if κ depends on pressure, this dependence can be factored out and absorbed into the definition of the depth z, which is then not the geometrical scale. The pressure structure of the atmosphere on the τ, z, and geometrical scales can be calculated by solving the hydrostatic equilibrium equation after $T(\tau)$ and $T(z)$ have been determined.

The transfer equation in terms of optical depth is given by

$$\pm\frac{dI^{\pm}}{d\tau} = I^{\pm} - B \quad , \tag{4}$$

and the formal solution for the specific intensity in the two directions in a semi-infinite atmosphere may be written as

$$I^{\pm}(\tau) = \Lambda^{\pm}(\tau,\tau')B(\tau') \tag{5}$$

(*cf.* Kalkofen 1985), with the half-range integral operators Λ^{+} and Λ^{-}. The equation describing the temperature structure of the atmosphere is

$$\Phi(\tau,\tau')B(\tau') = H(\tau) \quad , \tag{6}$$

where the integral operator Φ is defined by

$$\Phi(\tau,\tau') = \frac{1}{2}[\,\Lambda^{+}(\tau,\tau') - \Lambda^{-}(\tau,\tau')\,] \quad . \tag{7}$$

For constant net flux and zero incident radiation the solution for the Planck function is

$$B(\tau) = \mathcal{H}(1+\tau) \quad . \tag{8}$$

Note that the integral equation (6) for the Planck function is linear; thus the direct solution of the equation solves the problem.

Consider now the associated non-linear problem. Assume that the solution is to be determined on a geometrical depth grid, and ignore the short cut by which we could take the solution (8) just obtained and determine for it the opacity and integrate the equation (3) defining the optical depth; we would find the geometrical grid points corresponding to the optical depth grid points, and then interpolate to the geometrical grid points we want. — This is in fact what we do in order to calculate the exact solution of the problem on the z −grid.

Instead, we solve the problem now on the geometrical grid. We use the approach of Kalkofen (1974) to construct the operators for the linearization of the integral equation of transfer. In this method, the integral operators of the perturbed problem are developed from the first-order differential equation of transfer for the specific intensity. This is much simpler than the method chosen by Skumanich and Domenico (1971), where the linearization is taken in the integral expressions.

We first linearize the equation about a trial solution, assuming starting values for the temperature and hence for the opacity and the Planck function, *i.e.,*

$$\kappa = \kappa_0 + \delta\kappa \quad , \tag{9}$$

and

$$B = B_0 + \delta B \quad . \tag{10}$$

By solving the zeroth-order equation,

$$\pm\frac{dI_0^{\pm}}{dz} = \kappa_0(I_0^{\pm} - B_0) \quad , \tag{11}$$

we determine the specific intensity corresponding to these starting values on the zeroth-order optical depth scale,

$$d\tau_0 = \kappa_0 dz \quad , \tag{12}$$

using the formal integral (5). This implies the net flux, H_0,

$$H_0(\tau_0) = \frac{1}{2}(I^+(\tau_0) - I^-(\tau_0)) \quad . \tag{13}$$

The first-order transfer equation is

$$\pm\frac{d\delta I^{\pm}}{dz} = \kappa_0 \delta I^{\pm} + \delta\kappa I_0^{\pm} - \delta\kappa B_0 - \kappa_0 \delta B \quad . \tag{14}$$

It may be written in terms of the optical depth of the zeroth-order problem as

$$\begin{aligned} \pm\frac{d\delta I^{\pm}}{d\tau} &= \delta I^{\pm} - \left(\frac{\delta\kappa}{\kappa_0}(B_0 - I_0^{\pm}) + \delta B\right) \\ &= \delta I^{\pm} - \left(\frac{\partial \ln\kappa}{\partial T}(B_0 - I_0^{\pm}) + \frac{\partial B}{\partial T}\right)\delta T \end{aligned} \quad . \tag{15}$$

Note that the source function of the first-order transfer equation is anisotropic since it contains the specific intensity of the zeroth-order solution.

The formal integral of the transfer equation for the first-order intensity in the outward direction is

$$\begin{aligned} \delta I^+(\tau) &= \Lambda^+(\tau,\tau')\left(\frac{\partial \ln\kappa}{\partial T}(B_0 - I_0^+) + \frac{\partial B}{\partial T}\right)_{\tau'} \delta T(\tau') \\ &= \tilde{\Phi}^+(\tau,\tau')\delta T(\tau,\tau') \end{aligned} \quad ; \tag{16}$$

with the integral operators $\tilde{\Phi}^{\pm}(\tau,\tau')$ given by

$$\tilde{\Phi}^{\pm}(\tau,\tau') = \Lambda^{\pm}(\tau,\tau')\left(\frac{\partial \ln\kappa}{\partial T}(B_0 - I_0^{\pm}) + \frac{\partial B}{\partial T}\right)_{\tau'} \quad , \tag{17}$$

and the first-order flux operator by

$$\tilde{\Phi}(\tau,\tau') = \frac{1}{2}\left[\tilde{\Phi}^+(\tau,\tau') - \tilde{\Phi}^-(\tau,\tau')\right] \quad . \tag{18}$$

The temperature correction δT is then determined from the integral equation

$$\tilde{\Phi}(\tau,\tau')\delta T(\tau') = \delta H(\tau) \quad , \tag{19}$$

where δH is the flux error made by the zeroth-order solution,

$$\delta H(\tau) = \mathcal{H} - H_0(\tau) \quad . \tag{20}$$

Note that the first-order problem (19) has the same structure as the zeroth-order problem (6).

3. NUMERICAL SOLUTIONS

We have constructed model atmospheres in a depth space of 25 points covering the optical depth range from $\tau = 0$ to $\tau = 100$ with monotonically increasing interval length. The opacity had the temperature dependence

$$\kappa(T) = a + bT^n \quad . \tag{21}$$

We first determined the exact solution for $B(\tau)$ and hence $T(\tau)$ using equation (8), and $B(z)$ and $T(z)$ using the relation (3) between optical and geometrical depths. We then perturbed the temperature $T(z)$ by multiplying it by a constant factor, and computed for the perturbed temperature the corresponding opacity and the associated optical depths. In this depth space we evaluated the formal integral for the zeroth-order specific intensity and calculated the zeroth-order net flux. Given the flux error (20) we then solved the first-order integral equation (19) for the temperature perturbation, either with the operators (17) corresponding to complete linearization, or with the simpler isotropic operator

$$\tilde{\Phi}^{\pm}(\tau,\tau') = \Lambda^{\pm}(\tau,\tau')(\frac{\partial B}{\partial T})_{\tau'} \quad , \tag{22}$$

in which only the Planck function is perturbed and which corresponds to partial linearization. Note that the operators (17) and (22) are identical when the opacity does not depend on temperature; in that case the completely and the partially linearized equations are the same.

We discuss the case $a = b = 1$ and $n = 4$, where the opacity (21) depends on T^4; the exact surface value of the temperature is $T(0) = 1$. The initial perturbation was to double the temperature throughout. This led to a large flux error at the surface, but relatively minor errors at depth in spite of the large temperature perturbation because the flux there responds mainly to the gradient of the Planck function. Table 1 gives the maximal net flux relative to the exact solution; it shows rapid convergence when the equations are completely linearized; after five iterations the equations had converged to an error in the flux (*i.e.*, the numbers listed minus unity) below 1%, and a still smaller error in the temperature. Even though the maximal error, which is here the error at the surface, is positive, negative values are found in the interior. Clearly, the convergence shows the expected behavior of the Newton-Raphson method.

The partially linearized equations show completely different convergence properties. The flux error remained positive throughout and for all iterations and decayed only slowly. It had its maximum at first at the surface but then moved gradually inward. Once the maximal flux error occurred at the deepest depth point, which happened in the 9th iteration, convergence speeded up considerably and was much faster than one might have estimated on the basis of the earlier behavior. If the deepest depth point had been at a greater optical depth it clearly would have taken longer for the partially linearized equations to converge.

The pattern just described was found also when the opacity depended less strongly on temperature, but it was less pronounced. And when it depended on a higher power of the temperature than the fourth, it could happen that the completely linearized equations failed to converge whereas

TABLE 1: Relative Flux Extrema

	Linearization	
Iteration	Complete	Partial
0	8.9	8.9
1	2.7	4.0
2	1.45	2.4
3	1.11	2.0
4	1.04	1.79
5	1.009	1.67
6		1.58
7		1.52
8		1.46
9		1.42
...		..
15		1.008

the partially linearized equations maintained their poor convergence properties. By mixing the two operators, *i.e.*, by multiplying the term containing $\partial \ln \kappa / \partial T$ by a factor between zero and unity, convergence could be improved. We do not know whether this difficulty is intrinsic to the problem or is a consequence of the way in which it was solved. The case was not further studied since a large temperature perturbation throughout the atmosphere coupled with a strong temperature dependence in the opacity may have only limited practical interest.

4. CONCLUSIONS

We have discussed the construction of model atmospheres in radiative equilibrium with a grey opacity that depends only on temperature. The problem was formulated either as a linear integral equation in terms of optical depth, or as a non-linear integral equation in terms of geometrical depth. For the latter formulation the equations were solved either by means of complete or by means of partial linearization of the equations with respect to temperature. The convergence properties of the iterative solutions were compared by measuring their net fluxes relative to the known flux of the exact solution. The starting solution was obtained by multiplying the exact solution throughout the atmosphere by a constant factor.

If the opacity depends strongly on temperature, partial linearization leads to slow convergence. A strong temperature dependence in the opacity clearly calls for the complete linearization of the equations. This may complicate some approaches, but any difficulties are largely compensated if operator perturbation is used because in that case the integral operator for the first-order problem need not be the exact operator; an approximate operator incorporating the essential properties of the exact one will do.

In a practical case of model atmosphere construction, where the exact solution is not known, there is an additional hazard with partial linearization. When the iterative solutions change sufficiently slowly this might be misinterpreted as convergence. Thus, partial linearization may risk not only the inefficiency of slow convergence, it may also yield apparently converged solutions that still carry a large error.

REFERENCES

Auer, L. H. & Mihalas, D. 1969. *Ap. J.*, **156**, 157.
Avrett, E. H. & Krook, M. 1963. *Ap. J.*, **137**, 874.
Böhm-Vitense, E. 1964. *Proceedings of First Harvard-Smithsonian Ap. Obs., Spec. Rep.*, **167**, 99.
Kalkofen, W., 1974. *Astrophys. J.*, **188**, 105.
__________. 1984. *Methods in Radiative Transfer*, Cambridge University Press, Cambridge 427.
__________. 1985. This volume.
Schmid-Burgk, J. 1975. *Astron. & Astroph.*, **40**, 249.
Skumanich, A. & Domenico, B. A. 1971. *J. Quant. Spectrosc. Rad. Transfer*, **11**, 547.
Unsöld, A. 1955. *Physik der Sternatmosphären*, Springer Verl., Berlin.
Wehrse, R. 1977. *Astron. & Astroph.*, **59**, 283.

RADIATIVE TRANSFER DIAGNOSTICS: UNDERSTANDING MULTI-LEVEL TRANSFER CALCULATIONS

Andrew Skumanich
High Altitude Observatory
National Center for Atmospheric Research*

Bruce W. Lites
National Solar Observatory
National Optical Astronomy Observatories**

ABSTRACT. Efficient methods are now available for solving multi-level, multi-transition non-LTE transfer problems in stellar atmospheres in the approximation of complete redistribution. Once such a solution is obtained, however, it is frequently desirable to know more about the solution than simply the emergent intensity or the flux. We present a means by which one may easily understand how all the line and continuum transitions are formed. We cast each transition into an equivalent two-level form, consisting of a scattering and a source term. The behavior of these terms with height immediately reveals how the line excitation is driven. Numerical perturbations of the atomic rates, both collisional and radiative, about this solution reveal the sensitivity of the scattering and source terms to various atomic rate and helps one to identify the other interlocking radiative and collisional transitions important to the formation of a spectral line. We demonstrate this method through application to the formation of the solar hydrogen lines and the Lyman continuum.

I. INTRODUCTION

In multi-level, multi-transition radiative transfer problems the interactions between levels become important and the nature of the source function for the transition may differ radically from that based on a two-level approximation to the atom. In this paper we present methods that, once the solution to a multi-level problem has been found, reduce any transition to its <u>equivalent</u> two-level form. We make use of the fact that the source function for any transition has an explicit dependence upon its own (mean) radiation field (a scattering term), and also has a term representing the true sources of photons for the transition, a term with no explicit dependence upon the self radiation field.

---

*The National Center for Atmospheric Research is sponsored by the National Science Foundation.

** Operated by the Association of Universities for Research in Astronomy, Inc., under contract with the National Science Foundation.

J. E. Beckman and L. Crivellari (eds.), Progress in Stellar Spectral Line Formation Theory, 175–187.

Lites and Skumanich (1982, Paper I) have shown how this decomposition was useful not only in understanding the formation of resonance lines in a model of a sunspot chromosphere, but also in aiding the development of that model. Adjustment of the temperature structure of the model without the aid of these parameters was accomplished largely by trial and error. With the parameters, we were able to see which layers of the sunspot model contributed to the source of photons for the Ca II, Mg II, and H I resonance line, thereby allowing us to adjust the model to better fit all of these lines.

In this paper we present our decomposition of the photon source term and the scattering "albedo" into contributions representing the background opacity at the transition in question, collisional excitations, and excitations due to other paths, e.g. via absorption in interlocking transitions. This decomposition is useful, for example, in cases where it is desired to know whether radiative or collisional excitations are dominant at a given level in a model stellar atmosphere. We also show how we may guage the sensitivity of the source and scattering terms of this equivalent two-level representation to all other collisional and radiative rates within the model atom. This sensitivity analysis proves useful in isolating the particular interlocking transitions and collisional rates important to the transition under study. The information provided by the sensitivity analysis can lead to simplification of model atoms used in transfer problems, and it also helps to identify those atomic processes for which accurate atomic data must be known.

2. FORMULATION OF THE EQUIVALENT TWO-LEVEL PARAMETERS

Consider an atom with N levels including a continuum. The statistical equilibrium equations for an ensemble of such atoms can be put in the form

$$T\,\mathbf{n} = \mathbf{b} \tag{1}$$

Here $n = (n_1, \ldots, n_{N-1}, n_N)^t$ where t = transpose, and represents the state of the system with n_j being the number density state of atoms in level j and $b = (0, \ldots, 0, n_T)^t$ where n_T is the total number density. The matrix T is derived from the N - 1 individual rate equations which have the usual form

$$\frac{dn_j}{dt} = \sum_{l(\neq j)=1}^{N} n_l\,(R_{lj} + C_{lj}) - n_j \cdot \Gamma_j = 0, \qquad j = 2, N \tag{2a}$$

and the conservation equation

$$\sum_{J=1}^{N} n_j = n_T \tag{2b}$$

If one includes stimulated emissions as negative absorptions then one has that $R_{lj} = A_{lj}$ for downward radiative transitions ($l>j$) and $R_{lj} =$

$B_{lj}\,\bar{J}^{t}_{lj}$ for upward transitions (l<j), and $\Gamma_j = \sum_{l\neq j}^{N} (R_{lj} + C_{lj})$. The quantities A, B, C are the customary Milne form of the Einstein coefficients and collision rates respectively. Finally, $\bar{J}^{t}_{lj}$ represent a suitable frequency-angle average of the radiation field for the transition in question.

An explicit solution of Equation (1) is given by

$$\mathbf{n} = T^{-1}\,\mathbf{b}\,. \tag{3}$$

It is easy to convince oneself, ref. Cuny (1967), that if one considers the two states i, j (>i) with $n_i = (T^{-1}b)_i$ and $n_j = (T^{-1}b)_j$ and forms the frequency independent source function

$$S_{ij}^{sp} = \frac{2h\nu_0^3}{c^2}\,\frac{n_j}{n_i}\,\frac{g_i}{g_j} = \frac{2h\nu_0^3}{c^2}\,\frac{(T^{-1}b)_j}{(T^{-1}b)_i}\,\frac{g_i}{g_j} \tag{4}$$

where ν_0 is the central frequency of a line transition, or, for a ionizing transition, the frequency of the continuum edge (ionization frequency) then S^{sp} will have the form

$$S^{sp}_{ij} = (1-\varpi'_{ij})\,\bar{J}^{t}_{ij} + Q'_{ij}\,. \tag{5}$$

The quantities $(Q', \varpi)_{ij}$ are <u>independent</u> of $\bar{J}^{t}_{ij}$. We note that this is not the case if one solves, as does Athay (1972), the appropriate equation in (2a) for the ratio n_j/n_i. In this case (Q, ϖ) contain an implicit dependence on $\bar{J}^{t}_{ij}$ through the other occupation ratios that occur.

In general, one may not neglect the influence of a background opacity and associated radiation at the i->j transition, i.e. in $\bar{J}^{t}_{ij}$, and it is often useful to gauge the effect of this background opacity upon the formation of a <u>weak</u> line or a continuum transition. It is possible, to separate the influence of i->j radiative absorptions caused either by a photon previously emitted by a spontaneous j->i transition, or by a photon emitted by the background source. That is, one can write,

$$\bar{J}^{t}_{ij} = (1-\delta)\,\bar{J}^{sp}_{ij} + \delta_{ij}\,\bar{J}^{b}_{ij}\,. \tag{6}$$

where δ_{ij} is the probability that a radiative interaction is a background interaction while $\bar{J}^{sp}_{ij}$ is a suitable mean of the self or spontaneous radiation field while $\bar{J}^{b}$ is that for the background sources.

With decomposition (6) our basic equation (5) for the line source function becomes

$$S_{ij}^{sp} = (1-\tilde{\varpi}_{ij})\,\bar{J}_{ij}^{sp} + Q_{ij}\,. \tag{7a}$$

where

$$Q_{ij} \equiv Q'_{ij} + Q^b_{ij} \equiv Q'_{ij} + (1-\tilde{\varpi}'_{ij})\,\delta_{ij}\,\bar{J}^b\,. \tag{7b}$$

and

$$\tilde{\varpi}_{ij} \equiv \tilde{\varpi}'_{ij}(1-\delta_{ij}) + \delta_{ij}\,. \tag{7c}$$

This equation represents the equivalent two-level representation of the excitation of the i->j transition for our multi-level atom. Since $\bar{J}_{ij}^{sp}$ is a functional (through a generalized lambda transform) of S_{ij}^{sp}, equation (7a), given (Q_{ij}, ϖ_{ij}), yields a closed (integral) equation for S_{ij}. Indeed, the various scaling properties of the two-level transfer problem should hold for equation (7). Thus a check on the solutions to the multi-level problem, is that, when cast in the equivalent two-level form, the quantities S, Q, ϖ satisfy scaling laws similar to these for the pure two-level case.

We now formulate more specifically what we mean by S^{sp}, $\bar{J}^{sp}$, $\bar{J}^b$, Q, ϖ, A_{ji}, B_{ji}, and δ in Eqs. 4-7. The radiative rates for a transition i->j (i) are, following Mihalas (1978),

$$\textit{downward}:\; n_j R'_{ji} = 4\pi \int \frac{j_\nu^{sp}}{h\nu}\,d\nu = n_j\left[4\pi \frac{g_i}{g_j} \int \frac{\alpha_{ij}(\nu)}{h\nu}\frac{2h\nu^3}{c^2} e^{-x}\,d\nu\right] \equiv n_j A_{ji}\,. \tag{8}$$

$$\textit{upward}:\; n_i R'_{ij} = 4\pi \int \frac{\kappa_\nu^{sp}}{h\nu} J_\nu^t\,d\nu = n_i\, 4\pi \int \frac{\alpha_{ij}(\nu)}{h\nu} p_\nu J_\nu^t\,d\nu \equiv n_i B_{ij} \bar{J}^t. \tag{9}$$

where we have included stimulated emissions as negative absorptions via the term $p_\nu = (1 - \frac{n_j}{g_j}\frac{g_i}{n_i}e^{-x})$ with $x=h(\nu-\nu_0)/kT$. The Einstein coefficient of spontaneous decay A_{ji} as defined by Eq. (8) reduces to its usual meaning in the case of a line transition since $\nu=\nu_0$ and $x=0$. We treat continuum transitions in the same fashion as line transitions and require that the Einstein coefficient B_{ij} for absorption be given by

$$B_{ij} = \frac{g_j}{g_i}\left[\frac{2h\nu_0^3}{c^2}\right]^{-1} A_{ji} = 4\pi \int \frac{\alpha_{ij}(\nu)}{h\nu}\left[\frac{\nu}{\nu_0}\right]^3 e^{-x}\,d\nu\,. \tag{10}$$

For j = continuum we take $g_j = U_j \frac{2}{n_e}\left(\frac{2\pi m_e kT}{h^2}\right)^{3/2}$ where U_j is the ion statistical weight. Requirement (10) defines, via Eq. (9), the angular-frequency integrated intensity $\bar{J}^t$.

To determine the functional dependence of $\bar{J}^t$ on S^{sp} we proceed in terms of the integral solution to the transfer equation as represented by the familiar Λ-transform, $J_\nu^t = \Lambda_\nu^t\, S_\nu^t$. The Λ-operator can be expressed either as a "path" and solid angle integral (Kourganoff, 1963) or as a triple-space integral. The quantity S_ν^t is the total source function for a particular transition and is expressed in the usual way with respect to the self and background absorption and emission coefficients. Thus the mean intensity may be decomposed into a self and background term, viz.

$$J_\nu{}^t = \Lambda_\nu^t S_\nu{}^t = \Lambda_\nu{}^t\left[(1-\delta_\nu)\, S_\nu{}^{sp}\right] + \Lambda_\nu^t[\delta_\nu\, S_\nu{}^b]$$
$$\equiv (1-\delta)_\nu\, J_\nu{}^{sp} + \delta_\nu\, J_\nu{}^b \tag{11}$$

where $\delta_\nu = K_\nu^b/(K_\nu^{sp}+K_\nu^b)$. We now consider the defining equation for $\bar{J}^t$, Eq. 9, and express it in terms of λ-operators and source functions. We first decompose R'_{ij} as

$$R'_{ij} = R_{ij}^{sp} + R_{ij}^b$$

with

$$R_{ij}^{sp} = 4\pi \int \frac{\alpha_{ij}(\nu)}{h\nu}\, p_\nu \Lambda_\nu^t \left[(1-\delta_\nu)\, S_\nu^{sp}\right] d\nu\,. \tag{12}$$

R^b is given by a similar expression. Noting that the frequency-independent source function defined by Eq. (4) is related to the true source function by

$$S^{sp} = \frac{2h\nu_0^3}{c^2}\frac{n_j}{g_j}\frac{g_i}{n_i} = \left(\frac{\nu_0}{\nu}\right)^3 p_\nu\, e^{x}\, S_\nu^{sp}\,. \tag{13}$$

we find that the term in Eq. 12 due to the absorption of transition-generated photons may be written

$$\begin{aligned} R_{ij}^{sp} &= 4\pi \int \frac{\alpha_{ij}(\nu)}{h\nu}\, e^{-x}\left[\frac{\nu}{\nu_0}\right]^3 (1-\delta_\nu)\, \Lambda_\nu^{sp}\, S^{sp}\, d\nu\,. \\ &\equiv (1-\delta)\, B_{ij} \int \tilde{\Phi}(\nu)\, d\nu\, \Lambda_\nu^{sp}\, S^{sp} \\ &\equiv (1-\delta)\, B_{ij}\, \bar{J}^{sp} \end{aligned} \tag{14}$$

where the "new" Λ^{sp}-operator is defined in terms of the "old" operator by the similarity transformation

$$\Lambda_\nu^{sp} = \left[\frac{p_\nu e^{x}}{(1-\delta_\nu)}\right] \Lambda_\nu^t \left[\frac{p_\nu e^{x}}{(1-\delta_\nu)}\right]^{-1}. \tag{15a}$$

While the normalized profile function $\tilde{\Phi}(\nu)$ is given by

$$\tilde{\Phi}(\nu) = \frac{(1-\delta_\nu)}{(1-\delta)}\, \Phi(\nu) \tag{15b}$$

with $\Phi(\nu)$ as the standard profile function (for both lines and continua) which follows from the definition of B_{ij} in Eq. (17), viz.

$$\Phi(\nu) = \alpha_{ij}(\nu)\, \nu^2\, e^{-x} / \int \alpha_{ij}(\nu)\, \nu^2\, e^{-x}\, d\nu\,. \tag{15c}$$

We note that

$$\delta = \int \delta_\nu = \Phi(\nu)\, d\nu\,. \tag{16}$$

Equation (14) yields the decomposition we seek and shows the functional dependence of $\bar{J}^{sp}$ on S^{sp}. A similar relation holds for $\bar{J}^b$ but we shall not examine the explicit form for the background sources. We will simply calculate Q^b directly.

The reason for introducing this similarity transformation in Eq. 15 is that it preserves the normalization of the Λ-transformation for both lines and continua with (total) optical depth, and thus preserves, we believe, the appropriate transfer properties such as effective saturation scale, photon trapping and build-up as well as the scaling of surface and asymptotic values. It is this "scaling" insight that allows one to assess the accuracy and consistency of numerical solutions. The well known line case follows directly if one sets x=0 and $\nu = \nu_o$.

3. METHOD FOR COMPUTATION OF EQUIVALENT TWO-LEVEL PARAMETERS

We derive all of our radiative transfer diagnostic parameters, and the sensitivity of these parameters to the interlocking transitions, by alegbraic manipulation of the atomic and radiative rates in the statistical equilibrium equation. For all but the simplest atomic models the rate equations become too complex to allow an analytic solution for $(Q, \tilde{\omega})$. We have found the numerical methods for extraction of the equivalent two-level atom parameters of Eq. 7 to be very accurate and stable, most likely because the system of rate equations is well-conditioned for nearly all model atoms.

First, we determine the sum of all sources of photons for the transition, i.e. $Q = Q' + Q^b$. Note that if we take $\bar{j}_\nu^{sp} = 0$, then $\bar{J}^{sp} = 0$ in Eq. 14 and the solution (3) yields $S^{sp}\ (j_\nu^{sp} = 0) = Q$, cf. Eq. (7). Thus if we use the formal solution to the transfer equation with the solution populations, but use a modified total emissivity for the transition in which we include only the background emissivity, we will obtain an upward radiative rate from the resulting mean intensity that is representative of the background photon sources alone. We then use this (artificial) rate, in which $J^{sp} = 0$, to re-solve the system of statistical equilibrium equations with the given collisions and interlocking terms and find a set of populations, n_i^o from which we derive a source function via Eq. 4, which is in fact Q, thus

$$S^0 = \frac{2h\nu_0^3}{c^2} \frac{n_j^0}{g_j} \frac{g_i}{n_i^0} \equiv Q\ . \tag{17}$$

Similarly a solution of the statistical equilibrium equations with rates calculated with $j^{sp} = 0 = j^b$ i.e. $\bar{J}^t = 0 = R'_{ij}$ allows us to obtain another set of populations and hence another source function, in the same fashion as Eq. 17, which is $S^f = Q'$. We obtain the background source of photons by $Q^b = S^0 - S^f$.

The probability δ of branching into the background or thermal reservoir is obtained by direct integration using Eq. 16. There are several ways to obtain the probability $\tilde{\omega}$ of branching out of the transition, all of which are equally accurate. We have chosen to set the transition in question into detailed radiative balance, i.e. we require $\bar{J}^t = S \equiv S^d$, then we determine S^d from a subsequent solution of the statistical equilibrium equations, and obtain $\tilde{\omega}' = Q'/S^d$. Detailed radiative balance is achieved by removing all radiative rates for that transition from the statistical equilibrium equations.

The interlocking radiative rates and both direct and interlocking collisional rates generally enter Q' in a nonlinear fashion. A complete analysis would separate Q' into components from each of the paths within the model atom leading to excitation of the upper level of a transition. Such an analysis would be very cumbersome and, in general, not terribly instructive. One measure that is of interest, however, is the degree to which collisional excitations, Q_c , contribute to Q', since these measure the local thermal excitation of the line and, to some degree, specify the influence of the local temperature upon the source of photons for the transition. We cannot hope to separate pure collisional paths precisely from radiative paths since radiative absorptions do play some role in the local net collisional contribution, but we may often isolate the bulk of the influence of collisions as follows: We set all radiation fields to zero in the statistical equilibrium equations, but we retain the spontaneous emission rates. Solution of the statistical equilibrium equations then gives us populations, from which we obtain (as described earlier) a source function that we designate as Q_{csp}. This proves to be a useful measure of Q_c .

We often want to know which are the dominant processes that "quench" the excited state, i.e. that influence the branching fraction $\tilde{\omega}$. Frequently, it is the atomic branching ratios that set $\tilde{\omega}$, and thus it usually varies little throughout a stellar atmosphere. Spontaneous decays from interlocking transitions are the major contributor for all transitions except "raie ultime" resonance lines. We find the combined contribution of collisions and spontaneous decays to $\tilde{\omega}$ in a way similar to that described above for finding Q_{csp}. With all radiation fields set to zero, we set the transition in question into detailed radiative balance, then solve the statistical equilibrium equations for the resulting populations. The upper and lower level populations derived for that transition give a source function which, when divided into Q_{csp}, gives us the quantity $\tilde{\omega}_{csp}$. The differences $Q_{rad} = Q' - Q_{csp}$ and $\tilde{\omega}_{rad} = \tilde{\omega} - \tilde{\omega}_{csp}$ measure the remaining indirect and radiative absorption contributions to the parameters Q and $\tilde{\omega}$.

In order to carry our analysis further, we need to know which specific interlocking transitions have the dominant influence upon Q and ω for any given transition. For example, it might be necessary to know if an external radiation field pumping an interlocking transition would have a measureable influence on a spectrum line under study at some particular height in a stellar atmosphere. We perform a sensitivity analysis of Q and $\tilde{\omega}$ to all the other radiation fields and collisional rates within the model atom. This analysis immediately allows us to pinpoint the interlocking transitions that are crucial to the formation of any transition in the model atom. We accomplish this through a (small) numerical perturbation of each radiation rate and each collisional rate about its value at the solution. With the perturbed rates, we resolve the statistical equilibrium equations and with the new populations find a perturbed Q + d Q and $\tilde{\omega}$ + d$\tilde{\omega}$ from which we may calculate the sensitivities

$$\left(\frac{dQ}{dR_{kl}}\right)_{soln} \cdot R_{kl}{}^{soln} \text{ and } \left(\frac{d\tilde{\omega}}{d\,R_{kl}}\right)_{soln} \cdot R_{kl}{}^{soln} . \tag{18}$$

where R_{kl} is either a collisional rate, or a upward radiation rate for the interlocking transition k->l. Downward collisional rates are included in terms of upwards rates.

4. The Formation of the Quiet Sun Hydrogen Spectrum

We now apply the formalism and decomposition discussed above to the formation of the hydrogen spectrum in the solar atmosphere. For this purpose we use the Vernazza, Avrett and Loeser 1981 (here after VAL81) temperature-mass distribution for the average quiet-Sun, i.e. VAL3C. We consider a five-level plus continuum Hydrogen atom with eleven "free" transitions, whose transfer must be solved, and four "fixed" transitions, namely the Balmer (2-6), Paschen (3-6) Brackett (4-6), and (5-6) continuua. For these fixed transitions the "boundary" radiation fields were given by Planck functions with (radiation) temperatures fixed at 4940K, 4810K, 4716K and 4738K respectively. For depths in the photosphere where temperatures are above these radiation temperatures, these radiation fields were set to the Planck function at the local temperature. Such a behavior models the depth variation of the fixed fields in an approximate way and insures the appropriate asymptotic behavior for the free transitions. The atomic cross-sections, oscillator strengths, damping constants, etc. were the same as that used by VAL81. The Lyman α, β, γ, δ lines were treated in CRD and, for consistency, as Doppler broadened only. The wings of the Balmer lines were treated as in VAL81. The transfer-excitation (ionization) hydrostatic equilibrium was solved using HCODE2A.

4.1 The Formation of the Lyman α Line

The equivalent two-level transfer parameters, except for $\tilde{\omega}$, for the Lyman α (1-2) line in the VAL3C T(m) distribution is given in Fig. 1. These quantities are very similar to those of the spot umbral atmosphere of Lites and Skumanich (1982) and indicate that the line is formed under nearly the same conditions i.e. in a high temperature plateau (the so-called "Lyman α" plateau) of similar temperature and characteristic optical thickness. This thickness is seen from the values of Q in Figure 1 to be $\tau_0^* \simeq 200$. Since the quenching probability $\tilde{\omega}$ has the nearly constant value of $\simeq 4\text{x}10^{-4}$ throughout the line forming region, the Lyman α photon range or degradation length, Athay (1972), is $\tau_0^r \simeq (\sqrt{\pi}\tilde{\omega})^{-1} = 1.4\text{x}10^3$.

With $\tau_0^* < \tau_0^r$ one has an effectively-thin condition of line formation. Hence one expects to find the value $S_{max} \simeq \sqrt{\pi}\ \tau_0^*\ \overline{Q} \simeq \sqrt{\pi}$ $(200)(6\text{x}10^{-10}) = 2\text{x}10^{-7}$ ergs/cm^2 s sr. Hz which compares quite well with the 'exact' value in Figure 1. Because the line forming region is optically thick the source function is "built-up" by trapping to be above the sources Q but, since the situation is effectively-thin, S lies below the saturation function $Q/\tilde{\omega}$. As we shall see below, the region beneath the plateau, i.e. for $\tau_0 > 200$ is fed diffusively by photons penetrating inward from the plateau and undergoing "interlocking conversion" among the various subsidiary lines.

We next consider our sensitivity analyses in an effort to identify the "sources" and "sinks" in $(Q,\tilde{\omega})$. Figure 2 presents our results for Q. It is obvious that in the plateau region Q is essentially fixed by

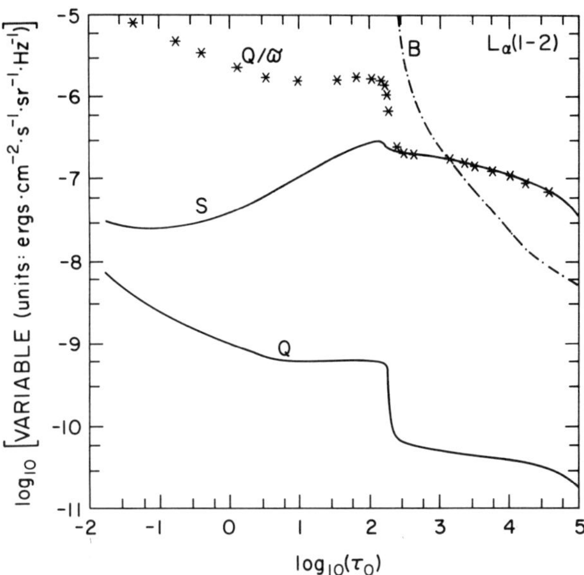

Figure 1. Transfer parameters for Lyman α vs. optical depth at line center. B' is the local Planck function (Wien Law). The saturation function Q/ϖ is ploted for each mass point. S is the spontaneous source function, Q, the source term (see text for definitions).

Q_{csp} while in the sub-plateau region, (or chromosphere), radiative interlocking, Q_{rad}, is the dominant source of Lyman α photons. In particular, we note that Q_{csp} is dominated by the (1-2) collisional excitation rate. This is the same as the εB source of two-level theory.

Thus as far as the excitation of Lyman α photons in the plateau is concerned Lyman α acts strictly as a two-level atom. It is only in the branching ratio, ϖ that a departure from the two-level atom, i.e. $\varpi \neq \varepsilon$, occurs. In the sub-plateau region the conversion of Lyman β photons, i.e. the (1-3) radiation term, dominates the interlocking conversion to Lyman α. Our results can be represented over the entire relevant range in τ_0 by the following approximation,

$$[\text{illegible}] \simeq \alpha C_{12} + \beta B_{13} J_{13} \equiv \hat{\alpha}(\varepsilon B)_{12} + \hat{\beta} J_{13} \tag{19}$$

with (α, β) given in terms of the calculated $(dQ/dR)_{soln}$. We note that our discussion of the excitation of Lyman α in the sub-plateau region (chromosphere) is incomplete without a consideration of the sources of the Lyman β radiation field. The quantity β should prove to be $\propto A_{32}$.

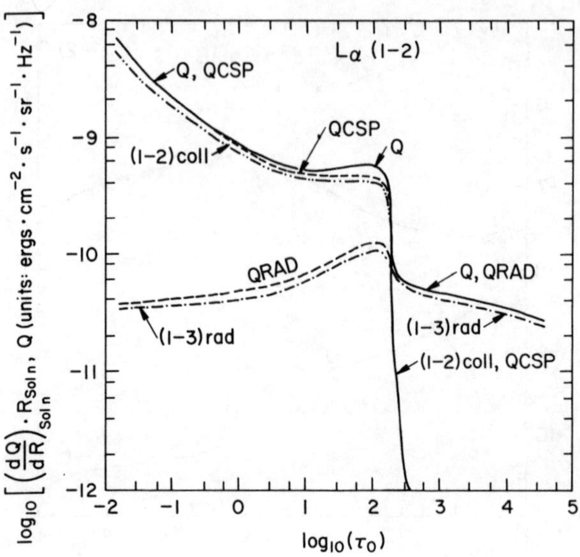

Figure 2. Decompostion of Q and sensitivity terms, (dQ/dR)R for Lyman α vs. optical depth at line center. $Q = Q_{csp} + Q_{rad}$. The numbers (1-u) identify the lower=1, and upper=u level index.

The sensitivity analysis of the quenching probability $\tilde{\omega}$ shows that it is fixed by radiative interlocking and, in particular, by Hα (2-3) radiation pumping out of the n=2 level. In a strict two-level Hydrogen atom we would expect $\tilde{\omega} \equiv \varepsilon \simeq C_{21}\ A_{21}\ (\equiv \varepsilon_{12})$. We find that the (2-1) collisional quenching, i.e. ε, plays a negligible role in $\tilde{\omega}$. Thus it is not the Planck function but $Q/\tilde{\omega} \simeq (\varepsilon/\tilde{\omega}) * B$, i.e. some fraction of the Planck function, that plays the role of the saturation function in the Lyman α plateau.

4.2 The Formation of the Lyman β line

The characteristic thickness of the "plateau" in Q for Lyman β is $\tau_0^* \simeq 20$. However here the quenching probability $\tilde{\omega} \simeq 0.5$ so that the photon degradation length $\tau_0^r \approx 1$. Thus we are in the <u>effective-thick</u> case and $S \simeq Q/\tilde{\omega}$ essentially everywhere except in the <u>optically thin</u> part of the plateau where $\bar{J}$ falls below Q here $S \simeq Q$.

Our sensitivity analysis of Q can be summarized by the approximate form

$$Q_{13} \simeq \alpha' C_{13} + \beta' J_{12} J_{23} \equiv (\varepsilon B)_{13} + \beta' J_{12} J_{23} \tag{20}$$

Once again direct (1-3) collisions, i.e. $(\varepsilon B)_{13}$, represent the true source of photons for Lyman β except in the sub-plateau (chromosphre), where pumping of the n=2 level by Lyman α (1-2) is followed directly by a Hα (2-3) absorption. This is our first instance of the role of higher order interlocking effects.

The quenching probability in a subsidiary line like Lyman β is essentially fixed by spontaneous decay via the subsidiary transition, in this case, via Hα branching. We do not consider its decomposition here since it is essentially a constant.

Our analysis shows that the excitation of Lyman α and β in the chromoshere may be driven by photons which diffuses downward from the plateau region. It is only to the extent to which Hα is excited by collisions in the chromosphere that photon diffusion may not be the only source.

4.3 The Formation of the Lyman Continuum

Here, as in Lyman β, the quenching probability $\tilde{\varpi}$ is dominated by spontaneous branching into other continuum transitions and remains at $\simeq$ 0.5 throughout the atmosphere. Once again the photon degradation length is $\simeq$ 1 so that Lyman continuum formation is effectively-thick and $S \simeq Q/\tilde{\varpi}$ through out the continuum forming region. The Lyman α plateau is now optically thin and since the upwelling Lyman continuum radiation, $\bar{J}$, is weak compared to $Q_{plateau}$ we have that $S \simeq Q$ in the plateau region. We note that the plateau irradiates the top of the Lyman continuum "photosphere" (i.e. chromosphere at $\tau_0\,(LyC) = 1$) with an intensity (at the head) of $\simeq Q_{plateau} \times \tau^*_{0\ plateau} = 8 \times 10^{-11} \times 3 \times 10^{-2} = 2.4 \times 10^{-12}$ ergs/cm^2 s sr Hz, which is small compared to the upwelling field.

Our sensitivity analysis shows that in the plateau, collisional ionization, viz., (1-6) coll, dominates the sources of Q while below the plateau the pumping of the n=2 level by Lyman α followed by a photoionization by the Balmer continuum is the dominant source. This is a similar interlocking process as in the Lyman β case. Thus we would represent our results in the form,

$$Q \simeq \alpha'' C_{16} + B'' J_{12} J_{26} \quad (21)$$

Below the plateau only the radiative term in important. The dominance of Balmer continuum pumping in this region is consistent with previous interpretations of the dominant ionization path for H in the solar chromosphere, refer Thomas and Athay (1961) and, Athay (1981). However, we find that it is Lyman α pumping and not "collisions to the n=2 level" that controls the H ionization, contrary to the quoted assertion of Athay (1981). Our result is consistent with the fact that Lyman α is saturated, i.e. in detailed radiative balance in the sub-plateau region (refer Fig. 1). In this case the population in level 2 is fixed by the Lyman α radiation field. Athay's assertion would require detailed collisional balance which we find does not occur.

4.4 The Formation of the Balmer α line

The analysis of this important line is currently in progress. Our preliminary sensitivity analysis indicates that the line is effectively thick, hence, saturated and that

$$Q \simeq \tilde{\alpha} C_{23} + \left[\frac{\tilde{\beta} J_{13}}{1 + \tilde{\alpha} J_{12}} \right] \tag{22}$$

Thus H has a collisional source, C_{23} as well as a radiative source due to pumping of the n=3 level by the Lyman β line. The Lyman β source is important both in the plateau and sub-plateau region (chromosphere). The new interlocking feature here is the role of Lyman α as a "sink". However the role of Lyman α is more complex than it appears from Eq. (22). We note from Eq. 21 that the electron density in the chromosphere is controlled by Lyman α (as well as by the Balmer continuum). Thus even the collisional contribution to Hα is radiatively controlled! It would appear that diffusion of Lyman α, β photons from the plateau into the chromosphere governs the source of chromospheric emissions.

V. DISCUSSION

We have presented here a method of interpreting the solution to a multi-level, multi-transition non-LTE transfer problem. The method represents the solutions in terms of equivalent two-level forms with a scattering and a source term. The resulting individual quenching probability, i.e. the difference of the scattering albedo from one, and source term are then decomposed by a perturbation method into their principal dependence on collisional and/or radiative rates. We have illustrated the method by considering the excitation and ionization of hydrogen in the VAL 3C model of the quiet sun chromosphere.

There are a variety of extant misconceptions concerning the nature of hydrogen excitation/ionization in the solar chromosphere. We expect our sensitivity analysis method to clarify these. The issue of the Lyman continuum (i.e. the H ionization problem) is an example. We note that Thomas and Athay (1961) assert (pg. 197) that in the region $0.1 \lesssim \tau_o \lesssim 10^2$ the Lyman continuum source function, i.e. b_1, "depends upon T_e, n_e, and the radiation field in the Lyman continuum." This was meant to imply that Lyman continuum photons escape from this region and that a deviation from saturation or detailed radiative balance occurs. In fact we find that the detailed radiative balance approximation holds here as well as for $\tau_o > 100$. Thus their use of "detailed balance in the Lyman continuum for $0.1 \lesssim \tau_o \lesssim 100$," in spite of their assertion, appears to be fully justified by our calculations.

From the point of view of computation, the method is fairly easily implemented. We have not explored here the use of our method as a guide in changing the model. The effects of perturbations in T_e (m) on the collisional rates, which represent the fundamental coupling to the thermal reservoir, or perturbations in the fixed rates, (e.g. the radiative temperature of the Balmer continuum) which represent a coupling to the underlying photosphere are fairly straightforward. However the effects of these perturbations on the "free" radiation fields, in particular

Lyman α, may require a resolution of the full problem. But even here we feel that scaling laws may help in designing an efficient scheme for iterating on the model.

5. References

Athay, R. G. 1972, Radiation Transport in Spectral Lines (Dordrecht-Holland: Reidel).

Athay, R. G. 1981, Ap. J. 250, 709.

Cuny, Y. 1967, Ann. d'Ap. 30, 143.

Jones, H. P. and Skumanich, A. 1980, Ap. J. Suppl., 42, 221.

Kourganoff, V. 1963, Basic Methods in Transfer Problems (New York: Dover).

Lites, B. W. and Skumanich, A. 1982, Ap. J. Suppl., 49, 293.

Mihalas D. 1978, Stellar Atmospheres (San Francisco: Freeman).

Thomas, R. N. and Athay, R. G. 1961, Physics of Solar Chromosphère (New York: Interscience).

Vernazza, J. E., Avrett E. H., and Loeser, R. 1981, Ap. J. Suppl., 45, 635.

A NEW METHOD FOR SOLVING MULTI-LEVEL NON-LTE PROBLEMS

G.B. Scharmer
Stockholm Observatory
S-133 00 SALTSJÖBADEN
Sweden

and

M. Carlsson
Uppsala Astronomical Observatory
Box 515
S-751 20 UPPSALA
Sweden
and
Institute of Theoretical Astrophysics
P.O.Box 1029 Blindern
N-0315 OSLO 3
Norway

ABSTRACT A new scheme for solving multi-level non-LTE problems is described. This method uses an approximate operator for the relation between the intensity and the source function. This operator results in a matrix equation for the population numbers which has a simple and characteristic structure. Solutions are obtained such that the results are "exact", irrespective of the choice of the approximate operator. The method has proven to be fast, stable and accurate and is suitable for vectorizing as well as non-vectorizing computers. This paper deals with the basic ideas of the method.

1. INTRODUCTION

The numerical expense of solving non-LTE problems comes from the large variation of the opacity with wavelength in spectral lines. These problems are often ill-conditioned numerically and numerical instabilities may occur with methods which are not carefully formulated.

In this paper we explain how the non-locality of spectral line formation problems may be understood and how this understanding can be used to create an efficient, accurate and numerically stable scheme for solving non-LTE problems. We will concentrate on the essential aspects of the method and refer the reader

J. E. Beckman and L. Crivellari (eds.), Progress in Stellar Spectral Line Formation Theory, 189–198.

to Scharmer and Carlsson (1984) and Carlsson (1984) for details.

2. FORMULATION OF THE PROBLEM

The solution of a multi-level non-LTE problem involves the simultaneous solution of the statistical equilibrium equations

$$n_i \sum_{j \neq i}^{N_l} P_{ij} - \sum_{j \neq i}^{N_l} n_j P_{ji} = 0 \tag{1}$$

the particle conservation equation

$$\sum_{j=1}^{N_l} n_j = n_{tot} \tag{2}$$

and the radiative transfer equation

$$\mu \frac{dI_{\nu\mu}}{dz} = -\kappa_{\nu\mu} I_{\nu\mu} + j_{\nu\mu} \tag{3}$$

c.f Mihalas (1978) for a derivation of these equations. In these equations n_i is the number of atoms per unit volume in level i, N_l is the total number of energy states considered and n_{tot} is the total number of atoms per unit volume (which is assumed to be a given quantity). P_{ij} is the total probability per unit time that an atom in level i will make a transition to level j. $I_{\nu\mu}$ is the specific intensity, $\kappa_{\nu\mu}$ is the absorption coefficient and $j_{\nu\mu}$ is the emission coefficient. For the purpose of discussing the basic numerical problems and the ideas behind the present method it is sufficient to assume that $\kappa_{\nu\mu} \propto n_i$ and that $j_{\nu\mu} \propto n_i$. μ is the cosine of the angle between the ray and the normal of the atmosphere.

The probabilities P_{ij} can be written as

$$P_{ij} = R_{ij} + C_{ij} \tag{4}$$

where R_{ij} is the contribution from radiative transitions and C_{ij} is the contribution from collisions with other particles. The main problem of solving non-LTE problems comes from the radiative transitions, the probabilities of which may be written as

$$R_{ij} = \begin{cases} A_{ij} + B_{ij} \bar{J}_{ij}, & \text{if } i > j; \\ B_{ij} \bar{J}_{ij}, & \text{if } i < j \end{cases} \tag{5}$$

for bound-bound transitions. In this equation A_{ij} is the Einstein spontaneous emission probability and B_{ij} is the coefficient of stimulated emission ($i > j$) and absorption ($i < j$) respectively.

The mean integrated intensity $\bar{J}_{ij}$ is defined as

$$\bar{J}_{ij} = \frac{1}{2} \int_{-1}^{1} \int_{0}^{\infty} \phi_{\nu\mu} I_{\nu\mu} d\nu d\mu \tag{6}$$

where $\phi_{\nu\mu}$ is the normalized absorption profile. The quantity $\bar{J}_{ij}$ couples the statistical equilibrium equations to the radiation field. The radiation field at any point in the gas depends in principle on emissions and absorptions at all other points in the gas i.e on the values of the population numbers of that transition at all other points in the gas. This is why non-LTE problems are generally non-local. The non-local character is particularly pronounced for line-transfer problems because of the rapid variation of the photon mean free path with wavelength in spectral lines. The small opacity in the wings allow photons emitted there to travel over very large distances before they are reabsorbed. Even though emissions in the far wings are very rare, they tend to be much more important in transferring radiation than emissions in the line core, since the latter process transfers radiation via a diffusion-like process which is very inefficient. Non-LTE problems are usually numerically ill-conditioned because of the fact that the bulk of radiation transported through the gas occurs as rare events in the wings whereas most emissions occur close to line center.

The dependence of the radiation intensity on the population numbers makes the statistical equilibrium equation non-linear. Even the transfer equation itself is non-linear, since the number of photons absorbed are proportional to the product of the specific intensity and the population density, both of which are unknown quantities.

In the line-core the photon mean free path is usually much shorter than the characteristic scale over which the source function, or the population numbers, vary appreciably. Photons which are emitted in the line core are therefore in practice absorbed locally which implies that

$$\kappa_{\nu\mu} I_{\nu\mu} \approx j_{\nu\mu} \tag{7}$$

i.e the non-linear absorption rate is balanced by a linear emission rate. If this were introduced as an approximation in the statistical equilibrium equations we would find that all non-linear terms drop out (small non-linear terms remain when overlapping continua are introduced). We conclude that the non-linearity of non-LTE problems primarily arises from the non-local part of the radiation field, i.e the part which is not emitted in the optically thick core of the line. Non-locality and non-linearity are thus strongly related concepts in spectral line formation problems.

3. LAMBDA ITERATION AND THE CORE SATURATION METHOD

In the lambda iteration scheme the radiation field and population numbers are iterated back and forth such that an initial estimate of the radiation field is used to calculate a set of population numbers, these are then used to calculate a new radiation field and so on. The reason why lambda iteration fails is that any perturbation in the population numbers at a particular depth generates a perturbation in intensity which propagates only over approximately one unit in optical depth per iteration. Therefore a perturbation in population numbers at large depth in the atmosphere propagates very slowly to the surface when lambda iteration is used.

Lambda iteration does not impose any joint constraint on the changes in population numbers and in the radiation fields. In particular it ignores an important limiting property of the optically thick part of the radiation: the balance between local absorption and emission rates at great optical depths. This property implies that whenever the population numbers are changed by a certain amount we should ensure that the optically thick part of the radiation field is also changed simultaneously so that Equation (7) remains valid during each step of the correction process. This is the basic idea of Rybicki's (1972) core saturation method. In his method a correction scheme is employed in which the radiation field and the population numbers are corrected simultaneously in the core such that detailed balance holds there, whereas the effects of the optically thin radiation field on the population numbers is corrected by lambda iteration. In this scheme the optically thick radiation field is correctly described at each iteration so that the remaining errors in the radiation field, which occur at small and intermediate optical depths, need to propagate only over rather small distances in optical depth. Therefore Rybicki's core saturation method converges much faster than the lambda iteration method.

An important advantage of using lambda iteration or Rybicki's core saturation method for solving non-LTE problems is that they eliminate the need for linearizing the transfer equation or the statistical equilibrium equation, since the statistical equilibrium equations are linear when the radiation field is known and the transfer equation is linear when the population numbers are known. The problem of linearization is that we can never be sure that the radiation fields implied in the linearization process are reasonable if the corrections are large. For example there is nothing which prevents negative radiation fields in a linearization approach, whereas with the lambda iteration method such things will not occur. Of course , the slow convergence of lambda iteration makes this method impractical. Unfortunately the core saturation method fails to describe the radiation field at intermediate optical depths sufficiently well for rapid convergence. Thus even though the convergence rate for the optically thin parts of the radiation field is very good, the over-all convergence rate of the core-saturation method is somewhat slow.

It should therefore be desirable to develop a hybrid method which uses lambda iteration for the optically thin wings and a more detailed description of the radiation field at intermediate optical depth, where much of the important radiative transfer occurs. To avoid slow convergence, it is necessary that the method is asymptotically correct at large optical depths.

4. AN APPROXIMATE LAMBDA OPERATOR

In previous papers (Scharmer 1981, Scharmer 1984) we have shown how an approximate radiative transfer operator can be constructed. This operator describes radiative transfer processes in spectral lines sufficiently well to permit for example a derivation of the so-called probabilistic equation (c.f Rybicki 1984 for a review). More important, however, is that this operator is quite general, allowing in principle the solution of multi-level non-LTE problems with velocity fields in three-dimensional media.

The operator is derived as a quadrature relation between the specific intensity and the source function. The exact relation between the intensity and the source function can be written as the usual integral

$$I^{+}_{\nu\mu}(\tau_{\nu\mu}) = \Lambda_{\nu\mu}[S_{\nu\mu}] \equiv e^{\tau_{\nu\mu}} \int_{\tau_{\nu\mu}}^{\infty} S_{\nu\mu}(\tau'_{\nu\mu}) e^{-\tau'_{\nu\mu}} d\tau'_{\nu\mu} \tag{8}$$

Most of the contributions to this integral come from a rather narrow interval in optical depth. By assuming that the source function varies linearly with optical depth

$$S_{\nu\mu} = a_{\nu\mu} + b_{\nu\mu}\tau_{\nu\mu} \tag{9}$$

we obtain an expression for the specific intensity

$$I^{+}_{\nu\mu}(\tau_{\nu\mu}) = a_{\nu\mu} + b_{\nu\mu}(\tau_{\nu\mu} + 1) \tag{10}$$

which is valid for the out-going radiation in a semi-infinite atmosphere. Comparing this solution to Equation (9) we can derive a quadrature relation of the type

$$I^{+}_{\nu\mu}(\tau_{\nu\mu}) = \omega^{+}(\tau_{\nu\mu}) S_{\nu\mu}(\tau^{+}_{\nu\mu}) \tag{11}$$

where the weight $\omega^{+}(\tau_{\nu\mu})$ and quadrature point $\tau^{+}_{\nu\mu}$ are given by

$$\omega^{+}(\tau_{\nu\mu}) = 1 \quad ; \quad \tau^{+}_{\nu\mu} = \tau_{\nu\mu} + 1 \tag{12}$$

To calculate the weight and the quadrature point for the incoming radiation ($\mu < 0$) we proceed in a similar way, obtaining (Scharmer 1981,1984)

$$\omega^{-}(\tau_{\nu\mu}) = 1 - e^{-\tau_{\nu\mu}} \quad ; \quad \tau^{-}_{\nu\mu} = \tau_{\nu\mu}/\omega^{-}(\tau_{\nu\mu}) - 1 \tag{13}$$

The asymmetry of these two expressions comes from the fact that the medium is assumed to be asymmetric, i.e having an upper boundary but no lower boundary. For a slab, which has a lower boundary, we obtain expressions which are similar for incoming and out-going rays (Scharmer 1984).

It can easily be verified that the above quadrature relations reduce to the core-saturation approximation in the limit of large optical depths, since then

$\omega^+ = \omega^- = 1$ and $\tau^+_{\nu\mu} = \tau^-_{\nu\mu} = \tau_{\nu\mu}$. In the limit of zero optical depth we obtain $\omega^- = 0$, which corresponds to zero incoming radiation at the surface and $\tau_{\nu\mu} = 1$, which corresponds to the Eddington-Barbier relation, for the out-going radiation. The quadrature relations derived above makes a gradual transition from a local relation between the intensity and the source function at large optical depths to a non-local relation at small optical depths. The quadrature relations can never be exact except for the trivial case when the source function is linear in optical depth. However, the quadrature relations give a sufficiently accurate description of the non-locally generated radiation field to enable fast convergence when used iteratively. In this respect the above approximations can be thought of as extensions of the core saturation approximation into the very important non-local regime. The fact that the quadrature relations can be used to derive a probabilistic equation which predicts the correct value for the source function at the surface, whereas the core saturation approximation fails badly, shows the importance of non-local transfer effects.

The approximations described above can be introduced into an iterative scheme such that the final solution converges to the "exact" solution independently of these approximations. This is done essentially by rewriting the relation between the intensity and the source function as

$$I^{(n)}_{\nu\mu} = \Lambda^{\dagger}_{\nu\mu}[S^{(n)}_{\nu\mu}] + (\Lambda_{\nu\mu} - \Lambda^{\dagger}_{\nu\mu})[S^{(n-1)}_{\nu\mu}] \qquad (14)$$

In this equation $\Lambda^{\dagger}_{\nu\mu}$ is an approximate operator corresponding to the quadrature relations derived above. $\Lambda_{\nu\mu}$ is an "exact" operator such that $\Lambda_{\nu\mu}[S_{\nu\mu}]$ is the "exact" value of the intensity as obtained from the source function. In the computations $\Lambda^{\dagger}_{\nu\mu}$ is represented as a matrix operator, but the quantity $\Lambda_{\nu\mu}[S_{\nu\mu}]$ is obtained by calculating the specific intensity using a known (from the previous iteration) source function and ordinary LTE techniques. Equation (14) is solved iteratively until the corrections are sufficiently small. The accuracy of the final solution depends only on the accuracy with which the intensity can be calculated from a given source function, i.e on the accuracy of the LTE-routine. The accuracy of the final solution is always independent of the accuracy of the approximate operator if the solution converges.

It is possible to formulate iterative schemes in different ways such as was done by Cannon (1973). We prefer to formulate the iterative method as a linearization scheme (Scharmer and Nordlund 1982), since this can most easily be used to solve non-linear problems.

5. A BRIEF DESCRIPTION OF THE METHOD

We now give a brief description of the method for solving non-LTE problems developed by the authors. It should be pointed out that for the purpose of clarifying the main ideas of this method we omit several important ingredients here. The interested reader can find a more detailed description of the method in Scharmer and Carlsson (1984) and in Carlsson (1984). The latter reference contains a complete listing of a code written by one of the authors and also a discussion of several test cases which has been run using this code.

The method consists of several steps: First the statistical equilibrium equations and the transfer equation are linearized with respect to the specific intensities and the population numbers. Secondly these equations are pre-conditioned in a way which removes the strong cancellation effects present in most non-LTE problems. This allows the solution of ill-conditioned non-LTE problems on 32-bit computers using single precision arithmetics. Then the approximate non-local statistical equilibrium equations are set up using the approximate lambda operator described in Section 4. To improve numerical stability, we set the approximate operator equal to zero for small optical depths, $\tau_{\nu\mu} < 0.1$. This corresponds to correcting the influence of the optically thin radiation field on the population numbers by lambda iteration. As explained in Section 3, this is a much safer procedure than linearization when the corrections are large. Lambda iteration of the optically thin radiation field does not reduce the computational speed since lambda iteration is very efficient at small optical depths. The matrix equation obtained for the population numbers has a very characteristic structure which is taken advantage of during the solution process. The last step of the solution method is an iterative procedure in which the population numbers are iterated to convergence.

It is of some interest to study the structure of a representative matrix. Figure 1 shows the matrix of a calcium II calculation. The general structure of the matrix is dominated by the approximate treatment of the non-local components of the radiation field, as formulated in terms of the approximate lambda operator. The elements to the left of the diagonal represents the approximate treatment of the incoming radiation field. The absence of matrix elements far to the left of the diagonal comes from the structure of the approximate lambda operator. The elements to the right of the diagonal contains the influence from the outgoing radiation field. The absence of elements far to the right of the diagonal implies that the influence from the distant lower layers is corrected by lambda iteration. Evidently these interactions come from the optically thin components of the radiation field. The most important interactions occur at optical depths for which the mean free path is of the order of a few grid points in depth. They give rise to bands of matrix elements fairly close to the main diagonal. This band structure is typical of non-LTE problems solved with the method described in this paper. It leads to important savings in computing time if taken advantage of during the solution process.

The fine structure of the matrix depends on the detailed collisional and radiative transitions in the model atom. This is described in more detail in the paper of Scharmer and Carlsson (1984).

6. EXPERIENCE WITH THE METHOD

A program based on this method has been written by Carlsson (1984). This program has been used to solve several "standard" non-LTE problems on dif-

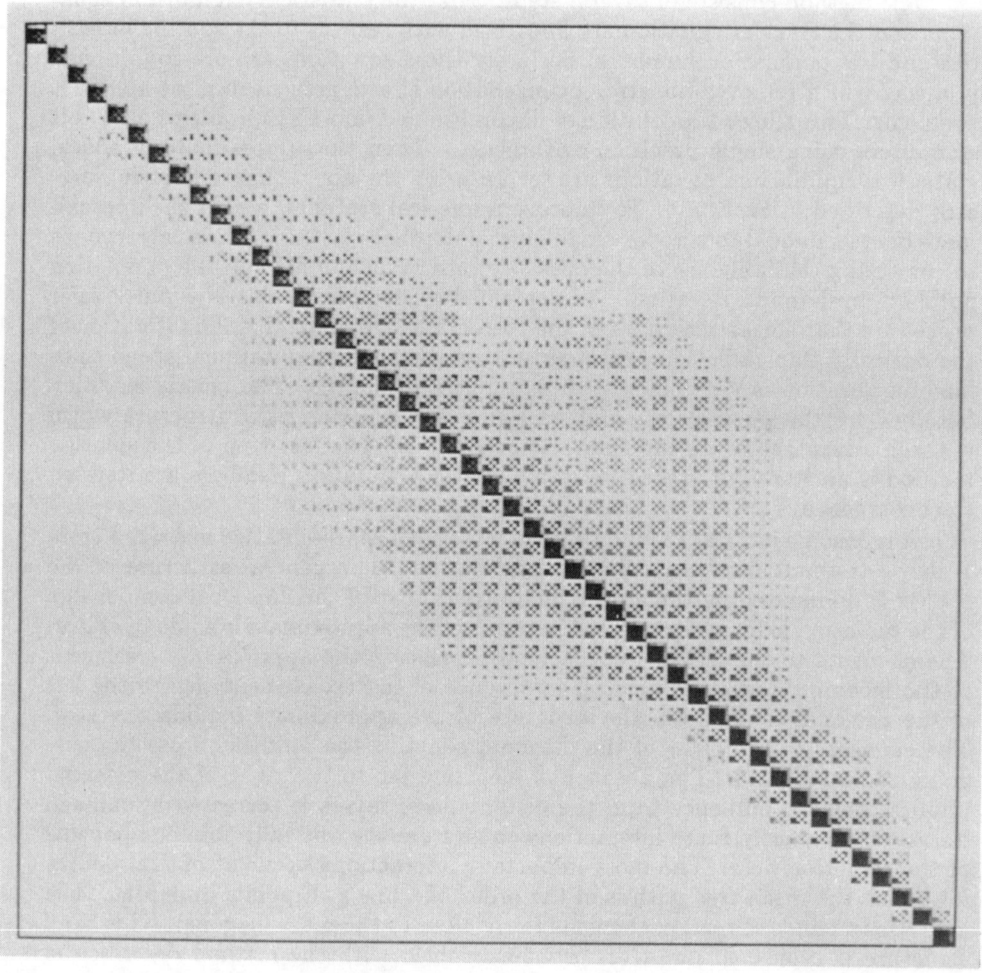

Figure 1. The grand matrix of a 5-level + continuum calcium II calculation corresponding to a solar model (VAL3-C). Each matrix element represents the interactions between two different levels at different depths. The absence of elements to the far right is due to the lambda iteration of the optically thin radiation field. The absence of elements to the far left is due to the approximate radiative transfer operator, $\Lambda^{\dagger}_{\nu\mu}$, itself. The non-zero matrix elements are clustered along the main diagonal showing that the most significant radiative interactions are of intermediate range. This relatively simple structure of the matrix allows efficient solution methods.

ferent computers and the results have been compared with those obtained from the often used code Linear-B (Auer and Heasley 1976). The test cases run were the following: A five-level plus continuum Ca II problem, a three-level plus continuum hydrogen problem and a 17-level plus continuum Fe I problem (Saxner 1985). All these problems were solved for static atmospheres. We also solved a very large number of three-level plus continuum Ca II problems with large-amplitude irregular velocity fields in an attempt to simulate mesoturbulence (Carlsson and Scharmer 1985a,1985b). Our experience with the method can be summarized as follows:

The computational speed of the present method is favorable. Many problems solved with the present method take only approximately a tenth of the time required by Linear-B.

The stability of the method is excellent. Problems which fail to converge with Linear-B converge easily with the present method. No artificial fix-ups are needed for convergence.

The method can be run in single precision on a 32-bit computer. Moreover, the internal accuracy of the present method is close to the machine accuracy thanks to the pre-conditioning of the equations. Fluxes calculated for the hydrogen problem on a VAX-11/750 (word length 32 bits) using single precision arithmetics agreed to within 5×10^{-6} of those obtained on a CYBER 170/835 (word length 60 bits).

The method treats incoming and out-going rays separately and can thus be used for problems where the usual Feautrier technique fails, such as problems involving blended lines in the presence of velocity fields.

The method permits straightforward programming and the "physical" approach to the numerical solution method makes the method rather appealing as well as easy to understand for the user.

The approach used is general and there should be no particular difficulty in generalizing the methods discussed in this paper to the solution of multi-dimensional multi-level non-LTE problems.

To summarize, we think that the method is a natural development of earlier methods and that it takes advantage of the numerical simplifications which are possible because of the physical nature of non-local radiative transfer in spectral lines. The numerical stability and high speed of the method and the fact that it can be used for moving atmospheres makes it suitable for a large class of radiative transfer problems.

7. REFERENCES

Auer, L.H., and Heasley, J.M.: 1976, *Astrophys. J.* **205**,165.

Cannon, C.J.: 1973, *J. Quant. Spectrosc. Radiat. Transfer* **13**, 627.

Carlsson, M.: 1984, *Uppsala Astron. Obs. Report*, in press.

Carlsson, M., and Scharmer, G.B.: 1985a, in Proceedings of National Solar

Observatory workshop on Chromospheric Diagnostics and Modeling, ed. B. Lites, Sacramento Peak Obs., Sunspot, in press.

Carlsson, M., and Scharmer, G.B.: 1985b, *Astron. Astrophys.*, in preparation.

Mihalas, D.: 1978, Stellar Atmospheres, 2nd ed., W.H. Freeman, San Fransisco.

Rybicki, G.B.: 1972, in *Line Formation in the Presence of Magnetic Fields*, ed. R.G. Athay, L.L. House, & G. Newkirk, Jr., High Altitude Obs., Boulder, p. 145.

Rybicki, G.B.: 1984, in *Methods in Radiative Transfer*, ed. W. Kalkofen, Cambridge University Press, p. 21.

Saxner, M.: 1985, *Astron. Astrophys.*, to be submitted.

Scharmer, G.B.: 1981, *Astrophys. J.* **249**, 720.

Scharmer, G.B.: 1984, in *Methods in Radiative Transfer*, ed. W. Kalkofen, Cambridge University Press, p. 173.

Scharmer, G.B., and Carlsson, M.: 1984, *J. Comput. Phys*, in press.

Scharmer, G.B., and Nordlund, Å.: 1982, *Stockholm Obs. Report* **19**.

ESCAPE PROBABILITY METHODS

George B. Rybicki
Harvard-Smithsonian Center for Astrophysics
60 Garden Street
Cambridge, MA 02138
USA

ABSTRACT. The physical foundations of escape probability methods, and methods derived from them, are briefly reviewed. First-order escape probability methods, the core saturation method, second-order escape probability methods, and Scharmer's method are discussed.

1. INTRODUCTION

Escape probability methods are relatively simple methods for obtaining solutions to spectral line formation problems. They are characterized by treating the transfer process solely through the escape probability $P_e \equiv$ the probability that an emitted photon will escape the medium in a *single step*, without absorption or scattering. Although escape probability methods in their simplest (first order) form are only approximate, they give deep insight into the process of line transfer, at least in the case of complete redistribution. This has, in turn, motivated several exact methods, such as the core saturation method, second-order escape probability methods, and Scharmer's method, which are outside the mainstream of direct discretization methods and which provide attractive alternatives to them.

Since the general subject of escape probability methods has recently been reviewed (Rybicki 1984), our main purpose here is to provide a condensed version that can serve as an introduction to that more extensive review. Consequently we simplify the physical discussion by assuming plane parallel geometry from the outset, and we omit the more mathematical aspects altogether. The interested reader can find more detailed discussions and additional references in Rybicki (1984).

2. REVIEW OF THE MULTILEVEL LINE PROBLEM

The simplest escape probability methods are resticted in their application to cases where complete frequency redistribution is a valid approximation. For convenience and simplicity we also assume here: a pure line problem with no associated continuum, plane parallel geometry, and no incident radiation field from outside the medium. With these assumptions the multilevel line problem can be formulated as the simultaneous solution of the equations of statistical equilibrium, and the equations of radiative transfer.

The equations of statistical equilibrium are

$$n_i \sum_j R_{ij} = \sum_j n_j R_{ji}, \tag{1}$$

where n_i are the level populations and the total rate coefficients for transitions between level i and

J. E. Beckman and L. Crivellari (eds.), Progress in Stellar Spectral Line Formation Theory, 199–206.

j are given by

$$R_{ij} = C_{ij} + B_{ij}\bar{J}_{ij} + A_{ij}, \tag{2}$$

The collisional rate coefficient C_{ij} is assumed to be due to thermal electrons. The spontaneous emission rate is given by the Einstein A-coefficient (zero for upward transitions), while the rates of stimulated emission and absorption are given by the product of the appropriate Einstein B-coefficients and the usual integrated mean intensity $\bar{J}_{ij}$.

The equation of radiative transfer can be formally solved to yield an expression for $\bar{J}_{ij}$ in terms of the line source function

$$S_{ij} = \frac{n_j A_{ji}}{n_i B_{ij} - n_j B_{ij}}. \tag{3}$$

In plane parallel geometry and with no incident radiation, the result is

$$\bar{J}_{ij}(\tau) = \int K_{ij}(\tau, \tau') S_{ij}(\tau')\, d\tau'. \tag{4}$$

Here we are using an integrated line optical depth variable τ and the kernel function $K_{ij}(\tau, \tau')$, which is given, for example, by Avrett & Hummer (1965). The integral operator on the right hand side is called the Λ-*operator*. Henceforth we shall omit the subscripts i and j, but it should be remembered that these quantities still refer to one of possibly many transitions within the multilevel atom.

An interpretation of the kernel function is that $K(\tau, \tau')\, d\tau'$ is the probability that a photon created at point τ will be absorbed in an interval $d\tau'$ about τ'. Thus the escape probability can be expressed as

$$P_e(\tau) = 1 - \int K(\tau, \tau')\, d\tau', \tag{5}$$

where the integration is over the entire medium.

3. FIRST-ORDER ESCAPE PROBABILITY METHODS

Suppose the source function $S(\tau)$ varies slowly over the scale of the kernel function $K(\tau, \tau')$. Then the source function can be removed from under the integration in equation (4), replacing τ' by τ. Using equation (5), we have the approximate result

$$\bar{J}(\tau) \approx [1 - P_e(\tau)] S(\tau), \tag{6}$$

a purely local equation involving the escape probability. This approximation can also be stated as

$$\rho(\tau) \approx P_e(\tau), \tag{7}$$

where the net radiative bracket is defined by

$$\rho \equiv 1 - \frac{\bar{J}}{S}. \tag{8}$$

Equation (6) or (7) is the fundamental relation for the simplest (first-order) escape probability method. Setting $\rho = P_e$ in each transition allows the complete solution of the multilevel line formation problem by iterating on purely local quantities. This is quite different from Λ-iteration, and typically convergence is quite rapid, although the end result is only approximate.

Since the criterion based on the variation of the source function is difficult to apply *a priori*, we now undertake to justify first-order escape probability methods by citing a relevant theorem and by showing their validity in a number of physically important cases:

a) Irons' Theorem. As first proved by Irons (1978) one has the general result

$$\int \rho(\tau)S(\tau)\,d\tau = \int P_e(\tau)S(\tau)\,d\tau, \tag{9}$$

which can also be written in the form

$$\langle\rho\rangle = \langle P_e\rangle, \tag{10}$$

where the brackets denote emission-weighted averages.

One can interpret Irons' theorem as stating that the otherwise approximate relation $\rho = P_e$ becomes ***exact*** when appropriately averaged over the entire medium. Of course, this does not imply that $\rho = P_e$ is valid at every point, but it does provide some overall constraint on its validity.

b) On-the Spot Approximation. Suppose that every emitted photon is absorbed at the same point. This might represent a good approximation deep in the interior of a very optically thick medium. The net radiative rate between the two levels is then zero,

$$n_jA_{ji} + n_jB_{ji}\bar{J} - n_iB_{ij}\bar{J} = 0. \tag{11}$$

This relation follows from the fact that all intensities become equal to the source function, $I_\nu = S$ under these circumstances. This case is also called ***detailed balance in the lines*** or ***complete saturation.*** It follows from this equation that $\rho = 0$. Since the escape probability is clearly also zero, we see that the relation $\rho = P_e$ is exact for this case.

c) The Dichotomous Model. As a simple extension of the on-the-spot model let us imagine that emitted photons fall into just two categories: A) Those absorbed at the same point; and B) Those that escape. Then the net radiative rate is just equal to that fraction of the spontaneous rate not compensated by absorptions, i.e.,

$$n_jA_{ji} + n_jB_{ji}\bar{J} - n_iB_{ij}\bar{J} = n_jA_{ji}P_e. \tag{12}$$

It follows immediately that relation $\rho = P_e$ is exact for the dichotomous model. Although the dichotomous model might seem artificial, we shall see presently that it applies to several physical cases of interest.

d) Sobolev Theory. In media with velocity gradients v', the ***Sobolev length*** is defined as $L = v_{th}/v'$, where v_{th} is the thermal velocity of the atom. The Sobolev length measures the distance over which parts of the medium cease to communicate radiatively due to the Doppler shift of the line profile. When L is small in comparison to other characteristic scales of the medium, this defines the regime of ***Sobolev theory,*** also known as ***large velocity gradient theory.*** In this case, if an emitted photon is going to be absorbed, it will be absorbed close to its point of emission, while photons that are not absorbed locally will be Doppler shifted out of the line profile and escape the medium entirely. [This latter statement depends on the line-of-sight velocity being monotonic in all directions, which may not always be true; see Rybicki (1984) for a fuller discussion.] This closely approximates the dichotomous model previously discussed, so that the relation $\rho = P_e$ is a good approximation in the Sobolev limit.

e) Longest Flight Property. In a static medium, scattering of line photons with complete frequency redistribution over Doppler or Voigt profiles gives rise to a random walk of a rather special kind (Rybicki & Hummer 1969). It is characterized by long sequences of scattering events confined to the line core, where the mean free path is very short, broken by very occasional emissions into the far wing, where the mean free path is very long. At least for scattering far from boundaries, it turns out that the greatest contribution to the net displacement of the photon occurs in a ***single longest flight*** in the wing, while all the scatterings in the core contribute only a small amount. This suggests that line scattering under these circumstances is approximately dichotomous, where escape is presumed to occur during one of the rare events of emission in the wings. Thus the relation $\rho = P_e$ again can be used as an approximation.

Having stated the case for the first-order escape probability methods in the most favorable light, let us now discuss some of the difficulties associated with them:

a) Boundaries. As indicated in the above discussion of the longest flight property for a static medium, the dichotomous model is really only justified deep in the medium. Near the boundaries scattering in the core of the line cannot be neglected, and the arguments for the validity of the relation $\rho = P_e$ become weak. It is in fact known that escape probability methods break down near the surface, predicting values for the source function that are much too small. It is this defect that largely motivated the development of the second-order escape probability methods, to be discussed below.

b) Negative ρ. Under certain circumstances, such as when a region of low excitation is bathed in radiation from a neighboring one of high excitation, the net radiative bracket ρ can actually be ***negative***, corresponding to a net upward radiative rate. Since the escape probability itself is manifestly positive, the relation $\rho = P_e$ is clearly a bad approximation.

c) Coherent Scattering. There are many cases, such as the treatment of types of continuum transfer, where the scattering is nearly coherent and is not well described by the complete redistribution we have considered above. The dichotomous model is then not applicable, since every scattering contributes about equally to the net displacement. The relation $\rho = P_e$ fails badly in these cases, not only near the surface, but at large depths as well.

4. CORE SATURATION METHOD

The on-the-spot approximation implies complete saturation of the specific intensity $I_\nu = S$ at all frequencies. This is a very restrictive assumption. In order to make it more generally applicable the method of ***core saturation*** was developed (Rybicki 1972; 1984). In this method the saturation assumption $I_\nu = S$ is made only for a set of ***core*** frequencies, for which the mean free paths are small in comparison to the characteristic scales for changes in the properties of the medium. In cases where the properties undergo no strong internal changes, such as might be associated with a shock, then the core frequencies are those for which the monochromatic optical depth τ_ν to the nearest boundary is of order unity or more. To be specific, in a semi-infinite medium one might define the core frequencies at a given spatial point to be those for which

$$\tau_\nu > \gamma, \tag{13}$$

where γ is a fixed parameter of order unity. The frequencies outside the core are called the ***wing*** frequencies, and no special assumption is made about them; they are calculated using the full transfer equation.

The integrated mean intensity, which is a frequency integral over the product of the mean intensity $J(\nu)$ and the line profile function $\varphi(\nu)$, can be written as a sum of contributions from the core and wing:

$$\bar{J} = \bar{J}_c + \bar{J}_w = \int_{core} \varphi(\nu)J(\nu)\,d\nu + \int_{wing} \varphi(\nu)J(\nu)\,d\nu. \tag{14}$$

Since $J(\nu) = S$ in the core, we have

$$\bar{J}_c = (1 - N_w)S, \tag{15}$$

where the ***wing normalization*** is defined by

$$N_w = \int_{wing} \varphi(\nu)\, d\nu. \tag{16}$$

It is known (Osterbrock 1962) that N_w is approximately related to the escape probability by

$$P_e = \frac{1}{2} N_w, \tag{17}$$

so equation (15) bears some resemblance to the escape probability result (6).

One now easily shows that the net radiative rate can be expressed as

$$n_j A_{ji} + n_j B_{ji} \bar{J} - n_i B_{ij} \bar{J} = n_j A_{ji} N_w + n_j B_{ji} \bar{J}_w - n_i B_{ij} \bar{J}_w. \tag{18}$$

The approximate expression on the right hand side has the same form as the exact one, except that there is a factor N_w multiplying A_{ji} and $\bar{J}$ is now replaced by $\bar{J}_w$. The core components no longer appear at all; they have been effectively eliminated from the problem. The numerical conditioning of the problem is very much improved, since the close cancellation of radiative rates due to scattering in the core has been analytically removed. However, the problem is of virtually the same form as before, so most common methods of solution can still be used. Thus we say that the equations of statistical equilibrium have been ***preconditioned.*** One consequence of preconditioning is that even Λ-iteration is now a practical method of solution, since only the relatively few scatterings in the wings need to be treated.

5. SECOND-ORDER ESCAPE PROBABILITY METHODS

As we have seen, the first-order escape probability methods are inaccurate near boundaries. In an attempt to overcome this failing Athay (1972) and Frisch & Frisch (1975) developed improved methods, which we call ***second-order*** escape probability methods. These methods still treat the transfer process solely through the escape probability P_e, but manage to take some account of non-local transfer by means of spatial derivative terms acting on $\bar{J}$ and S. Such methods have been extensively used by Canfield and his collaborators [see Canfield *et al.* (1984) for references].

The basic equation of the second-order methods can be expressed in a variety of ways. A convenient form for multilevel problems is

$$\frac{d\bar{J}}{d\tau} = \frac{dS}{d\tau} - S\frac{dP_e}{d\tau} - 2P_e\frac{dS}{d\tau}, \tag{19}$$

which is to be regarded as a differential approximation to the exact Λ-operator (4). When applied to the simple "standard" problem for two-level atoms, equation (19) yields the correct surface value of the source function (Frisch & Frisch 1975), showing its superiority to the first-order methods near boundaries.

The derivation of equation (19) by Canfield *et al.* (1981) was based on the mathematical approach devised by Frisch & Frisch (1975). We sketch here a more approximate, but more physically motivated, derivation due to Scharmer (1981), as modified by Rybicki (1984), for the simple case of a semi-infinite static medium. This derivation is based on the separation into core and wing components, with the core components again being determined by the saturation condition (15).

However, the wing components are now assumed to be completely optically transparent. At a certain frequency in the wings, the inward intensity is just equal to the zero incident field, while the outward intensities are given by the value of the source function S^* at the optical depth where the core begins at that frequency. In the two-stream approximation the mean intensity in the wing is equal to $\frac{1}{2}S^*$, so the wing integrated mean intensity is

$$\bar{J}_w = \frac{1}{2}\int_{wing} S^* \varphi\, d\nu = \frac{1}{2}\int_{wing} S^*\, dN_w. \tag{20}$$

Now if this integral were over *fixed* frequency limits, it would be independent of depth, since it involves a constant profile function and constant intensities. Thus the change in $\bar{J}_w$ with depth is due solely to the change in the wing frequency range, and we can write

$$d\bar{J}_w = \frac{1}{2} S\, dN_w = S\, dP_e, \tag{21}$$

where we have used equation (17). The replacement of S^* by S is valid because the new wing components are determined by the local source function. [See Rybicki (1984) for more details.] Differentiating equation (15) and adding equation (21) then yields equation (19).

The preceding derivation clearly points out the physical origin of the various terms in equation (19) for the simple case treated. It should be pointed out, however, that the derivation of Frisch & Frisch (1975) does apply to a wider variety of problems, for example, a medium with a constant velocity gradient.

The second-order escape probability methods are an improvement on the first-order ones in many respects, but not all. They cannot treat problems in which the profile function varies with depth, nor can they treat general nonconstant velocity gradients. In addition these methods involve first-order differential operators, so that the solution at a certain depth can depend on properties of the medium only to one side of that point (usually at larger depths). A related difficulty is that the appropriate boundary condition to be used with this differential equation is not always obvious, especially in a finite medium.

6. FINAL REMARKS

A general method of solving transfer problems is to use a very simple approximation for the Λ-operator, followed by iterative improvement to the exact solution (Cannon 1973). The success of such a method depends on choosing an approximate operator that has the essential physical properties of the exact one. There already exist exact iterative methods of this type based on the core saturation method [Rybicki (1972, 1984) and Flannery *et al.* (1979, 1980)] and second-order escape probability methods [Frisch & Froeschle (1977)]. In these methods the choice of approximate operator was clearly motivated by the physical ideas of escape probability.

Another example of such of method is that of Scharmer (1981), who uses a single-point quadrature scheme for expressing the formal solution along a ray, a choice originally motivated by his analysis of the second-order escape probability methods. The approximate Λ-operator so obtained is numerically represented by a matrix of strongly upper triangular structure, which can be inverted in order N^2 operations instead of the usual N^3. Scharmer's method has recently been developed considerably [see the reviews by Scharmer (1984; this volume)], and it has been shown to be capable of solving quite general transfer problems.

Although the simplest escape probability methods are restricted to the assumption of complete redistribution, there are now several related methods that can treat partial redistribution, for example, Frisch (this volume) and Scharmer (1983).

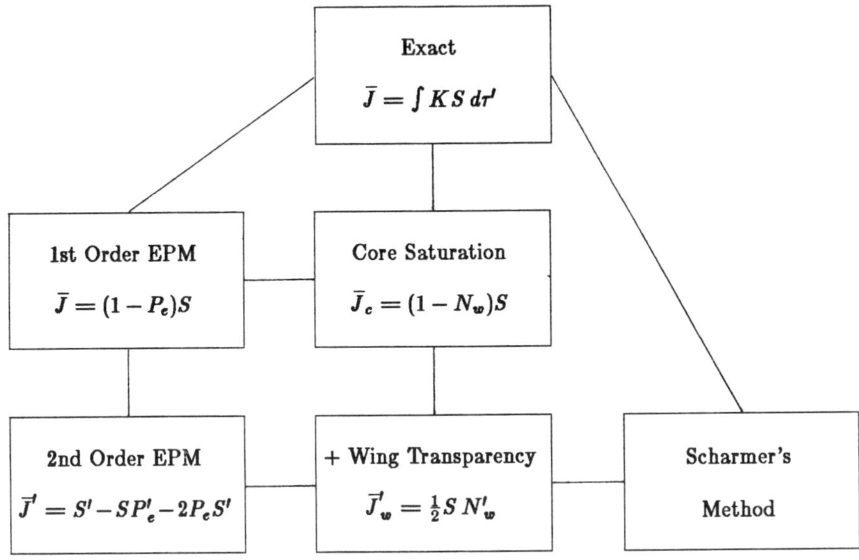

Fig. 1 – A schematic diagram suggesting the logical connections between the different methods discussed in the text. Note that the method denoted "+ Wing Transparency" implies that the core saturation approximation is also made.

The uses of escape probability theory are clearly varied, but we are beginning to see that one of its major uses may simply be to motivate physically realistic choices for approximate operators to be used in iterative improvement schemes. It appears that a variety of methods based on escape probability concepts can already provide attractive alternatives to conventional discretization techniques.

By way of a final summary, we have attempted in figure 1 to characterize the various methods discussed in this paper by their treatment of the relationship between $\bar{J}$ and S, and to suggest the logical connections between them.

REFERENCES

Athay, R.G. 1972, *Astrophys. J.*, **176**, 659.
Avrett, E.H., & Hummer, D.G. 1965, *Mon. Not. Roy. Astron. Soc.*, **130**, 295.
Canfield, R.C., Puetter, R.C., & Ricchiazzi, P.J. 1981, *Astrophys. J.*, **248**, 82.
Canfield, R.C., McClymont, A.N., & Puetter, R.C., 1984, MRT p 101.
Cannon, C.J. 1973, *J. Quant. Spectrosc. Rad. Transf*, **13**, 627.
Flannery, B.P., Rybicki, G.B., & Sarazin, C.L. 1979, *Astrophys. J.*, **229**, 1057.
________. 1980, *Astrophys. J. Suppl.*, **44**, 539.
Frisch, U., & Frisch, H. 1975, *Mon. Not. Roy. Astron. Soc.*, **173**, 167 (FF).
Frisch, H., This volume.
Frisch, H., & Froeschle, Ch. 1977, *Mon. Not. Roy. Astron. Soc.*, **181**, 281.

Irons, F.E. 1978, *Mon. Not. Roy. Astron. Soc.*, **182**, 705.
Kalkofen, W., ed., *Methods in Radiative Transfer*, (Cambridge: Cambridge Univ. Press), (MRT).
Osterbrock, D.E. 1962, *Astrophys. J.*, **135**, 195.
Rybicki, G.B. 1972, in *Line Formation in the Presence of Magnetic Fields*, ed. R.G. Athay, L.L. House, & G. Newkirk, Jr. (Boulder: High Altitude Observatory), p. 145.
———. 1984, MRT, p. 21.
Rybicki, G.B., & Hummer, D.G. 1969, *Mon. Not. Roy. Astron. Soc.*, **144**, 313.
Scharmer, G.B. 1981, *Astrophys. J.*, **249**, 720
———. 1983, *Astron. Astrophys.*, **117**, 83.
———. 1984, MRT, p. 173.
———. This volume.

NUMERICALLY STABLE DISCRETE ORDINATE SOLUTIONS OF THE RADIATIVE TRANSFER EQUATION

R. Wehrse
Institut für Theoretische Astrophysik
Im Neuenheimer Feld 294
D-6900 Heidelberg
Federal Republic of Germany

ABSTRACT. A new numerically stable and highly accurate method for the computation of the radiation field in a spectral line is presented. It involves the following three main steps: First by means of a discretisation of the angle and frequency space the transfer equation is transformed into a system of coupled first order differential equations. Next the system is solved analytically and finally it is cast into a form, which gives the emergent intensities as a function of the inflowing radiation and which contains only negative eigenvalues (assuring a numerically benign behaviour). We give the condition for closed solutions and discuss the versatility, accuracy and numerical performance of this approach.

1. INTRODUCTION

The first method employed to solve the radiative transfer equation was the discrete ordinate method (Schuster, 1905, Schwarzschild, 1906, see also e.g. Chandrasekhar, 1960), in which the continous frequency and angle dependence of the specific intensity is replaced by a discrete representation. By this the transfer equation is transformed from an integro-differential equation into a system of differential equations. Such a system can easily be solved for initial value or equivalent problems by standard methods. However, in radiative transfer one generally has to deal with two-point-boundary-value problems. In these cases a direct determination of the integration constants is practically impossible for total optical depths larger than about 10 (the exact number depends on the size of the system and the word-length of the computer), because the system of linear algebraic equations for the calculation is very ill-conditioned as a consequence of exponentially increasing terms (comp. e.g. Mihalas, 1978). Hence up to now such problems had to be solved piece-wise in optical depth space by difference, integral or iterative methods (see e.g. several contributions in Kalkofen, 1984, Kalkofen, 1974, Mihalas, 1978).
But they have the general disadvantage that the accuracy of a numerical result is often very difficult to assess. In addition, they are fre-

J. E. Beckman and L. Crivellari (eds.), Progress in Stellar Spectral Line Formation Theory, 207–213.

quently numerically slow and/or difficult to program.

In this paper we present a rather general way to eliminate the increasing exponentials of the discrete ordinate method so that the difficulties mentioned above no longer occur and a highly accurate and versatile algorithm is obtained. We derive the corresponding equations in the next chapter. In Section 3 the performance and some first applications are discussed.

2. DERIVATION OF THE STABLE SOLUTIONS

We consider in this chapter the radiative transfer equation for a static, un-polarized plane-parallel medium

$$\mu \frac{dI}{d\tau} = -I + \frac{\sigma}{\kappa+\sigma} \iint R I d\mu' d\nu' + \frac{\kappa}{\kappa+\sigma} B \tag{1}$$

where in the usual way μ = cosine of the angle between the ray and the normal direction, I = specific intensity, τ = optical depth, σ = scattering coefficient, κ = absorption coefficient, R = redistribution function, ν = frequency and B = Planck-function. Now we discretize the frequency and μ-space so that eq. (1) reads

$$\mu_m \frac{dI_{mn}}{d\tau} = - I_{mn} + \frac{\sigma}{\kappa+\sigma} \sum_{m'=1}^{M} \sum_{n'=1}^{N} R_{mnm'n'} I_{m'n'} a_{m'} b_{n'} + \frac{\sigma}{\kappa+\sigma} B \tag{2}$$

$$m = 1,2,\ldots\ldots,M$$
$$n = 1,2,\ldots\ldots,N$$

When we consider the I_{mn} as components of a vector $\bar{I}=(\bar{I}^+,\bar{I}^-)^T$ ($\bar{I}^+$ representing the intensities in the positive, $\bar{I}^-$ those in the negative direction) and collect all coefficients in the matrix $\bar{\Phi}$ eqs. (2) can be written very concisely in matrix notation

$$\frac{d\bar{I}}{d\tau} = \Phi\bar{I} + \bar{B} \quad . \tag{3}$$

The vector $\bar{B}$ contains the quantities $B\kappa/(\kappa+\sigma)/\mu_m$.
If we assume that

$$\bar{\Phi},\bar{\Phi}' = \bar{\Phi}\bar{\Phi}' - \bar{\Phi}'\bar{\Phi} = 0 \tag{4}$$

and furthermore that all eigenvalues of $\bar{\Phi}$ are distinct the matric differential equation (3) has the solution

$$\bar{I}(\tau) = \exp(\int_0^\tau \bar{\Phi} dt) \{\bar{I}(0) + \int_0^\tau \exp(-\int_0^t \bar{\Phi} ds)\bar{B}(t)dt\} \tag{5}$$

see e.g. Arnol'd (1970)

where the matrix exponential function $\exp(\bar{\Omega})$ is given by

$$e^{\bar{\Omega}} = \sum_{i=0}^{\infty} \frac{1}{n!} \bar{\Omega}^n = \bar{T} \begin{pmatrix} e^{\lambda_1} & & \\ & \ddots & \\ & & e^{\lambda_K} \end{pmatrix} \bar{T}^{-1} \tag{6}$$

with $\lambda_1, \lambda_2, \ldots \lambda_K$ being the eigenvalues of $\bar{\Omega}$ and T representing the modal matrix of $\bar{\Omega}$. The assumption (4) is necessary to assure that

$$\frac{d}{d\tau} e^{\bar{\Omega}} = \frac{d\Omega}{d\tau} e^{\bar{\Omega}} \quad . \tag{7}$$

We use the boundary conditions

$$\begin{aligned} \bar{\bar{I}}^-(0) &= \bar{I}^-_{bc} \\ \bar{I}^+(T) &= \bar{I}^+_{bc} \end{aligned} \tag{8}$$

where $\bar{\bar{I}}^-_{bc}$ and $\bar{I}^+_{bc}$ are considered to be given.

In order to introduce these boundary conditions into eq. (5) we write

$$\exp\left(\pm \int_0^\tau \bar{\Phi} dt\right) = \begin{pmatrix} \bar{u}_{\pm\tau} & \bar{v}_{\pm\tau} \\ \bar{w}_{\pm\tau} & \bar{x}_{\pm\tau} \end{pmatrix} \tag{9}$$

and

$$\left(\bar{\lambda}_{\pm\tau}\right)_{ij} = \exp\{\pm\lambda_i\}\delta_{ij} \tag{10}$$

(λ_i = positive eigenvalues of $\int_0^\tau \bar{\Phi}dt$; if the same angle set is employed for the positive and the negative direction there is symmetry of the λ values with respect to 0).

Now eq. (5) reads

$$\begin{pmatrix} \bar{I}^+(\tau) \\ \bar{\bar{I}}^-(\tau) \end{pmatrix} = \begin{pmatrix} \bar{u}_\tau & \bar{v}_\tau \\ \bar{w}_\tau & \bar{x}_\tau \end{pmatrix} \left\{ \begin{pmatrix} \bar{I}^+(0) \\ \bar{\bar{I}}^-(0) \end{pmatrix} + \int_0^\tau \begin{pmatrix} \bar{u}_{-t} & \bar{v}_{-t} \\ \bar{w}_{-t} & \bar{x}_{-t} \end{pmatrix} \bar{B}(t)dt \right\} \tag{11}$$

or with the abbreviation for the source term

$$\begin{pmatrix} \bar{\Gamma}^+_\tau \\ \bar{\Gamma}^-_\tau \end{pmatrix} = \begin{pmatrix} \bar{u}_\tau & \bar{v}_\tau \\ \bar{w}_\tau & \bar{x}_\tau \end{pmatrix} \int_0^\tau \begin{pmatrix} \bar{u}_{-t} & \bar{v}_{-t} \\ \bar{w}_{-t} & \bar{x}_{-t} \end{pmatrix} \bar{B}(t)dt \tag{12}$$

$$\begin{pmatrix} \bar{I}^+(\tau) \\ \bar{I}^-(\tau) \end{pmatrix} = \begin{pmatrix} \bar{u}_\tau & \bar{v}_\tau \\ \bar{w}_\tau & \bar{x}_\tau \end{pmatrix} \begin{pmatrix} \bar{I}^+(0) \\ \bar{\bar{I}}^-(0) \end{pmatrix} + \begin{pmatrix} \bar{\Gamma}^+_\tau \\ \bar{\Gamma}^+_\tau \end{pmatrix} \quad . \tag{13}$$

If $\bar{I}^+(0)$ is expressed in terms of the boundary values (by means of the first group of eqs. of 13)

$$\bar{I}^+(0) = \bar{\bar{u}}^{-1}_T \, \bar{I}^+_{bc} - \bar{\bar{u}}^{-1}_T \, \bar{v}_T \, \bar{\bar{I}}^-_{bc} - \bar{\bar{u}}^{-1}_T \, \bar{\Gamma}^+_T \tag{14}$$

and inserted we find

$$\begin{pmatrix} \bar{I}^+(\tau) \\ \bar{\bar{I}}^-(\tau) \end{pmatrix} = \begin{pmatrix} \bar{u}_\tau \bar{\bar{u}}^{-1}_T & \bar{v}_\tau - \bar{u}_\tau \bar{\bar{u}}^{-1}_T \bar{v}_T \\ \bar{w}_\tau \bar{\bar{u}}^{-1}_T & \bar{x}_\tau - \bar{w}_\tau \bar{\bar{u}}^{-1}_T \bar{v}_T \end{pmatrix} \begin{pmatrix} \bar{I}^+_{bc} \\ \bar{\bar{I}}^-_{bc} \end{pmatrix} - \begin{pmatrix} \bar{u}_\tau \bar{\bar{u}}^{-1}_T \bar{\Gamma}^+_T \\ \bar{w}_\tau \bar{\bar{u}}^{-1} \bar{\Gamma}^+_T \end{pmatrix} + \begin{pmatrix} \bar{\Gamma}^+_\tau \\ \bar{\Gamma}^-_\tau \end{pmatrix} \tag{15}$$

$$= \begin{pmatrix} \bar{u}_\tau \bar{u}_T^{-1} & \bar{v}_\tau - \bar{u}_\tau \bar{u}_T^{-1} \bar{v}_T \\ \bar{w}_\tau \bar{u}_T^{-1} & \bar{x}_\tau - \bar{w}_\tau \bar{u}_T^{-1} \bar{v}_T \end{pmatrix} \begin{pmatrix} \bar{I}^+_{bc} \\ \bar{I}^-_{bc} \end{pmatrix} + \begin{pmatrix} 0 & \bar{v}_\tau - \bar{u}_\tau \bar{u}_T^{-1} \bar{v}_T \\ 0 & \bar{x}_\tau - \bar{w}_\tau \bar{u}_T^{-1} \bar{v}_T \end{pmatrix} \int_0^\tau \begin{pmatrix} \bar{u}_{-s} & \bar{v}_{-s} \\ \bar{w}_{-s} & \bar{x}_{-s} \end{pmatrix} \bar{B}(s)ds$$

$$- \begin{pmatrix} \bar{u}_\tau & \bar{u}_\tau \bar{u}_T^{-1} \bar{v}_T \\ \bar{w}_\tau & \bar{w}_\tau \bar{u}_T^{-1} \bar{v}_T \end{pmatrix} \int_\tau^T \begin{pmatrix} \bar{u}_{-s} & \bar{v}_{-s} \\ \bar{w}_{-s} & \bar{x}_{-s} \end{pmatrix} \bar{B}(s)ds \qquad (16)$$

This is the final expression giving the specific intensities at optical depth τ as a function of the boundary values. In the appendix we show that no term contains an increasing exponential any longer and therefore the expression is numerically benign for all total optical thicknesses T.

3. DISCUSSION

We have programmed the solution eq. 16 for the case of a line with complete redistribution and a Planck function that is approximated by a polynomial. The absorption and scattering coefficients are assumed not to depend on the depth. For 24 frequency-angle points and an internal relative accuracy of about 10^{-16} the calculation of the specific intensities at some optical depth τ takes about 250 msec on an IBM 3081 D computer. For some examples see Fig. 1. The CPU time is essentially determined by the calculation of the eigenvalues and the subsequent matrix operations and therefore scales as the cube of the number of frequency and angle points. The storage requirement scales as the square of these numbers. We note as additional properties of this method that

(i) the use of partial redistribution functions or anisotropic phase functions does not increase the CPU times (assuming that the calculation of R itself is fast) or the storage requirements.

(ii) the CPU times are independent of the total optical depth of the medium.

(iii) a depth dependence of $\bar{B}$, which allows an analytic evaluation of integrals $\int \exp(-\psi(s))\bar{B}(s)ds$ (see eq. 16) reduces the computer time drastically.

Since the system 3 is integrated exactly (i.e. up to machine accuracy) and all linear equations are well conditioned, errors result only from the frequency and angle discretisation. However, test calculations showed that they can be made very small (e.g. $< 10^{-5}$) already for few meshpoints by proper selection of the discretisation.

Limits on the applicability of this method are posed by the two requirements stated in Section 2:

(i) The condition that all eigenvalues must be distinct excludes the pure scattering cases $\kappa/\sigma = 0$, where there are two roots $\lambda_{1,2} = 0$. For such a problem the diagonal λ matrices have to be replaced by the corresponding Jordan matrices.

(ii) More severe is the condition $[\bar{\Phi},\bar{\Phi}']=0$ which inhibits eg. direct and closed solutions for depth dependent redistribution functions and spherical media (except for special cases). Possible ways out are here the division of the medium into subslabs or subshells, for which $\bar{\Phi}$ can be approximated by a constant and which subsequently are put together (see e.g. Peraiah, 1984, Kalkofen and Wehrse, 1985), or by a perturbation technique similar to that used in quantum electrodynamics (see e.g. Louissell, 1973).

On the other hand, because of the analytical character of this method it offers a number of easy generalisations, e.g.

(i) Polarisation phenomena can be incorporated by simply modifying $\bar{\Phi}$, the formalism itself remains totally unchanged.

(ii) Velocity fields can be taken into account both in the observer's and comoving frame.

(iii) The equations 3 can be transformed into stochastic differential equations for the (stochastic) variables I and B. For a solution only the source term in eq. 16 has to be interpreted as a sum of stochastic integrals.

Acknowledgement. I would like to thank M. Schmidt, Drs. W. Kalkofen and G. Traving for many helpful discussions. This work was supported by the Deutsche Forschungsgemeinschaft (SFB 132).

APPENDIX

In order to show that the terms of eq. 16 can be written without increasing exponentials we put

$$\begin{pmatrix} \bar{u}_\tau \bar{v}_\tau \\ \bar{w}_\tau \bar{x}_\tau \end{pmatrix} = \bar{T} \begin{pmatrix} \bar{\lambda}^+_\tau & 0 \\ 0 & \bar{\lambda}^-_\tau \end{pmatrix} \bar{T}^{-1} \equiv \begin{pmatrix} \bar{A} & \bar{B} \\ \bar{C} & \bar{D} \end{pmatrix} \begin{pmatrix} \bar{\lambda}^+_\tau & 0 \\ 0 & \bar{\lambda}^-_\tau \end{pmatrix} \begin{pmatrix} \bar{a} & \bar{b} \\ \bar{c} & \bar{d} \end{pmatrix} \quad \text{(A1)}$$

(Note that $\bar{B}$ here is not the Planck-function of Section 2, but a block of the modal matrix).

Since for a plane parallel problem

$$\bar{\lambda}^+_\alpha \cdot \bar{\lambda}^-_\alpha = \bar{1} \quad \text{(A2)}$$

if the same angle set is used for both hemispheres, we find for the submatrix

$$\begin{aligned} \bar{u}_\tau \bar{u}_T^{-1} &= (\bar{A}\,\bar{\lambda}^+_\tau\,\bar{a} + \bar{B}\,\bar{\lambda}^-_\tau\,\bar{c})(\bar{A}\,\bar{\lambda}^+_T\,\bar{a} + \bar{B}\,\bar{\lambda}^-_T\,\bar{c})^{-1} \\ &= (\bar{A}\,\bar{\lambda}^+_\tau + \bar{B}\,\bar{\lambda}^-_\tau\,\bar{c}\,\bar{a}^{-1})(\bar{A}\,\bar{\lambda}^+_T + \bar{B}\,\bar{\lambda}^-_T\,\bar{c}\,\bar{a}^{-1})^{-1} \\ &= (\bar{A} + \bar{B}\,\bar{\lambda}^-_\tau\,\bar{c}\,\bar{a}^{-1}\,\bar{\lambda}^-_\tau)\bar{\lambda}^+_\tau\bar{\lambda}^-_T(\bar{A} + \bar{B}\,\bar{\lambda}^-_T\,\bar{c}\,\bar{a}^{-1}\,\bar{\lambda}^-_T)^{-1} \end{aligned} \quad \text{(A3)}$$

and similarly

$$\bar{w}_\tau \bar{u}_T^{-1} = (\bar{c} + \bar{D}\,\bar{\lambda}^-_\tau\,\bar{c}\,\bar{a}^{-1}\,\bar{\lambda}^-_\tau)\,\bar{\lambda}^+_\tau\,\bar{\lambda}^-_T\,(\bar{A} + \bar{B}\,\bar{\lambda}^-_T\,\bar{c}\,\bar{a}^{-1}\,\bar{\lambda}^-_T)^{-1} \quad \text{(A4)}$$

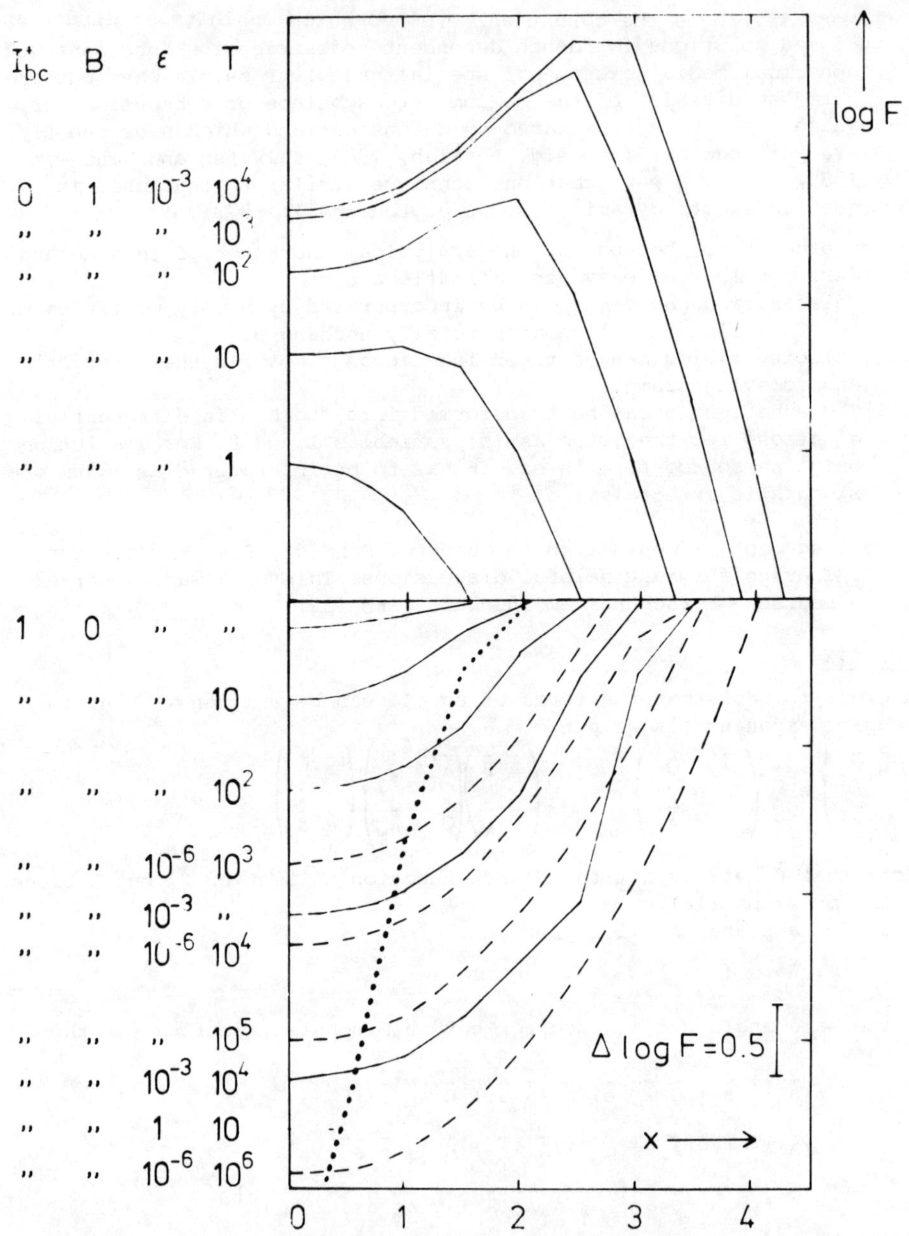

Fig. 1. Some examples of emergent fluxes as a function of the normalized frequency x for lines of total optical depth T. The boundary condition at τ=0 is kept $\bar{I}_{bc}$=0, whereas the other parameters ($\bar{B}$=Planck function, $\bar{I}^+_{bc}$= boundary condition at T, $\varepsilon=\kappa/(\kappa+\sigma)$) take the specified values. The profiles are calculated for 2 angle and 12 frequency points. Complete redistribution and Gaussian profile functions are assumed.

$$\bar{v}_\tau - \bar{u}_\tau \bar{u}_T^{-1} \bar{v}_T = \bar{A}\, \bar{\lambda}^+_\tau\, \bar{\lambda}^-_T (\bar{1} + \bar{A}^{-1} \bar{B} \bar{\lambda}^-_T\, \bar{c}\bar{a}^{-1} \bar{\lambda}^-_T)^{-1} \bar{A}^{-1} \bar{B}\bar{\lambda}^-_T (\bar{c}\bar{a}^{-1}\bar{b} - \bar{d})$$

$$+ \bar{B}\bar{\lambda}^-_\tau \left[\bar{d} - \bar{c}(\bar{a} + \bar{\lambda}^-_T \bar{A}^{-1} \bar{B}\bar{\lambda}^-_T \bar{c})^{-1} (\bar{b} + \bar{\lambda}^-_T \bar{A}^{-1} \bar{B}\bar{\lambda}^-_T \bar{d}) \right] \tag{A5}$$

$$\bar{x}_\tau - \bar{w}_\tau \bar{u}_T^{-1} \bar{v}_T = \bar{c}\bar{\lambda}^+_\tau \bar{\lambda}^-_T (\bar{1} + \bar{A}^{-1} \bar{B}\bar{\lambda}^-_T \bar{c}\bar{a}^{-1} \bar{\lambda}^-_T)^{-1} \bar{A}^{-1} \bar{B}\bar{\lambda}^-_T (\bar{c}\bar{a}^{-1}\bar{b} - \bar{d})$$

$$+ \bar{D}\bar{\lambda}^-_\tau \left[\bar{d} - \bar{c}(\bar{a} + \bar{\lambda}^-_T \bar{A}^{-1} \bar{B}\bar{\lambda}^-_T \bar{c})^{-1} (\bar{b} + \bar{\lambda}^-_T \bar{A}^{-1} \bar{B}\bar{\lambda}^-_T \bar{d}) \right] \tag{A6}$$

It is seen that all λ^+ terms are eliminated in eqs. A3 to A6 except for the combination $\bar{\lambda}^+_\tau \bar{\lambda}^-_T$, which however is smaller than or equal to unity because of $\tau \leq T$. For the source terms no such general derivation can be given, because it depends on the actual depth dependence of $\bar{B}$, but e.g. for polynomials it proceeds exactly along the same lines and also results in expressions containing only $\bar{\lambda}^-$ and $\bar{\lambda}^+_\tau \bar{\lambda}$.

REFERENCES

Arnol'd, V.I., 1980, Gewöhnliche Differentialgleichungen, Springer Verlag Berlin, Heidelberg

Chandrasekhar, S., 1960, Radiative Transfer, Dover-Publications, New York

Kalkofen, W., 1984, Ed., Numerical Methods in Radiative Transfer, Cambridge University Press, Cambridge

Kalkofen, W., Wehrse, R., 1985, submitted

Louissell, W.H., 1973, Quantum Statistical Properties of Radiation, J. Wiley, New York

Mihalas, D., 1978, Stellar Atmospheres, Freeman, San Francisco

Peraiah, A., 1984, in Numerical Methods in Radiative Transfer, Kalkofen, W., Ed. Cambridge University Press, Cambridge

Schuster, A., 1905, Astrophys. J. 21,1

Schwarzschild, K., 1906, Göttinger Nachrichten, p. 41

NLTE SPECTRAL LINE FORMATION IN A THREE-DIMENSIONAL ATMOSPHERE WITH VELOCITY FIELDS.

Åke Nordlund
Copenhagen University Observatory
Øster Voldgade 3
Dk-1350 Copenhagen K
Denmark

ABSTRACT: A method to solve the "two-level-atom-with-overlapping-continuum" problem in a three-dimensional atmosphere is presented. The method is based on treating the radiative transfer along a number of rays through the model as separate sub-problems. In each iteration, the error in the source function is evaluated along all the rays through the model, and an estimate of the necessary correction is obtained for each ray. The converged solution is an exact solution to the problem. The method requires a computer time per iteration equivalent to the solution of N one-dimensional two-level atom problems, plus 2N solutions of the radiative transfer problem with given source function, where N is the number of rays = $N_\Omega N_x N_y$. As an application, the method is used on the case of a neutral iron line in the solar photosphere. The solar photosphere is represented by a snapshot from three-dimensional hydrodynamic simulations of the solar granulation. The presence of the granular velocity field has a substantial influence on the behavior of the line source function in the upper photosphere. The resulting line source function exceeds the local Planck function in most of the upper photosphere, thus decreasing the central depth of synthesized spectral line profiles, relative to the LTE case.

1. INTRODUCTION

The presence of a velocity field with a geometrical scale of variation of the order of the optical depth scale height is known to have a strong influence on the spectral line formation process. The major cause for this influence is the Doppler shift of the (local) line absorption profile, relative to the (non-local) radiation intensity field profile. In LTE, one result is to increase the number of photons absorbed in the line; i.e., a strengthening of the line absorption ("micro-turbulence"). In non-LTE, the increased photon absorption in the line leads to an increase in the line source function (Shine, 1975).

J. E. Beckman and L. Crivellari (eds.), Progress in Stellar Spectral Line Formation Theory, 215–224.

In contrast to the excitation equilibrium problem, the ionization equilibrium (at least in the case of the solar iron ionization equilibrium) may be treated as a problem with a given radiation field, and calculations are relatively straight-forward (if costly). The results of such calculations for the ionization equilibrium of iron in the solar photosphere show that iron may possibly be overionized by as much as 0.2 dex relative to the LTE case (Nordlund, 1984). This systematic effect in the average iron ionization is related to the quite low temperatures (< 4000 K) of the upper photospheric layers found in the hydrodynamical model. These low temperatures are caused by the weakness of the radiative heating (mainly in spectral lines) relative to the expansion cooling of the ascending gas in the upper photosphere. Observations of very dark cores of infrared CO-lines by Ayres & Testerman (1981) indicate that the upper layers of the photosphere are indeed quite cool. Because of these low temperatures, there could also be a significant over-excitation, caused by the non-local radiation field in spectral lines, just as the over-ionization is caused by the non-local UV ionizing radiation.

Synthetic spectral lines calculated (in LTE) from numerical simulations of the hydrodynamics of granular convection in the photosphere (Nordlund 1982, 1984), show good agreement with observed shapes, widths, and strengths of observed photospheric spectral lines (Dravins, Lindegren & Nordlund; 1981). However, there is a systematic tendency for the synthetic spectral lines to have deeper line cores than the observed ones.

To investigate quantitatively to what extent this discrepancy is caused by departures from LTE in the excitation equilibrium, one needs to solve the NLTE line formation problem in three spatial dimensions. The purpose of the present contribution is to present a method that is sufficiently rapid to allow practical computations in 3D.

2. THE TWO-LEVEL ATOM WITH OVERLAPPING CONTINUUM

The line source function S_L is related to the Planck function B, and to the mean intensity in the line J_L by (cf. Mihalas, 1978, p. 336ff)

$$S_L = \frac{\varepsilon' B + J_L}{1 + \varepsilon'} \tag{1}$$

or

$$S_L = \varepsilon B + (1-\varepsilon) J_L \quad . \tag{2}$$

In complete redistribution, J_L is an average over angle and frequency:

$$J_L = \int_\nu \int_\Omega I(\nu,\Omega)\ \phi(\nu,\Omega)\ d\nu\ d\Omega \quad , \tag{3}$$

where $\phi(\nu,\Omega)$ is the line absorption and emission profile (which depends on Ω when there is a macroscopic velocity field present). [In order not to overburden the notation, the dependence of everything on spatial coordinates (x,y,z) is implicit throughout the paper].

Along each ray in the medium, the specific intensity $I_\nu = I(\nu,\Omega)_{[\text{along the ray}]}$ is governed by the transfer equation

$$\frac{dI_\nu}{d\tau_\nu} = I_\nu - S_\nu \tag{4}$$

where

$$d\tau_\nu = \rho\,(\kappa_{cont}+\phi\kappa_L)\,ds_{[\text{along the ray}]} \tag{5}$$

$$S_\nu = r_\nu B + (1-r_\nu)S_L \tag{6}$$

and

$$r_\nu = \frac{\kappa_{cont}}{\kappa_{cont}+\phi\kappa_L} \quad . \tag{7}$$

For numerical reasons, it is convenient to work in the average of in-coming and out-going intensities (Feautrier transformation),

$$P_\Omega = \int_\nu \phi_\nu \tfrac{1}{2}[I(\nu,\Omega) + I(-\nu,-\Omega)]\,d\nu \quad . \tag{8}$$

Then

$$J_L = \int_\Omega P_\Omega \frac{d\Omega}{4\pi} \tag{9}$$

and

$$S_L = (1-\varepsilon)\int_\Omega P_\Omega \frac{d\Omega}{4\pi} + \varepsilon B \quad . \tag{10}$$

By the transfer equation (4), and Eqs. (5) - (7), P_Ω is a linear function of S_L and B,

$$P_\Omega = \Lambda_\Omega^S[S_L] + \Lambda_\Omega^B[B] \quad . \tag{11}$$

The two equations (10) and (11) summarize the line transfer problem. All complexities of the radiative transfer and frequency redistribution are hidden in the operators $\Lambda_\Omega^S[\,]$ and $\Lambda_\Omega^B[\,]$.

2.1 The correction procedure.

Let us denote by $S_L^{(n)}$ a current estimate of the line source function S_L, and let

$$P_\Omega^{(n)} = \Lambda_\Omega^S[S_L^{(n)}] + \Lambda_\Omega^B[B] \quad . \tag{12}$$

Now let

$$E_\Omega^{(n)} = \Lambda_\Omega^S[(1-\varepsilon)\int_{\Omega'} P_{\Omega'}^{(n)}\frac{d\Omega'}{4\pi}+\varepsilon B] + \Lambda_\Omega^B[B] - P_\Omega^{(n)} \quad . \tag{13}$$

By equations (10) and (11), $E_\Omega^{(n)}$ vanishes when $S_L^{(n)} \to S_L$; i.e., $E_\Omega^{(n)}$ is a measure of the current error in $S_L^{(n)}$.

Variations $\delta P_\Omega^{(n)}$ cause variations in $E_\Omega^{(n)}$;

$$\delta E_\Omega^{(n)} = \Lambda_\Omega^S[(1-\varepsilon)\int_{\Omega'} \delta P_{\Omega'}^{(n)}\frac{d\Omega'}{4\pi}] - \delta P_\Omega^{(n)} \quad . \tag{14}$$

Here, the first term represents the transfer effects through the δJ_L term in the source function. The requirement that perturbations of the radiation field should be isotropic at large depths suggests the approximation

$$\int_{\Omega'} \delta P_{\Omega'}\frac{d\Omega'}{4\pi} \equiv \delta J_L \simeq \delta P_\Omega^{(n)} \quad . \tag{15}$$

With this approximation, the coupling to other rays vanishes, and

$$\delta E_\Omega^{(n)} = \Lambda_\Omega^S[(1-\varepsilon)\delta P_\Omega^{(n)}] - \delta P_\Omega^{(n)} = \Gamma_\Omega[\delta P_\Omega^{(n)}] \quad . \tag{16}$$

The requirement

$$E_\Omega^{(n)} + \delta E_\Omega^{(n)} = 0 \tag{17}$$

leads to

$$\delta P_\Omega^{(n)} = -\Gamma_\Omega^{(-1)}[E_\Omega^{(n)}] \quad , \tag{18}$$

and the next estimate of S_L is

$$S_L^{(n+1)} = (1-\varepsilon) \int_\Omega (P_\Omega^{(n)} + \delta P_\Omega^{(n)}) \frac{d\Omega}{4\pi} + \varepsilon B \quad . \tag{19}$$

Equations (12) through (19) define the correction procedure. Each iteration requires (per ray) two evaluations of the radiative transfer [Λ_Ω^S[], Eqs. (12) and (13)], and a solution of a problem equivalent to a one-dimensional complete redistribution problem [$\Gamma_\Omega^{(-1)}$[], Eq. (18)].

2.2 Numerical methods.

To calculate the error $E_\Omega^{(n)}$, one must solve the radiative transfer equation with a given line source function $S_L^{(n)}$. This is done by first selecting a set of rays through the model. An arbitrary horizontal plane is chosen as a reference plane, and rays of certain inclinations θ and orientations Φ are passed through all the (x,y) points of this plane. This produces specific intensities at all the points of intersection with the other horizontal planes. To perform the space angle averaging, these intensities must all refer to the same (x,y) points in the planes. This is achieved by Fourier interpolation back onto the Cartesian mesh points of the model, from the values supplied by the transfer solutions at the intersection points. The use of Fourier interpolation is essential, in that it guarantees that successive interpolations back and forth do not lead to an "erosion" of the (x,y) dependence of S_L. The solution of the one-dimensional two-level atom problem along each ray is performed simultaneously with one of the solutions of the radiative transfer with given source function, as described in Scharmer & Nordlund (1982) and Scharmer (1984).

3. RESULTS AND DISCUSSION

The numerical input data for the present analysis consist of values for the temperature, pressure, and velocity from a three-dimensional hydro-

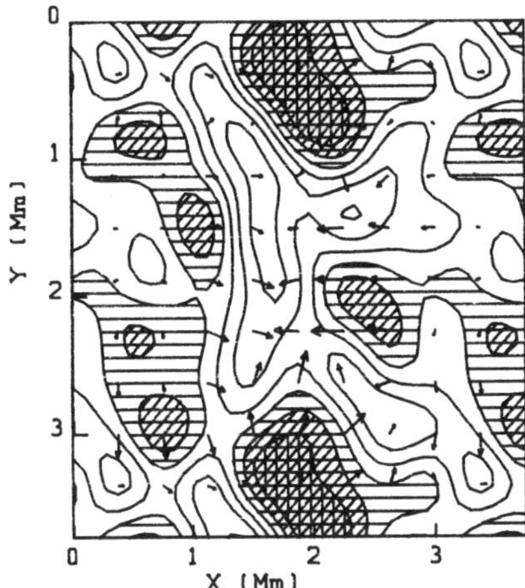

Fig. 1: The continuum radiation intensity (at 522.5 nm) for the snapshot from the hydrodynamical simulations used in the spectral line calculation. The contour levels correspond to 50, 70, 90, 110, 130, and 150 % of the average intensity. Levels above 110 % are shaded. The corresponding horizontal velocity field is also shown, with arrows corresponding to the distance traveled in 40 s.

dynamical model of the solar photosphere (Nordlund 1982, 1984). To reduce the computational requirements sufficiently for a readily available SPERRY 1100/82 computer, only a single snapshot from the hydrodynamical simulations were used, and these data were further reduced to sets of 8x8x32 points, by dropping higher wave-number components in the horizontal Fourier representation. Opacities and ionization equilibria were supplied from tables calculated with a standard model atmosphere subroutine package (Gustafsson, 1973). Line strength parameters, profile functions, and line absorption coefficients were taken from an LTE line synthesis program, with the FeI line at 522.55 nm as a "mock up" for the two-level atom problem. Values for the collision parameter ε were taken from an estimate due to Böhm [1960, cited in Mihalas 1978, Eq. (5-47)].

The continuum intensity pattern and the horizontal velocity at zero height (average optical depth unity) are shown in Fig. 1. This particular snapshot is characterized by somewhat smaller than average vertical velocities, and by somewhat lower than average temperatures in the upper layers. Apart from the larger than average line depth due to these fluctuations, the spectral line bisector happens to agree quite well with the bisector of the time averaged spectral line profile (cf. Fig. 5). Note that bisectors vary strongly as a function of time, and only occasionally resemble the bisector of the time averaged profile (cf. Dravins et al. 1981, Fig. 11).

3.1 Convergence properties.

The correction procedure was tried out on schematic test cases with prescribed variations of the source function, the line and continuum absorption coefficients and the velocity. (In particular, with no variations; i.e., the standard 1D test case). On these test cases, the method converged without modifications, with typically 3-4 iterations

Table 1: Maximum values [over all (x,y,z)] of 1) the error term $E_\Omega^{(n)}$ relative to $P_\Omega^{(n)}$ [cf. Eq. (13)], 2) the error in the line source function radiation temperature after n iterations, and 3) the error in the spectral line shift (as measured by the average spectral line bisector).

iteration n	$\lvert E_\Omega^{(n)}/P_\Omega^{(n)}\rvert_{max}^{1)}$	$\lvert T_S^{(n)}-T_S\rvert_{max}^{2)}$ (K)	$\lvert v_b^{(n)}-v_b\rvert_{max}^{3)}$ (m/s)
0	–	1409	29
1	1.00	277	5
2	4.04	99	3
3	0.157	48	1
4	0.061	23	0
5	0.021	13	0
6	0.0084	6.3	0
7	0.0038	3.6	0
8	0.0016	1.7	0

needed for less than 1 % error in the source function. However, when the actual physical data from the granulation simulation were used, the correction procedure overestimated the necessary corrections, and the introduction of an under-relaxation parameter was necessary to obtain convergence. For the particular case reported here, an under-relaxation with a factor of 0.3 was applied in the layers above a height of 100 km relative to the continuum forming layers. Table 1 shows the convergence properties of the method, with this modification of the correction procedure. Note that the spectral line shape has converged to within 1 m/s in three iterations.

3.2 Properties of the converged solution.

Fig. 2 shows the horizontally averaged values of the temperature and the line source function radiation temperature as a function of height.

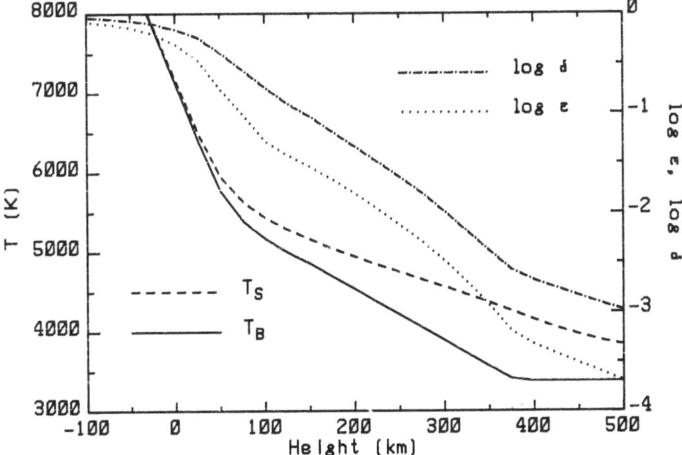

Fig. 2: The horizontal averages of the kinetic temperature T_B, the line source function radiation temperature T_S, the collision parameter ε, and the relative background continuum strength δ.

Note that the line source function exceeds the local temperature by over 500 K (on the average) over a heigh interval from 250 to 500 km above the continuum layers. Fig. 2 also shows the average values of the collision parameter ε and the background continuum strength measure

$$\delta = \int_{\nu} \int_{\Omega} r_{\nu} \, \phi_{\nu} \, d\nu \, \frac{d\Omega}{4\pi} \quad , \tag{20}$$

(cf. Athay, 1972). δ is significantly larger than ε over the entire photosphere (at least with the estimate of ε used here). This is important, in that it implies that photon destruction through absorption in the background continuum is much more likely than photon destruction through collisions, and that uncertainties in collision cross sections are less important.

Fig. 3 illustrates the correlation of the temperature and the line source function with the continuum intensity fluctuation pattern. The

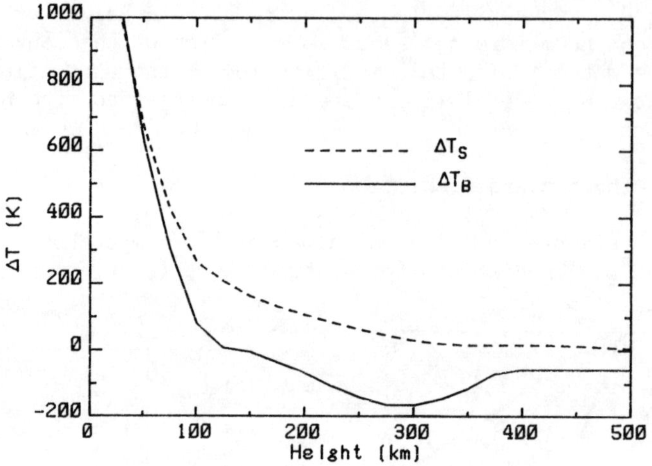

Fig. 3: The fluctuation amplitudes of the kinetic temperature (ΔT_B) and line source function radiation temperature (ΔT_S) with height. The amplitudes are defined by taking the scalar products of the quantities with the continuum radiation intensity pattern shown in Fig. 1.

figure shows the scalar product of the quantities with the continuum intensity pattern, normalized with the rms intensity fluctuation (29 %). The temperature fluctuation associated with the driving, convectively unstable layers vanishes rapidly above the continuum forming layers, and the temperature fluctuations in the convectively stable layers of the upper photosphere are negatively correlated with the continuum intensity fluctuations. This negative correlation carries over to a negative correlation between the continuum intensity and the line center intensity of (not too weak) photospheric spectral lines formed in LTE. The properties of the temperature and pressure fluctuations in the upper photosphere have been discussed at length elsewhere (Nordlund, 1984). The horizontal fluctuations of the NLTE line source function differ qualitatively from those of the LTE source (Planck) function: The NLTE line source function, because of its close coupling to the continuum intensity, is positively correlated with the continuum intensity at all heights. Despite this, it turns out that the correlation between the NLTE line center intensity and the continuum intensity remains negative. This is possible because the line center opacity is strongly influenced by the local temperature (also with NLTE ionization, see Nordlund 1984).

Figs. 4 and 5 show the resulting effects on the spectral line profile. As could be expected from the increase of the line source function, the NLTE line center intensity is increased relative to the LTE

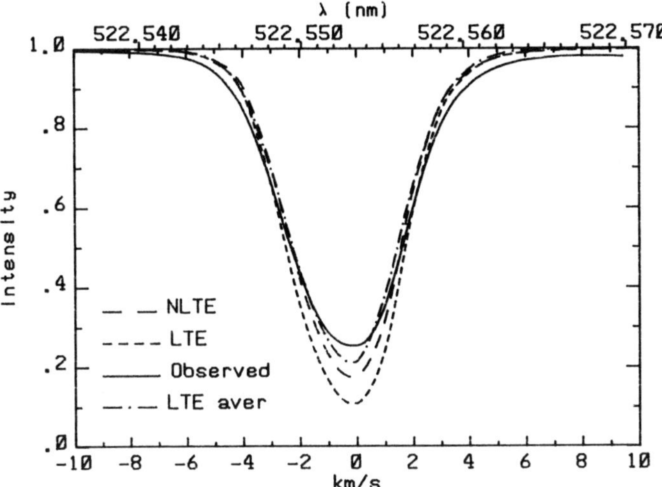

Fig. 4: The observed (full drawn, Delbouille et al. 1973) spectral line profile of the FeI line 522.55, together with synthetic spectral line profiles in LTE (dashed) and NLTE (long dashed) for the snapshot, and time averaged LTE spectral line profile (dot-dashed; Nordlund 1984).

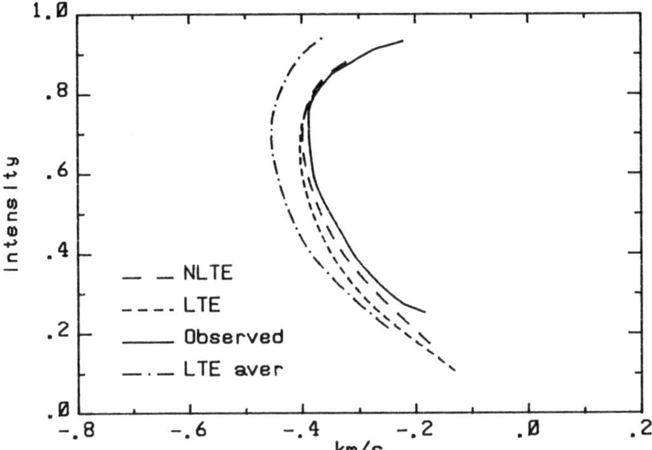

Fig. 5: The spectral line bisectors corresponding to Fig. 4.

case. Whether or not this is enough to bring the depths of the synthetic line profiles into agreement with the depths of the Jungfraujoch line profiles cannot be decided at the present time; this requires time averages of a time sequence of synthetic NLTE line profiles.

4. CONCLUSIONS

The method presented in this contribution has proven to be able to provide useful solutions to the two-level-atom-with-overlapping-continuum problem in just a few iterations. Although the convergence properties of the present method probably gets progressively worse with increasing line strength, the method is useful for investigations of quantitative effects on photospheric spectral lines.

In the case studied here, two significant properties of the problem and its solution should be stressed:

1) The departure from LTE in the excitation equilibrium causes a significant increase in the line source function, which actually *exceeds* the local Planck function over a large part of the photosphere.

2) Thermalization through the overlapping continuum may well be more important than collisional thermalization for typical photospheric spectral lines.

Further work is necessary in order to draw quantitative conclusions regarding the temperature structure and the energy equilibrium of the upper photosphere.

ACKNOWLEDGEMENTS

Financial support by the Danish Space Board and the Danish Natural Science Research Council is gratefully acknowledged.

REFERENCES

Athay, R.G. 1972, *Radiation Transport in Spectral Lines*, Reidel, Dordrecht.

Ayres, T.R., Testerman, L. 1981, *Ap.J.* **245**, 1124.

Böhm, K.H. 1960, in *Stellar Atmospheres*, ed. J.L. Greenstein, University of Chicago Press.

Delbouille, L., Neven, N., Roland, G. 1973, *Photometric Atlas of the Solar Spectrum from λ 3,000 to λ 10,000*, Liege.

Dravins, D., Lindegren, L., Nordlund, Å. 1981, *Astron. Astrophys.* **96**, 345.

Gustafsson, B. 1973, *Uppsala Astron. Obs. Ann.* **5**, No. 6.

Mihalas, D. 1978, *Stellar Atmospheres*, 2nd ed., Freeman & Co., San Francisco.

Nordlund, Å. 1982, *Astron. Astrophys.* **107**, 1.

Nordlund, Å. 1984, "Modeling of Small Scale Dynamical Processes: Convection and Wave Generation", in *Small Scale Dynamical Processes in Quiet Stellar Atmospheres*, Sacramento Peak Observatory conference, ed. S. Keil, Sac-Peak Publication (in press).

Scharmer, G.B., Nordlund, Å. 1982, *Stockholm Obs. Rep.* **19**.

Scharmer, G.B. 1984, in *Methods in Radiative Transfer*, ed. W. Kalkofen, Cambridge University Press, Cambridge.

Shine, R.A. 1975, *Ap. J.* **202**, 543.

A CODE FOR LINE BLANKETING WITHOUT LOCAL THERMODYNAMIC EQUILIBRIUM

Lawrence Sven Anderson
Ritter Observatory
The University of Toledo
Toledo, Ohio 43606
USA

ABSTRACT. A numerical code has been written which is designed to calculate radiation transport and atmospheric structure under the constraints of statistical equilibrium in atomic transitions and radiative and hydrostatic equilibrium in the medium. In addition to the complete linearization and variable Eddington factor techniques of Auer and Mihalas, it uses a multi-frequency / multi-grey algorithm which admits the inclusion of many spectral lines in full statistical equilibrium. The program can comfortably accept up to about 300 specific lines arising from about 30 lower states and any number of continua. Cleverly constructed artificial model atoms can extend the number of lines to 3000 or more, where opacity sampling techniques can begin to approximate the blanketing accomplished by Kurucz in LTE. By way of example, I present a model of a stellar atmosphere with effective temperature 35,000 K and surface gravity 10^4 cm s^{-2}. The calculation includes 98 bound-free transitions and 93 bound-bound transitions (57 with radiative rates) between 91 states in 36 ions of 9 cosmically abundant species.

1. INTRODUCTION

Our ability to calculate completely self-consistent stellar atmospheres under the constraints of plane parallel geometry and hydrostatic and radiative equilibrium has stood at an impass for the last dozen years. On the one hand, if we wish to include the thermodynamic effect of all the atomic transitions present in the real medium we must resort to statistical methods such as opacity distribution functions (Kurucz 1979) or opacity sampling (Sneden, et al. 1976), and worse, local thermodynamic equilibrium (LTE) for the atomic statistics. On the other hand, if we wish to remove the LTE constraint we can include only a few transitions since we must follow the radiation distribution in detail. Unfortunately, when radiative rates exceed collisional rates in the transitions, the LTE approximation introduces "errors" as large or larger than the differences between models with and without thousands of lines. It is neither satisfying nor correct to assume that such

J. E. Beckman and L. Crivellari (eds.), Progress in Stellar Spectral Line Formation Theory, 225–232.

non-linear processes as radiative excitation and radiative blocking add linearly. Thus we cannot "correct" one kind of model for the effects generated in another kind of model.

The impass must be overcome before we can successfully relax the classical constraints in the first sentence. To that end I have developed a new computer code to solve for the transfer of radiation through a medium subject to the rate equations for many atomic transitions. In the next sections I briefly describe the features of the code which set it apart from previous codes and present some models which demonstrate its power.

2. PAM

PAM (Plane parallel Atmosphere Modeling) solves the coupled differential transfer and constraint equations as a tridiagonal block of linearized matrix equations in standard form (cf. Mihalas 1978, p. 230f). Unlike Mihalas, et al. (1975) I have analytically eliminated the statistical equilibrium equations from the linearization. Atomic populations affect only three coefficients in the atmospheric structure equations: the opacity, the source function, and the ratio of gas to electron pressure. The code calculates these coefficients and their partial derivatives with respect to the dependent variables temperature, electron pressure, and radiation field on each iteration cycle.

The most significant departure from other codes of this type is the treatment of the radiation field. Integral properties of the radiation distribution determine the transport of radiation and the structure of a radiative atmosphere. Thus the state variables are not particularly sensitive to the details of the distribution of radiation over various photon energy ranges (e.g. the Lyman continuum, the cores of all Lyman lines taken together, etc.). PAM takes advantage of this insensitivity by dividing the photon spectrum into a relatively small number of blocks (less than 100) and considering the total energy density of radiation in a block at one depth as one dependent variable. It iteratively corrects (via the linearized matrix) these densities assuming that the spectral distribution within blocks remains unchanged. The treatment is similar to the variable Eddington factor formalism (Auer and Mihalas 1970). The internal distributions are updated at the same time as the Eddington factors by a formal solution of the equation of transfer for each individual photon energy. This technique is a logical extention of the multi-frequency / grey algorithm developed by Freeman, et al. (1968). Those authors placed the entire spectrum in one block and assumed LTE. Full statistical equilibrium cannot be treated with only one block since photon redistribution would not be accounted for. PAM explicitly follows redistribution through the partial derivatives of the transfer coefficients in one block with respect to the radiation density in other blocks. Thus individual transitions may require more than one block but related transitions can be grouped together into the same blocks. An appropriate name for the method is multi-frequency / multi-grey (MF/MG).

The MF/MG algorithm greatly reduces the number of dependent variables in a problem, making possible the inclusion of many more

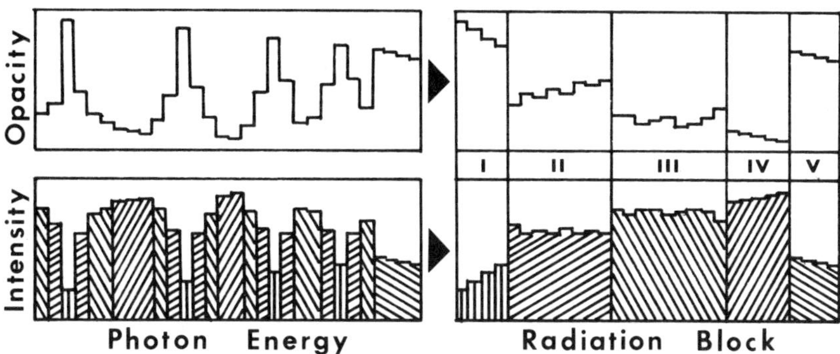

Figure 1. The sorting of individual photon energies into blocks for the MF/MG algorithm as described in the text.

individual photon energies and atomic transitions. Figure 1 shows an example block structure for a converging series of transitions from the same ground state. There, 30 photon energies have been sorted into five blocks; one for strong line cores, two for line wings, and two for continua. Notice that block members need not be contiguous in the energy spectrum. In this example the dependent variables are electron temperature and pressure and the radiation energy density in each of the five blocks. Thus the linearized correction matrices are dimensioned 7x7. If all photon energies and atomic populations were treated explicitly the matrices would be 38x38 (depending on the source of the background continuum). In MF/MG higher resolution in line profiles and more transitions from the same lower state can be included at (almost) no extra cost. The spectrum blocking is done "by hand" such that the photons in each block experience roughly the same physics; e.g. they scatter from the same lower state with similar probabilities for redistribution to other blocks.

I wish to stress that, while the updating of internal distributions is by lambda iteration, the overall solution is not. MF/MG makes approximations in the correction matrix by assuming certain photons behave similarly and can be grouped together. The lambda iteration used to determine distributions only sorts out the (small) differences between these "similar" photons. Since these differences do not propagate in the same way as individual photons, the lambda iteration is sufficient. To illustrate, consider an artificial source of monochromatic photons within a bound-bound transition profile at a specific depth. If the transition suffers complete redistribution, the anomalous radiation profile will persist for only a few optical depths whereas the photons themselves will persist for a thermalization depth. The converged solution is that of the <u>full</u> set of transfer, statistical, and constraint equations.

The user specifies all transitions and atomic levels in input files. PAM itself contains no data on transitions except quantities that depend on state variables (e.g. line profiles and free-free Gaunt factors). Thus the physics of model atoms is entirely up to the user, making PAM very flexible. The code sees little distinction between LTE and non-LTE (running times are virtually the same) or between bound-free and bound-bound transitions (any number of each may be included depending on the resources of the user; running time increases roughly in proportion). PAM reflects the philosophy that approximations should be made in model atoms, which can be changed and refined, and not in global properties of the statistics and radiative transport.

3. MODELS

Here are some results from two calculations at (T_{eff}, log g) = (35000,4), both constrained by radiative and hydrostatic equilibrium. A more complete discussion of these and other models has been submitted to the Astrophysical Journal. Both models employ 63 depth points, 3+ and 3- zenith angle cosines, and 370 photon energies. Nine elements are present with solar abundances (Kurucz 1979; in parentheses are the number of ions and excitation states): H(2,10), He(3,23), C(4,14), N(5,9), O(5,10), Ne(5,7), Mg(4,5), Si(3,5), and S(5,9). There are 98 bound-free transitions. Of 93 bound-bound transitions, 57 (21H, 34He, 2C) include radiative rates with complete redistribution over fully state dependent pressure*Doppler profiles, while the remainder (15He, 21C) see only collisional rates. Model RCS includes radiative-collisional statistics for all states. Model BSS assumes Boltzmann-Saha statistics (LTE).

Figure 2 shows the depth dependence of the electron temperature in RCS, BSS, Mihalas (1972; hereafter M), and Kurucz (1979; hereafter K). The small differences between BSS and K are due to the large number of lesser cooling contributions in the latter. The dramatic difference between RCS and M is due to the carbon resonance transitions C III λ 977 Å and C IV λλ 1548,1551 Å in the former. Models without carbon

Figure 2. Temperature vs. column density of atomic nuclei.

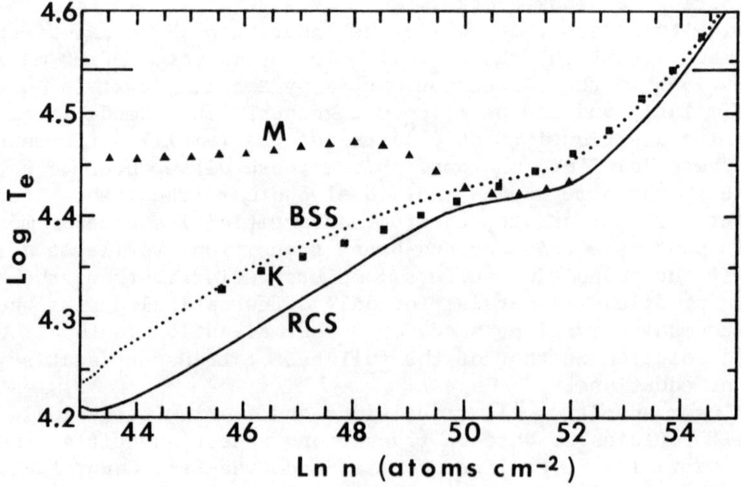

but with more H and He transitions than M used raise the boundary temperature by providing more electron cascade channels, decreasing source functions and cooling efficiency (Anderson 1982). However, the abundance of C IV in RCS is about 4000x the abundance of H I, and the collisional excitation rate of the C IV doublet is about 10x that of Lyman alpha. This one doublet dominates the thermodynamics of the outer atmosphere (in BSS, the C III singlet dominates; see below).

The C IV ion is well represented as a collisionally dominated two level atom (PAM assumes complete redistribution over the doublet). Thus one may express the line source function as $S = (1-\delta)\bar{J} + \delta B$, where $\delta \approx C_{u\ell}/A_{u\ell}$ is the photon destruction probability and $\bar{J}$ is the mean intensity averaged over the line opacity profile (cf. Mihalas 1978, p. 336f). Then, dominated by this transition, radiative equilibrium leads to $\chi_{\ell u}\,\delta(\bar{J} - B) = 0$ or $\bar{J} = S = B$. One expects and finds a Boltzmann population ratio and LTE-like behavior (however, the ionization is far from LTE).

The transition should control the thermodynamics down to its thermalization depth, beyond which δ exceeds the escape probability. For a Lorentz profile thermalization occurs when the core optical depth reaches about δ^{-2}. Model diagnostics show that that happens at a column density of 4.2×10^{21} atoms cm^{-2} ($\ln \underline{n} = 49.8$ and $\tau = 5.4 \times 10^4$). In RCS the temperature begins its downward slope under the influence of λλ 1548, 1551 Å right at this depth. Now, since $S = B$, B must be the solution of the integral equation of transfer $B(\tau) = \int_0^\infty B(t)K|t-\tau|\,dt$ (see Mihalas 1978, Eq. 11-18). If we assume B has a form $B_0\tau^\alpha$ for $\tau < 5 \times 10^4$, a simple numerical exercise leads to $\alpha \approx 0.4$. We can understand this value as follows. Without radiative equilibrium, a B constant with depth and a Lorentz profile produce a source function proportional to $\tau^{0.25}$ (Avrett and Hummer 1965). Replacing B with this S can only produce a steeper S. However, S cannot be steeper than $\tau^{0.5}$ without the integral over $S \times K$ becoming unbounded for a Lorentzian K. Finally, at $T = 20{,}000$ K and $h\nu = 8.0$ eV, B is proportional to $T^{4.7}$ so we should find $T \propto \tau^{0.085}$. The C IV fraction is roughly constant with depth so τ is proportional to the column density. Indeed, in RCS $T \propto \underline{n}^{0.081}$ for $\ln \underline{n} < 50$.

Figure 3 contains the Eddington fluxes emergent from RCS, BSS, M, and K' (Kurucz 1984 private communication; the 1979 data are in error). The two LTE models differ insignificantly. The difference between RCS and M above the He I 1s photoionization limit is due to the more realistic He atom used in RCS and M's use of a fictitious oxygen-like atom to approximate all atoms heavier than helium.

Even though consistently cooler than BSS, RCS emits from 10 to 100x more radiation at all energies above 24.6 eV. This increase is because the statistical equilibrium is driven toward higher ionization. Radiative rates dominate all of these continua. For example, the effective photon destruction probability $C_{u\ell}/R_{u\ell}$ for the O III $2s^22p^2\ ^3P$ continuum above 51.45 eV is 3×10^{-7}. Thus photons tend to scatter within a given continuum, although they get redistributed by the thermalized recombining electrons. These continuua also emerge from regions with steep gradients of $B(\tau)$. Thus $\bar{J}$, and therefore S and the emergent flux, influenced by B one thermalization depth inward, are high. In BSS the O III ^3P continuum reaches optical depth 2/3 at $\ln \underline{n} = 47.0$ where $T = 23{,}450$ K. In RCS it

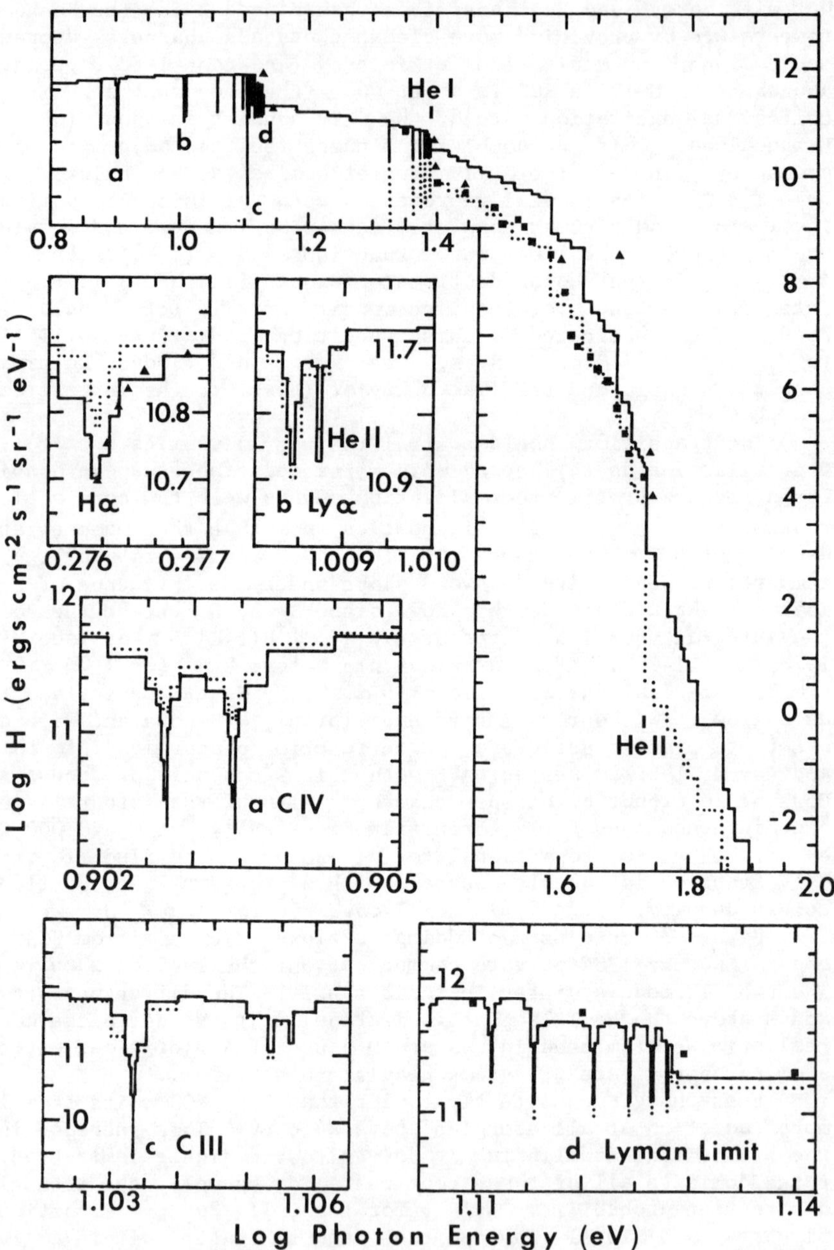

Figure 3. Emergent Eddington flux vs. photon energy. The data symbols are the same as found in Figure 2. a-d: Locations of expanded labeled segments.

reaches 2/3 at $\ln \underline{n} = 47.5$ where T = 21,618 K. However, the source function is 54 x B, which more than compensates for the 8.6 reduction in B.

The RCS ionization accounts for the shift in dominance from C III to C IV. The shift is most dramatic near the upper boundary where the declining temperature arbitrarily forces recombination in BSS. At $\ln \underline{n} = 44.0$ the BSS C IV 2s fraction drops to 43% while C III $2s^2$ climbs to 48%. In RCS, however, C IV 2s remains above 95% and C III $2s^2$ below 0.3% throughout most of the atmosphere. There is a significant difference in the emergent flux profiles of the two resonance features as a result. The enhancement of C IV will be even more pronounced at cooler effective temperatures where it has more to gain.

Figure 3 also contains expanded portions of the emergent flux showing the profiles of Hα, Lyman α, the carbon lines, and the Lyman limit. These profiles illustrate the high spectral resolution that PAM achieves directly in the structure calculations. The statistics of hydrogen is dominated by radiation, so none of its emergent profiles respond very much to changes in the temperature. Thus the RCS profiles are almost identical to those of M. Not shown in Figure 3 are the infrared lines. Despite the monotonic temperature structure, in RCS they appear in emission as they do in M. Lower levels drain to the ground state more easily, tending to invert population ratios. Interestingly, He λ 10830 Å also appears in emission in RCS although I remain skeptical until more subordinate transitions in the helium atom are accounted for.

4. DISCUSSION

The two major conclusions are a) radiative-collisional equilibrium in late O-stars leads to higher ionization relative to LTE, which in turn enhances the contribution of the C IV 2s-2p transition, and b) this transition determines the thermodynamics of the outer atmosphere above its thermalization depth. Unfortunately, the presence of winds in luminosity class V stars at 30,000 K masks the comparison of RCS with real stars. If one applies Abbott's (1982) relation between mass loss rate and luminosity, from the equation of continuity one may estimate the depth at which the wind speed becomes equal to the isothermal sound speed. In RCS this depth is at $\ln \underline{n} = 45.75$, where the core of C III λ 977 Å is just reaching $\tau = 2$. Therefore, one expects to find little evidence for a wind in that transition. The observed profile from μ Col (Olson and Castor 1981) does appear symmetric. In BSS the line is much more opaque. It definately would be asymmetric unless the estimate for $\dot{M}$ is 100x too large. Better comparisons should be possible for cooler stars where winds are weaker and C IV $\lambda\lambda$ 1548,1551 Å becomes a better diagnostic.

Will the inclusion of more lines affect the atmospheric structure? It is hard to imagine any transition or group of transitions being more effective than the C IV resonance doublet when C IV is the dominant carbon ion. The resonance transitions of He, N, and O will have little additional effect since collision rates scale approximately as $\exp(-h\nu/kT)$. However, if there is significant line <u>blocking</u> in the

C III continuum, the ionization balance will shift back toward C III.

I thank the University of Toledo Computer Center and the National Center for Atmospheric Research in Boulder, Colorado for many hours of computer time and the High Altitude Observatory division of NCAR for a summer visiting appointment. Special appreciation goes to D. Mihalas for his support and discussion. Other useful discussion came with D. Abbott, D. Hummer, and P. Kunasz. The International Travel Grant Program of the American Astronomical Society and the Physics and Astronomy Department of the University of Toledo provided travel funding.

REFERENCES

Abbott, D. C. 1982, Ap. J., 259, 282.
Anderson, L. 1982, Bull. AAS, 14, 921.
Auer, L. H., and Mihalas, D. 1970, M. N. R. A. S., 149, 65.
Avrett, E. H., and Hummer, D. G. 1965, M. N. R. A. S., 130, 295.
Freeman, B. E., Hauser, L. E., Palmer, J. T., Pickard, S. O., Simmons, G. M., Williston, D. G., and Zerkle, J. E. 1968, DASA Report No. 2135, Vol. I (La Jolla: Systems, Science, and Software, Inc.).
Kurucz, R. L. 1979, Ap. J. Suppl., 40, 1.
Mihalas, D. 1972, Non-LTE Model Atmospheres for B and O Stars, NCAR Technical Note NCAR-TN/STR-76, (M).
Mihalas, D. 1978, Stellar Atmospheres, (San Francisco: Freeman).
Mihalas, D., Auer, L., and Heasley, J. 1975, A Non-LTE Model Stellar Atmosphere Computer Program, NCAR Technical Note NCAR-TN/STR-104.
Olson, G. L., and Castor, J. I. 1981, Ap. J., 224, 179.
Sneden, C., Johnson, H. R., and Krupp, B. M. 1976, Ap. J., 204, 281.

PANNEL DISCUSSION ON RADIATIVE TRANSFER METHODS

Pannel members: W. Kalkofen, J. Linsky, G. Rybicki, G. Scharmer, R. Weherse.

The second pannel discussion dealt with the application of physical principles to real stellar atmospheric situations. The ball was set rolling by a question to the pannel to define the limit of usefulness of approximations in radiative transfer methodology. Here there were two distinct views. Those who thought that approximate methods should be used whenever possible gave the following reasons:

1) physical simplification leads to computational simplification and therefore to major saving in computer time;

2) more fundamentally approximation methods which allow analytical treatment can give a very clear insight into the way the physics of a problem develops;

3) frequently the observations are not detailed enough to warrant a "complete" treatment.

Examples of useful approximative methods such as that of the escape porbability technique were brought up. Not only for a set of stellar atmospheric problems, but also in situations where the radiative transfer is of limited importance compared with the methods of energy transport such as hydrodynamics, approximative methods are very worthwhile.

However limits to their general applicability were pointed out by those whose basic stance was that modern computers make approximations outmoded. They emphasized that in case of high velocity gradient, for example, escape probability is not appropriate and Sobolev method is sufficient, while for static cases "complete" non-approximate methods are preferred. For the case of solar plages, 2nd order escape probability methods have been applied, but fall

J. E. Beckman and L. Crivellari (eds.), Progress in Stellar Spectral Line Formation Theory, 233–237.

short because of the large velocity fields. Even in the case of QSO spectra, where the line profiles are not well enough measured to demand a full treatment, the possible presence of velocity gradients makes the use of particular approximations a somewhat problematical enterprise.

On the subject of computer time, a rule of thumb was proferred, which said that an approximate method which introduces errors in the 20% range but saves time by an order of magnitude is clearly a preferred approach. This was countered by the claim that a full solution in which 3-dimensional geometry was taken into account could be of supreme importance in some cases, and no approximation in 2-dimensions would be good enough.

As far as the pedagogical aspects of approximation were concerned, those who were involved in "full" solutions admitted that much work was needed to understand the details of the output of computer codes in terms of distinct physical effects which could be used for diagnostics.

It was difficult to find quantitative statements about the limits to approximation, because different types of transfer problems are more or less susceptible to approximate treatment. Although CRD may be a practical approximation to PRD, one must be aware that PRD not only affects the use of lines as diagnostics, but can be important in the interaction of radiation with the mechanism of transfer, and can affect basic physical quantities such as the thermalization length. This is clearly important when the wings play a major role in the transfer. Another case where an apparently good approximation can be inadequate is where non-linear effects apply. A 20% margin in the initial assumption can produce instability and a large error in the final output.

Moving from attempts to grapple with the limits of approximation to attempts to inject realism into models, the importance of three-dimensional modelling was brought out. Although horizontal transfer, when considered quantitatively is only a perturbation on vertical transfer, the significance of the ability to handle a third dimension is the possibility to take velocity fields adequately into account.

Although specialized types of three-dimensional codes have been in use since the mid 1970's, these have comprised essentially two-dimensional models plus a set of periodical boundary conditions. They differ from earlier one- and two- dimensional models in the structure of the emergent radiation, but the source function does not suffer much impact. On the contrary a three-dimensional model which incorporates velocity fields can entail the shifting of photons from core to wings or from wings to core, and this necessarily implies an important role for PRD, with a significant effect on the intrinsic profile. Similar effects can in fact occur in two-dimensional and even one-dimensional models, given suitable formulations of velocity fields, but clearly the three-dimensional case should be more realistic. The ultimate goal will be an "all dancing all singing" multi-transition, multi-dimensional model, but it was made clear in the discussion that so far models of this kind have not been run with PRD, and so comparison with observations are not yet "à propos".

It was pointed out, furthermore, that a single spectral line could not be an adequate test for any model, and great stress was laid on the need for observations of many lines within a given atmosphere. These should include a number of lines from the same species, and could entail simultaneous observations (discount variability) within different wavelength ranges. This touched a responsive chord in one observationalist, who stated that the greatest unsolved problem in line formation is a mechanism to persuade observatory committees of the need for scheduling such observations in parallel. In similar lightheartred vein, a theorist noted wrily that the strongest requirement for approximate methods was the difficulty of obtaining computer time on large machine.

At the end of the session, pannel members were asked to pick out the key advantages and disadvantages of their own particular method.

The advantages of the multi-level method were held to be that it can be applied to a wide variety of atomic species, and can easily incorporate velocity fields, including (countrary rumours not withstanding) high velocities. On the other hand it cannot readily cope with more than 20 levels, and it would run into time problems when dealing with two- or three-dimensional transfer. It is of interest that although it is not easy to assign immediate physical significance to each element of the transfer matricies, some tentative steps have been already made in that direction. In the uppermost part of certain trial atmospheres negative populations were produced, which were the effects of the linearization technique, and could be removed by abstracting certain distant off-diagonal matrix elements.

Escape probability methods are useful in offering clear physical insight into the behaviour of photons within an atmosphere, and are economical in computer time. The benefit of physical clarity has been bequeathed to subsequent, more complete, techniques. The drawbacks of escape probability are that it requires CRD, at least in the simplest formulation, and that on the other hand it is overelaborate to handle high velocity fields.

An alternative approach to the problem of global radiative transfer within an atmosphere in a part of the spectrum where there are many lines is to handle the radiation in "blocks" of photons which see the same physics as they proceed through the atmosphere. Such a block might comprise the wings of all lines originating in a common state, or alternatively the cores of such lines. The advantages of such an approach for numerical handling of this stage of building atmospheric models are very clear. All photons within a block have equivalent probability of absorption, scattering or redistribution into another block. This type of code is designed to handle statistical equilibrium in a large number of species, but is not so effective for a single species with many different lower states. It is not possible to see how the multi-level method (given frequency as a function of depth) and the present multi-block method (given population as a function of depth) can be effectively combined, and indeed the two approaches are not directly compatible. Both will be of major value.

Another general line transfer technique, that of operator perturbation methods, is in wide use. It has given rise to computer-based multi-level atom approach. It can be very economical because, relying on a differential formulation and not on a set of integral equations, the matrix produced for a transfer problem with many depth points and many levels scales only linearly with the number of depth points, which makes computer handling possible. It is also useful in time-dependent situations. In fact the major point of difference between practicioners nowadays is not whether to use such methods, but how to choose the initial simple operator to perturb. Because it is iterative it is useful in hydrodynamics problems.

In conclusion it now appears that we have entered the epoch where large computers are making a major impact on what is possible in several ways. Handling of the very large matrices produced by multi-dimensional, multi-stream, multi-level models is a feature of the new generation of parallel-processing machines. In any applications of this type there is always the danger that the physics is lost sight of, but as long as there are practicioners of transfer problems for whom elegance and insight are as important as realism, we will be able to maintain a clear vision of the physical wood through the numerical trees.

Aknowledgment.

The editors are happy to thank Dr. Giampapa, whose accurate notes proved very helpful when writing the summary of this discussion.

LINE FORMATION IN LABORATORY PLASMAS

P. Jaeglé, G. Jamelot, A. Carillon
Laboratoire de Spectroscopie Atomique et Ionique, associé au C.N.R.S., Bat. 350, Université Paris-Sud
91405 Orsay
and
GRECO "Interaction Laser-Matière"
Ecole Polytechnique
91128 Palaiseau
France

ABSTRACT. Spectral line formation in laboratory plasmas has concern with plasmas of high density, such that photon escape is not free, though the plasma size may be small. Plasmas produced by laser impact on solid targets, like in inertial confinement experiments are typical examples ot the kind. Given the high density range under consideration, the effect of electron-ion and ion-ion collisions upon optically thin profiles is important. Then, radiation transfer brings out the relation between the thin and the thick profiles. In order to build computable models, it is necessary to consider the role of various physical factors, like frequency redistribution, plasma inhomogeneity, coupling of radiation with electron level populations. Moreover fast time-variation is a common feature of laboratory plasmas. This leads to departure from equilibrium which modifies the intensity of peculiar spectral lines, mainly in recombining plasmas. The evidence of population inversions has been given.

1. INTRODUCTION

For being retained on decided density and temperature, laboratory plasmas need external means of confinement and heating. Regarding hot plasmas, i. e. plasmas in which atoms are stripped so that there are only multicharged ions surrounded with a free electron gas, magnetic confinement has been for a long time the only way to hold the plasma within the walls of a machine. Technics differ substantially according to the way of producing the magnetic field and the characterisctics of its space and time configuration. In the "pinch" devices, the confinement comes from an intrinsic magnetic field induced by axial (Z-pinch) or azimuthal (Θ-pinch) current flow. Considerable efforts are made since thirty years, in view of thermonuclear fusion, for developping Tokamaks in which the lines of magnetic field are helicoïdal because of the conjunction of a poloïdal intrinsic field with a toroïdal one produced by circular coils around the plasma chamber. In Tokamaks, like in mirror machines - another type

J. E. Beckman and L. Crivellari (eds.), Progress in Stellar Spectral Line Formation Theory, 239–263.

of fusion device - the plasma density is 10^{13}-10^{14} particles/cm^3. From the point of view of radiative transfer, this is much too small for being likely to reabsorb significantly the photons emitted by the ions, even in the largest existing machines. Then, line profiles represent merely the thermal motion of the emitters, i.e. they exhibit the familiar Gaussian shape due to Doppler shift averaged over Maxwellian distribution.

A new class of laboratory plasmas has come in more recently, following the growth of inertial confinement research. In inertial confinement devices a strong energy pulse heats very steeply the external layer of a dense matter target. A hot plasma is created, the fast expansion of which gives rise to a ten or hundred megabar pressure against the target surface. The plasma produced in the small volume between the energy deposition layer and the central cold matter is hard squeezed and may reach a density much larger than the density of solid. For thermonuclear fusion it would be necessary to confine the plasma during a few nanoseconds with a density about 1000 times the solid density. In the present state of art these conditions are seeked after by lighting a ≃100μ spherical target with several laser beams focused so that the incoming radiation is as close as possible to a spherical converging wave. The sketch of such an experiment is shown in fig. 1. The minimum number of beams is two, a con-

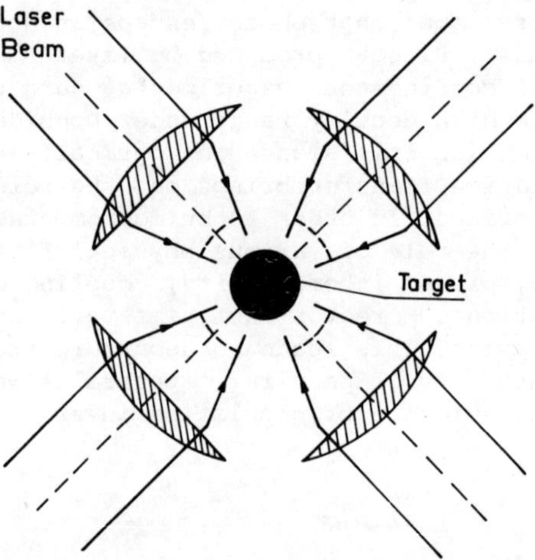

Fig. 1. Sketch of an inertial confinement experiment. The diameter of the target is ≃100μ. The ablation of matter on the target surface, owing to heating by laser radiation, generates a pressure shock wave towards the target center.

figuration which requires a very large aperture of focusing optics. Several tens beams are used in some experiments.

Although the plasma volumes produced by inertial confinement are small, the density is so large that reabsorption of emitted photons has generally a large probability. Electron density comprised between 10^{24} and 10^{25} cm^{-3} has been reached in implosion experiments [1] . Given the extended density scale under consideration, it is useful to have a simple criterion to estimate if reabsorption is important or not. For a two level transition, 1 and 2 labelling respectively the lower and the upper level, the absorption coefficient can be written:

$$\alpha_\nu = \frac{h\nu}{c} B_{12}(N_1 - \frac{g_1}{g_2} N_2)\ \Phi(\nu) \tag{1}$$

where $h\nu$ is the transition energy, B_{12}, Einstein's absorption coefficient, $\Phi(\nu)$, the absorption profile function, N_1 and N_2, the level populations, g_1 and g_2, the statistical weights. Assuming Boltzmann's distribution at electron temperature kT_e, replacing the true profile function by the inverse line width $1/\delta\nu$, restricting the estimation to the lines whose the energy lies in the range of electron temperature ($h\nu \simeq kT_e$) and using CGS units and electronvolts give:

$$\alpha_{cm^{-1}} = 1.58\text{x}10^{-24}.\frac{A_{21\ sec^{-1}}}{E^2_{\ ev}}.\frac{1}{\delta E_{\ ev}}.\frac{g_1}{g_2}.N_1\ _{cm^{-3}} \tag{2}$$

where A_{21} is the radiative transition probability, E, the transition energy, δE, the line width in energy units. We lay down that the photoabsorption has a significant effect on line shape only when the photon escape probability is smaller than 0.99. In other words, L being the length travelled by the radiation in the plasma, the line formation problem comes ut when:

$$\alpha L \simeq > 10^{-2}$$

i.e., for a typical value,

$$L = 100\mu$$

in a laser-produced plasma:

$$\alpha \simeq > 1\ \text{cm}^{-1}$$

From (2) this condition means that:

$$\frac{g_2}{g_1} N_{1\,cm^{-3}} \simeq > 6.3\text{x}10^{23}.\frac{E^2\ _{ev}}{A_{21\,sec^{-1}}}.\delta E_{ev} \tag{3}$$

Various processes, discussed in section 2, contribute to line broadening in dense hot plasmas. Here, for the simplicity, we assume that δE results from thermal Doppler broadening, that is:

$$\frac{\delta E}{E} = 7.67\text{x}10^{-5}.\frac{(kT_{ev})^{1/2}}{M_g} \tag{4}$$

where M is the atomic mass of the plasma component. Then the electron level density must satisfy the condition:

$$\frac{g_2}{g_1} N_{1\,cm^{-3}} \simeq > \frac{4.8\text{x}10^{19}}{M^{1/2}_{\ g}}.\frac{E^3\ _{ev}}{A_{21\,sec^{-1}}}.(kT_{ev})^{1/2} \tag{5}$$

For numerical applications, let us remember that $1\,ev \leftrightarrow 8066\ cm^{-1}$ for atomic transitions and $1\,ev \leftrightarrow 11605\,^{o}K$ for plasma temperatures. Suppose an aluminum plasma at 100 *ev* temperature; we have M= 27 *g*; for a transition energy about 100 *ev* ($\lambda \simeq 120$ Å), corresponding to the XUV spectrum of the plasma, and a typical value of radiative probability in this spectral range $A_{21}=10^{11}$ sec^{-1}, condition (5) gives:

$$\frac{g_2}{g_1} N_1 \simeq > 10^{15} \quad cm^{-3}$$

This order of magnitude is frequently achieved in laser plasmas.

An important feature of these plasmas is that electron density and temperature exhibit steep gradients, on a few micron length scale. In order to introduce some consequences of this for spectroscopic studies, we will consider one-beam plane-target experiments, which don't produce plamas as much compressed as in multi-beam experiments but which are performed for studying many interaction and transport processes. Fig. 2 is a scheme of temperature and density distributions corresponding to this case. Along laser axis an important separation is set up between two plasma regions by the critical density surface, at few tens microns from the target. The fact is that the laser beam cannot propagate beyond

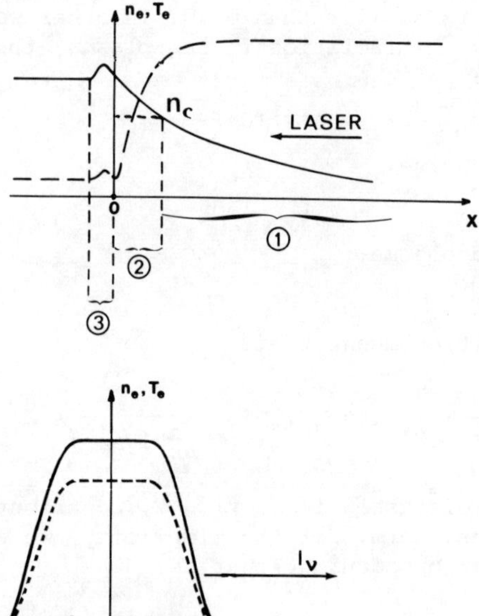

Fig. 2. Plasma density (solid curve) and temperature (dashed curve) distribution in plan target experiment. Along the laser axis OX, there are three regions: 1) plasma corona below the critical density, 2) ablation zone between critical density and target, 3) compressed plasma between unperturbed solid and ablated plasma. Perpendiculary to the laser axis (OY-axis), a hot core inside the ≃100µ focal spot is surrounded by a cooling expanding plasma;(real distributions are generally more complex).

the critical density for which the plasma frequency and the laser frequency are equal. Thus direct heating by laser beam takes place only in the corona, below the critical density. The plasma of larger density, near the target, is heated by electronic conduction from the corona; this leads to gradients of opposite signs for temperature and density. Perpendiculary to laser axis, the central hot plasma volume is shown to be surrounded by expanding plasma shells; I_ν represents an emerging plasma radiation of frequency ν in this direction which is frequently used for spectral observations.

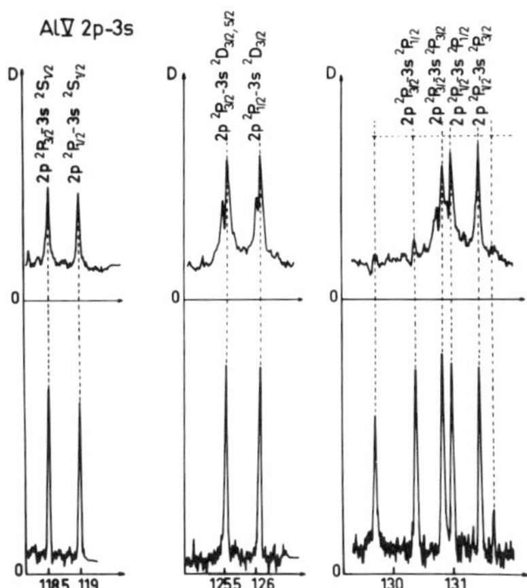

Fig. 3. Example of spectral recordings at two different distances from the target, perpendiculary to laser axis, showing line shape modifications at large plasma density (upper curves).

Keeping in mind these characteristics, let us consider now the former result presented in fig. 3 (2-5). The figure shows the densitometer trace of Al^{4+} spectrum emitted by a plasma produced by laser impact with an illumination power density around 10^{12} W/cm^2 and a 30 ns pulse duration. The spectrograph was perpendicular to laser axis. For the lower spectra in the figure, the observed zone was in the far corona (region 1 in fig.2) and there is no trace of reabsorption on line shapes due to instrumental function. In contrast, the upper spectra, which correspond to the ablation and critical density region of the plasma (region 2 in fig. 2) exhibit strongly modified line profiles. We notice especially the splitting which appears on the 125-126 Å lines at the center of the figure. Assumptions of very different contents could be made for interpreting this result at the time of the experiment. For instance, similar results for He-like resonance lines were suggested to be due to peculiar Stark effect raising the prohibition of $\Delta l=0$ transitions and shifting the line by the plasma frequency (6). In fact, radiation transfer trough inhomogeneous plasma, like shown in fig. 2, accounts fairly well for such results which have been therefore the sarting-point of line formation study in laboratory plasmas. We will present below four aspects of actual interest of this study: optically thin profiles, frequency redistribution, plasma modelling and stimulated emission.

2. OPTICALLY THIN LINE PROFILES

As seen on expression (1), the coefficients entering into the treatment of radiation transfer need the knowledge of optically thin line profiles. Although drastic approximations are generally made about the profile functions before integrating transfer equations, a general view of thin profile theory is necessary to make an appropriate choice between Gaussian, Lorentzian, Voigt or other possible shapes as well as to use a correct estimation of optically thin line width. On the one hand, one has to consider the motion of the emitters due to plasma temperature or produced by plasma expansion. On the other hand, since the interaction between radiation and emitters takes place only at large density, electron-ion and ion-ion collisions occur generally at a rate large enough to cause broadening and shift of lines by Stark effect.

A view of the line profile change along temperature and density gradients is given on the photography of fig. 4. A spatial resolution system(7) reproduces into two equivalent symetric patterns the distribution of plasma emittance along laser axis. The most intense line on the photography is due to the 3d-4f transition of Al^{10+} ions at λ = 154 Å. It has been chosen as an example because the reabsorption has practically no effect on its profile owing to a pecularity of the 3d/4f population ratio(8). Going from the end of the line to the intense continuous spectrum, one sees first the broadening produced by a varying Doppler shift due to plasma spherical expansion. Nearer the target, line wings are

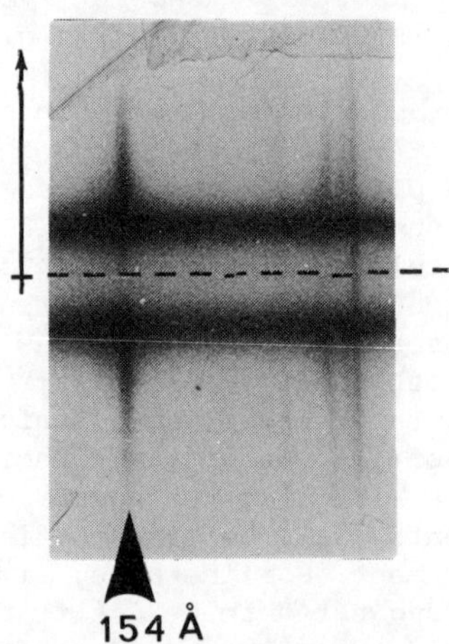

Fig. 4. Photographic recording of the Al^{10+} soft X-ray spectrum in a laser plasma. The spatial resolution (direction of the arrow) shows the change of line profile along the laser axis. The middle horizontal dashed line points the plasma region near the narget surface; the ends of the lines correspond to plasma shells lying at more than 100µ from surface.

growing more and more as a result of increasing quasistatic perturbation by near-by ions. The break-down of emission very near the target, i.e. along the middle axis of the photography, reproduces the fall of temperature shown in region 2 of fig. 2.

As far as many calculations have been devoted to optically thin profiles, we are going to survey some important results of the works. Let us mention that all these results are obtained in assuming Maxwell's distribution of ion and electron velocities.

2.1 Doppler Broadening

The effect of expansion velocity Doppler shift on line profiles is often important but it can be calculated only by numerical modelling of the plasma, with the possible help of hydrodynamical calculations or, if it be not, in using experimental results. Simple examples will be given in section 4. Here we just recall the familiar Gaussian profile corresponding to ion thermal motion at temperature T:

$$\Phi(\nu) = \frac{1}{\sqrt{\pi}\xi\nu_0} \exp\left(-\left(\frac{\nu - \nu_0}{\xi\nu_0}\right)^2\right) \tag{6}$$

where:

$$\xi = \left(\frac{2kT}{Mc^2}\right)^{1/2}$$

M being the atomic mass of the ions. It is convenient to calculate the full-width at half-maximum of this profile in using the formula:

$$\frac{\delta\nu}{\nu} = 7.67\text{x}10^{-5} \left(\frac{kT_{ev}}{M_g}\right)^{1/2} \tag{7}$$

Farther we will make some comparisons of the widths given by this formula with values provided by Stark broadening calculation.

2.2 Stark Broadening

In the plasma, the emitters are perturbed by the electrical microfield due to neighbouring particles. Fig. 5 is a schematic view of the action of an external electric field F, in Oz direction, on the atomic potential. For calculations, the lack of central symetry makes necessary to use new parabolic coordinates in the wave equation in order to solve two separate equations. Qualitatively on sees that the discrete levels are more or less shifted; in the case of hydrogenic ions a separation in sublevels occur. Moreover a potential barrier appears in the direction of electric field; this affords bound electrons to escape either by ionization at an energy smaller than the unperturbed limit or by tunnel effect through the barrier. In the plasma, perturber motion produces a statistical distribution of perturbing microfield and these effects are averaged over the distribution, what results in line broadening. The general methods

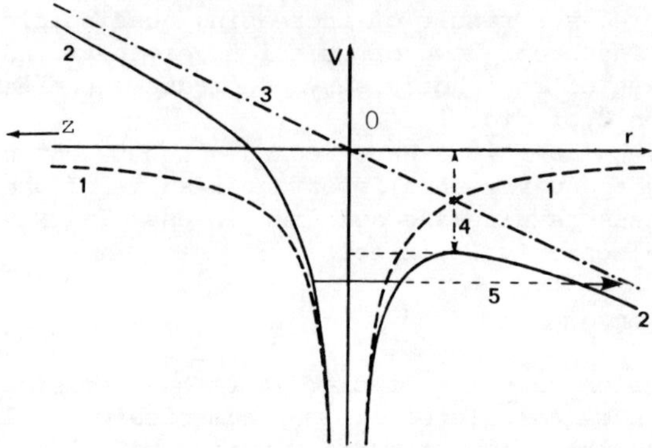

Fig. 5. Atomic system perturbed by an external electric field; 1. unperturbed atomic potential; 2. perturbed potential; 3. perturbing field; 4. lowering of ionization limit; 5. tunnelling through the potential barrier.

of calculation are presented in the litterature[9,10]. Here, as in the next paragraphs, we summarized the most important features of these density dependent perturbations according to the present state of art.

As regards line broadening, it is usual to consider separately the high and low frequency components of the microfield. The effect of high frequency microfield, which comes mainly of the fast moving free electrons, is calculated in the impact approximation. It gives a Lorentzian profile; let us assume the two levels to be degenarate, the indices α and β numbering the states belonging respectively to the initial and final levels. We have:

$$\Phi(\nu) = \sum_{\alpha\beta} n_\alpha |d_{\alpha\beta}|^2 \frac{1}{2\pi} \frac{\gamma_{\alpha\beta}}{(\nu - \nu_{\alpha\beta} - \Delta_{\alpha\beta})^2 + (\gamma_{\alpha\beta})^2} \tag{8}$$

where n_α is the population of the state α, $d_{\alpha\beta}$ is the matrix element of dipole moment, $\Delta_{\alpha\beta}$ is the shift of the α-β transition, $\gamma_{\alpha\beta}$ the broadening. The quantitative calculation of (8) needs to determine the collision matrix. For the perturbation being proportional to the number of impacts per time unit, it is easily shown that $\Delta_{\alpha\beta}$ and $\gamma_{\alpha\beta}$ are proportional to the electron density N_e. Let us mention that, in practical transfer calculations, optically thin profiles are often obtained by convoluting (8) and (6).

The low frequency component of the microfield comes mainly from the slowly moving ions. If the field variation is slow enough, it is possible to assume that the perturbation is constant during the emission. If, in addition, electrons and ions are considered as statistically independent, this quasi-static approximation allows to write the profile in the form:

$$\Phi(\nu) = \int d\beta \; W(\beta) \; J(\nu,\beta) \tag{9}$$

Here β is the ratio F/F_0, where the so-called normal field is given by:

$$F_0 = 2.603 x Z e N_i^{2/3}$$

or:

$$F_0 = 2.603 x e Z^{1/3} N_e^{2/3} \tag{10}$$

$W(\beta)$ describes the statistical distribution of the microfield and $J(\nu,\beta)$ is the general spectral profile[11]:

$$J(\nu,\beta) = \frac{1}{\pi} \text{Re Tr}\{d \frac{1}{i\Delta\nu(\beta) - \Gamma_e(\Delta\nu(\beta))} \rho d\} \tag{11}$$

ρ is the density operator; the trace is taken over the electronic states contributing to the line under consideration; $\Delta\nu(\beta)$ is the shift and Γ_e the electron broadening operator in presence of the field βF_0. Electric microfield distribution functions have been calculated for hot dense plasmas[12]. The quasi - static approximation is valid in the wings of the lines. Between wings and center, where electron collisions are dominant, intermediate approximations, including ion dynamic effects, may be necessary.

Most of the Stark profiles, which have been calculated to-day, correspond to Lyman's lines of hydrogenic ions[12-16]. As an example of these calculations, fig. 6 illustrates the various contributions written in (6), (8) and (9) for Ly-α line of Ne X at T_e = 800ev and $N_e = 10^{23}$ cm^{-3} [12].

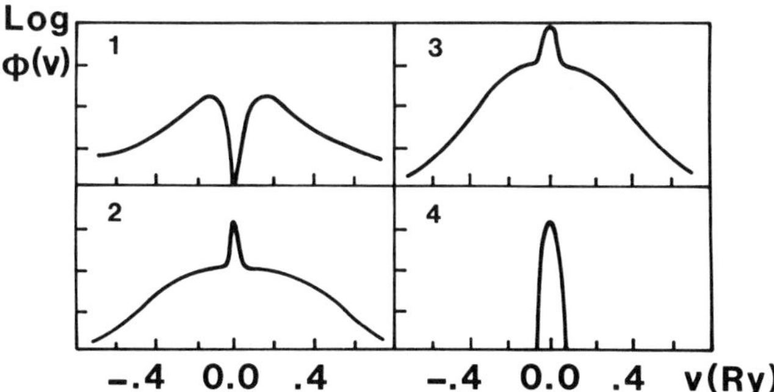

Fig. 6. Ly-α line profile of Ne X in dense hot plasma (ref. 12); (1) static-ion profile; (2) electron-impact plus static-ion profile; (3) convolution of profile (2) with Doppler effect; (4) Doppler profile.

In numerical calculations of radiation transfer, it is often possible to spare time by choosing between Doppler and Stark profiles rather than convoluting both. This would need a simple criterion about the dominance of one of these effects. General approximate expressions can be used for line widths only in the case of hydrogenic ions; even in this case, a difficulty arises from the fact that the estimation of the width is very different for perturbations by electrons and by ions. For the Lyman series of H-like ions we can write, for Doppler broadening:

$$\delta\nu_{ev} = 1.04 x 10^{-3} \left(\frac{kT_{ev}}{M_g}\right)^{1/2} Z^2 \frac{n^2 - 1}{n^2} \tag{12}$$

for Stark broadening by electron collisions[17]:

$$\delta\nu_{ev} = 1.58\text{x}10^{-22}.\frac{N_{e\ cm^{-3}}}{Z^2(kT_{e\ ev})^{1/2}}\ n^2\ (n^2-1)Q \tag{13}$$

and we may assume for the static-ion broadening the Holtsmark width:

$$\delta\nu_{ev} = 8.25\text{x}10^{-15}.\frac{n^2-1}{Z^{2/3}}\ N_{e\ cm^{-3}}^{2/3} \tag{14}$$

Z is the ion charge, n, the main quantum number of the upper level, Q, a Gaunt factor. In (13), (n^2-1) is to be replaced by (n^2-l^2-l-1) for the levels of non-zero orbital momentum. From these expressions we deduce that Stark width will dominate the line profile when:

$$N_{e\ cm^{-3}} \simeq> 6.58\text{x}10^{18}.\frac{Z^4}{n^4}\frac{kT_{ev}}{M_g^{1/2}}\frac{1}{Q} \tag{15}$$

for electron-collision broadening, and:

$$N_{e\ cm^{-3}} \simeq> 4.48\text{x}10^{16}.\frac{Z^4}{n^3}\left(\frac{kT_{ev}}{M_g}\right)^{3/4} \tag{16}$$

for static-ion broadening. Considering again an aluminum plasma (M=27) of 100ev temperature and the Ly-γ line (n=4, Q≃1), the previous conditions give respectively:

$$N_e \simeq> 1.4\text{x}10^{22}\ \text{cm}^{-3}$$

and:

$$N_e \simeq> 5.3\text{x}10^{19}\ \text{cm}^{-3}$$

A more detailed model, producing the line shapes of Lymannlines emitted by an aluminum plasma, leads to a similar estimation[18]. This can be seen in fig. 7 which shows the variation of Ly-γ line width versus the electron density, as given by this model. The static-ion line width is measured at the "shoulders" of the profile (see fig. 6). The Doppler widths at 100ev and 1Kev are reported on the figure (dashed lines).

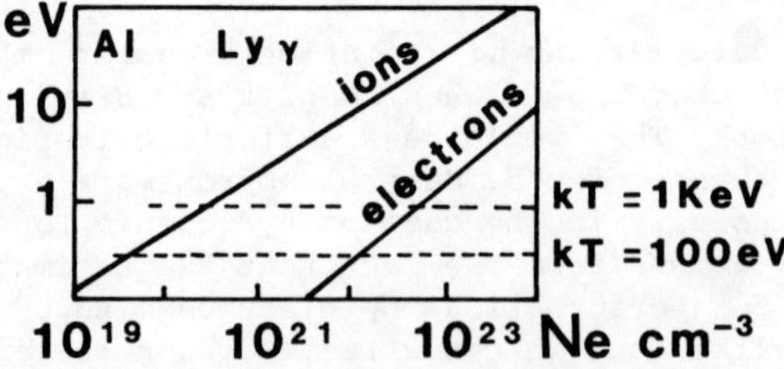

Fig. 7. Calculated full width at half maximum of the Ly-γ line of aluminum broadened by electron collisions (T_e= 1Kev) and distance between the shoulders of the static-ion profile; Doppler widths are shown (ref. 18).

For the line wings being affected by Stark effect well before the center, it is clear that the Gaussian shape of Doppler profile is inappropriate and must be given up as soon as the electron density gets the value fixed by the condition (16).

2.3 Series-limit

It is known that Stark broadening does mix the high members of the spectral series in such a way that discrete structures disappear above a maximum value of the main quantum number. This is the Inglis-Teller limit which is estimated, in H-like ions, by equating the separation between to successive levels with the static linear shift produced by the normal field F_0 (10). In assuming one and the same charge for the emitters and the perturbers, this leads to:

$$\text{Log } n_{max} = 3.11 - 0.133 \text{ Log } N_e + 0.533 \text{ Log } Z \qquad (17)$$

For instance, in an aluminum plasma with 10^{22} cm^{-3} density, the highest member of the series given by this relation corresponds to n=6.

However we have mentioned other effects as able to change the spectrum near the ionization limit (fig. 4). On the one hand, the ionization of the high levels by tunnel effect through the potential barrier gives an attenuation of the emission coefficients[19]. Let us $\Gamma(n,F)$ be the ionization rate, depending on the electric field F [20] and $A(n,n',F)$, the radiative transition probability between the levels n and n'. The attenuation factor δ is to be calculated from:

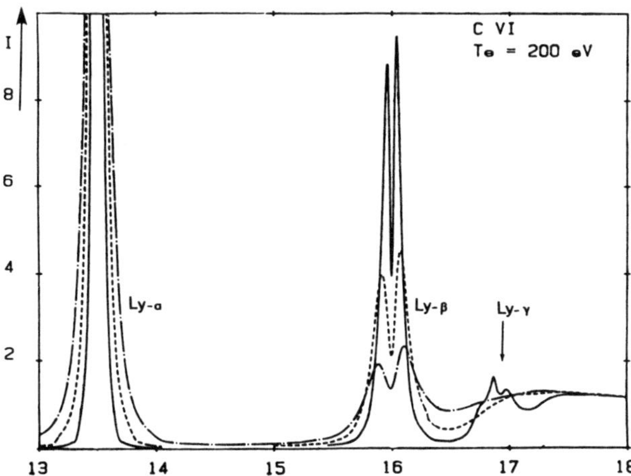

Fig. 8. Calculated spectrum of C VI showing the effect of plasma density in the region of the series-limit; solid line: $N_e=10^{22}$ cm^{-3}; dashed line: $N_e=3x10^{22}$ cm^{-3}; dashed-dotted line: $N_e=7x10^{22}$ cm^{-3}. The calculation includes line-broadening, attenuation due to ionization by tunnel effect and lowering of radiative recombination limit. One sees the Ly-γ line to disappear in the continuous spectrum (ref. 19).

$$\delta = \frac{\Sigma_{n'} A(n,n',F)}{\Sigma_{n'} A(n,n',F) + \Gamma(n,F)} \tag{18}$$

On the other hand, the ionization potential is lowered and the recombination spectrum moves forward to the low frequencies, overlapping the highest members of the series. Bringing together these effects makes it possible to calculate the spectral profile of the series as a whole. An example is given, for the H-like ions of carbon, in fig. 8. It shows the Ly-γ line to disappear in the continuum when the electron density increases. This provides a valuable method of density diagnostic in laser plasmas, as shown in fig. 9 where an experimental spectrum of Boron is compared to a calculated one, the density yielding the best fitting being about 1.9×10^{22} cm^{-3} (21).

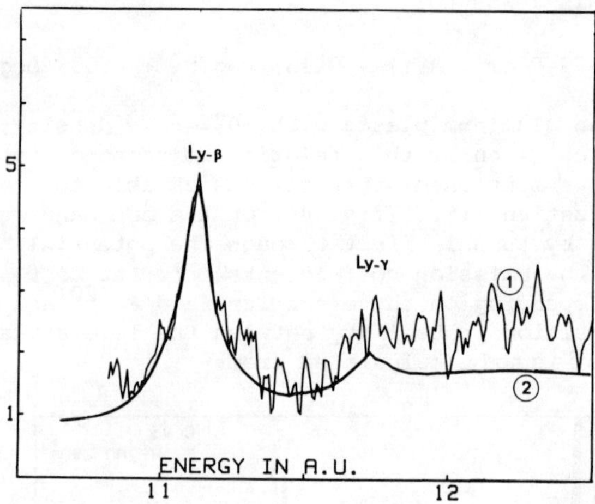

Fig. 9. H-like lines of Boron in a laser-produced plasma. 1) Experimental spectrum at less than 15μ from the target; 2) Calculated spectrum with assuming 1.9×10^{22} cm^{-3} electron density (ref. 21).

2.4 Line Shifts due to Electron Impact

The first-order shifts of the kind produced by static fields, averaged over the low frequency component of the microfield distribution, contribute to the static-ion profiles like shown in fig. 6. Another question, subject to controversial discussions during the last years (22-29), regards the line shifts caused by electron impact, namely the magnitude and the sign of the term $\Delta_{\alpha\beta}$ in expression (8). The interest for this question is related to its importance for plasma diagnostics, for radiative transfer calculations and, more precisely, for the attempts of optical pumping performed in view of X-ray laser production.

A first difficulty to succeed in theoretical predictions, that could be confronted to experiments, comes from the fact that a line shift does involve generally two different level shifts and possibly additional crossed terms (24). Thus the sign of the observable line shift can be different from the one calculated for the individual levels and there is no *a priori* reason for the magnitude of the shift being the same for both line and levels. On the other hand, accurate measurements are difficult in experiments[23,30-32]. In plasmas with density large enough for producing observable effects fast space- and time-variations of density tends to substitute an apparent asymetric broadening to the physical shifts. Few hundreds picosecond time-resolution combined to $\simeq 5\,\mu$ space-resolution should be necessary to achieve a good separation between shift and broadening. Moreover experimental shifts can be occasioned by various processes such as reabsorption in the expanding plasma (see section 4). For the lines including two or more components, a change of relative level populations may also shift the center of the unresolved profile. Therefore the calculations of electron-impact effect on spectral line position have hitherto few experimental support. The simplest case for investigation is the Lyman series of hydrogenic ions because it makes it possible to assume line shifts to be merely determined by the shifts of the transition upper-levels.

The frequency shift affecting the levels is due to shielding of the nucleus by free electrons. As a matter of fact, in dense plasmas, there is a non-negligible probability to find free electrons at distance from nucleus shorter than the radius of excited orbitals. A simple way of taking into account the perturbation produced by these electrons is to replace the usual Coulomb potential:

$$V_o = \frac{e(Z-1)}{r}$$

by the Debye-Hückel screened potential:

$$V_D = \frac{e(Z-1)}{r}\, e^{-r/r_D} \tag{19}$$

where r_D is the Debye radius for electrons:

$$r_D = \left(\frac{kT}{4\pi e^2 N_e}\right)^{1/2} \tag{20}$$

The new level positions required by the perturbed potential may be easily calculated for hydrogenic ions[23]; for Lyman series it gives a red shift expressed as:

$$\delta E_{ev} = -8.59\times10^{-22}\,\frac{Z-1}{Z}\,(n^2-1)\,\frac{N_e\ cm^{-3}}{kT_{ev}} \tag{21}$$

for the first-order perturbation energy. This model proves to be unsuitable for large densities, that is just when it should be used. Therefore more sophisticated methods have been carried out for calculating the potential. They generally consist in self-consistent calculations taking into account neighbouring particles[17,22,33]. In contrast with the simple Debye-Hückel model, the electron density is no longer uniform through the emitter volume but its distribution is modified by attraction to the nucleus. The energy of the levels is obtained by solving Schrödinger equation including the new potential. Several iterations are necessary. Recently Nguyen et al.(17) calculated i) the broadening and the shift of Lyman lines as given by quantum impact theory, Coulomb-Born-Oppenheimer collision cross-sections being introduced in the numerical calculation of the broadening operator, ii) the shifts predicted by self-consistent field calculation, taking into account the free electron distribution near the nucleus. They found a surprinsingly good agreement between the two methods for densities as large as $\simeq 10^{25}$ cm^{-3}, the temperatures being comprised between 200 ev and 1 Kev. The results are summarized in the expression:

$$\delta E_{ev} = -10^{-22} \frac{N_{e\,cm^{-3}}}{Z^2} \Delta(n,l,T_e) \qquad (22)$$

where the values of $\Delta(n,l,T_e)$ are tabulated (see Table 1).

$T_{e\,ev} / Z^2$	Δ(2p)	Δ(3p)	Δ(4p)
1	8.93	38.05	98.82
2	6.53	29.40	79.99
5	4.52	22.26	65.42
10	3.50	18.41	56.76
13	3.26	17.55	54.80

Table 1

Both relations (21) and (22) provide red line shifts. However the scaling of the shift with n,Z and T_e changes strongly from the approximate relation (21) to the "exact" calculation (22). Taking again Al^{12+} as an example, the two methods give approximately the same shift for Ly-β and Ly-γ lines at intermediate temperatures, but the Debye-Hückel model underestimates all the shifts by a factor 2-5 at large temperature. Regarding the scaling law with respect to Z, Skupsky(22) has shown that, if δE_o is the shift for an ion of charge Z_o, in a plasma of temperature T_o and density N_o, the line shift for an ion of charge Z is

$$\delta E = \delta E_o (Z/Z_o)^2 \qquad (23)$$

at a density $N = N_o(Z/Z_o)^4$ and temperature $T = T_o(Z/Z_o)^2$.

3. FREQUENCY REDISTRIBUTION

Let ν_A be the frequency of a photon absorbed in a transition between two levels of an emitter and ν_S, the frequency of the photon "scattered" when the emitter falls back on its initial state. The frequency redistribution problem consists to find the profile function $\Phi(\nu_A,\nu_S)$. Relatively few efforts have been devoted so far to redistribution calculation in laboratory plasmas.

In radiative transfer studies, the redistribution is assumed to be complete, i.e. no coherence at all is introduced between absorbed and scattered photons (see section 4). This assumption is no questionable when radiation has only a weak part in level populations, that is in very small plasma volumes of high density plasmas. The situation may be different if plasma dimensions get larger, as it does for instance in X-ray laser experiments, especially in the underdense region ($N_e < 10^{21}$ cm^{-3}). Therefore it is interesting to consider shortly the problem of frequency redistribution in these plasmas. We won't discuss all the processes which might contribute to redistribution; for instance, the emitter thermal motion(34,35) may play a part but it is not specific of high density calculation.

The first question is to know which part of the line profile may be affected by redistribution. A qualitative answer is given by comparing the frequency separation to line center:

$$\Delta\nu = \nu_o - \nu$$

with the plasma frequency ν_P. Remembering that ν_P is related to the Debye length r_D by:

$$\nu_P = \frac{1}{2\pi r_D}\left(\frac{kT}{m}\right)^{1/2} \tag{24}$$

and considering the thermal velocity $V_{th} = (2kT/m)^{1/2}$, we see that the plasma frequency introduces a time:

$$T_P = \frac{1}{\nu_P} \propto \frac{r_D}{V_{th}} \tag{25}$$

during which, on an average, a particle remains in the Debye sphere of an emitter. Thus, if:

$$\Delta\nu \ll \nu_P \tag{25}$$

the time of interest for the line profile involves many independent perturbingprocesses and the effect of collisions will be randomized.In contrast with this, if:

$$\Delta\nu \simeq> \nu_P \tag{26}$$

each fluctuation may affect the emitter and change its frequency between absorption and emission in such a way that a function $\Phi(\nu_A,\nu_S)$ takes place. Now we have, in energy units:

$$h\nu_{Pe_{ev}} = 3.7\text{x}10^{-11}\ N_{e\,cm^{-3}}^{1/2} \tag{26}$$

for electrons, and:

$$h\nu_{Pi} = \sqrt{Zm_e/m_i}\ h\nu_{Pe} \tag{27}$$

for ions, what gives for current plasma conditions:

$$h\nu_{Pe} = 1\text{ev} - 10\text{ev}$$
$$h\nu_{Pi} = 10^{-2}\ \text{ev} - 10^{-1}\text{ev}$$

Thus, for electron collisions, condition (25) will be generally satisfied over all the observable line profile, in ion X-ray spectra; but a large part of the profile, except the very center, is such as condition (26) is verified for the ion plasma frequency ν_{Pi}. For frequency redistribution, we conclude that electron collisions will be negligible, while ion collisions may be of importance for the largest part of the profile.

Let us remark that, in the quasi-static treatment of ion perturbation presented in section 2, the profiles result mainly from a statistical distribution of constant shifts (9). There is no change of level position between absorption and emission. Therefore the only effect of redistribution results from the broadening of each peak. In fact, true effect of frequency redistribution is a typical result from ions dynamics.

Achieving collisional redistribution function require generally difficult calculations[36-38]. We present below the result of a recent study[39] which shows that taking into account the statistical properties of perturbers in the plasma affords substantial simplification. The authors, instead of considering a large number of perturbers with customary approximations, used the so-called Microfield Method Model (MMM) for modelling the collective effect of the perturbing bath. The spectrum of an emitter immersed in the bath being described by the Fourier transform of the radiation interaction operator averaged over the system and bath variables, the general expression of the redistribution function has the form:

$$\Phi(\nu_S,\nu_A) \propto \lim_{\eta\to 0} \mathrm{Tr}(\{V_S G(i\eta) V_A\}_{av}\ \rho) \tag{28}$$

V_S and V_A are the radiator-field interaction operators at the frequencies ν_S and ν_A, $\{\cdots\}_{av}$ indicates an average over the stochastic bath variables and $G(i\eta)$ is the one-sided Fourier transform of the Liouville time propagator $U(t)$:

$$G(Z) = -i\int_t^\infty U(t,t')\ e^{iZ(t - t')}dt'$$
$$U(t) = e^{iL(t - t')}$$

MMM enables to define a rate of change of the bath state, Γ, which is equivalent to the collision rate of the ordinary picture. However a simplification with respect to the collisional treatment occurs from the fact that the rate of plasma microfield fluctuation introduced by MMM is independent of the state of the emitter. These fluctuations represent ion motion effects and the redistribution by a fluctuating microfield can be calculated. Numerical calculations have been performed in using a Holtsmark distribution of fluctuating ion microfield in order to illustrate the effect of ion motion on redistribution.

For these calculations, the simple case of Lyman line of hydrogenic ions has been chosen. Fig. 10 shows two examples of results for two values of the bath fluctuation rate, $\Gamma=2\gamma_e$ and $\Gamma=6\gamma_e$, where γ_e is the homogenous width due to electron impact. In the second case, the rate of fluctuations is large enough for the impact limit being almost achieved in the one-photon spectrum given in inset. The redistribution function is seen in a two dimensional representetion, for $\Delta\nu_A<0$; the redistribution is symetric about $\Delta\nu_A$. For comparaison, the static limit, $\Gamma=0$, is presented in fig. 11. The one-photon profiles represented for the three cases in inset should be the optically thin profiles for radiative transfer calculation

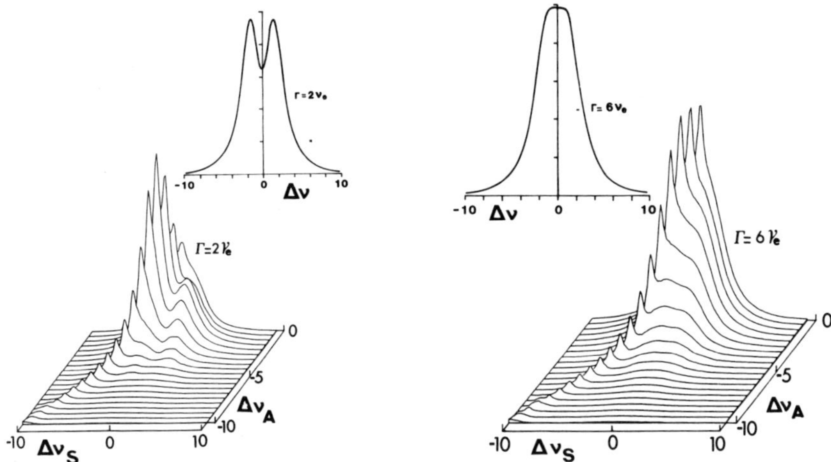

Fig. 10. Graphical representation of the frequency redistribution function for the Ly-β line and two fluctuation rates, Γ , of the Holtsmark microfield distribution; γ_e is the electron broadening; in inset, one-photon profiles. These results are reported from ref. 39.

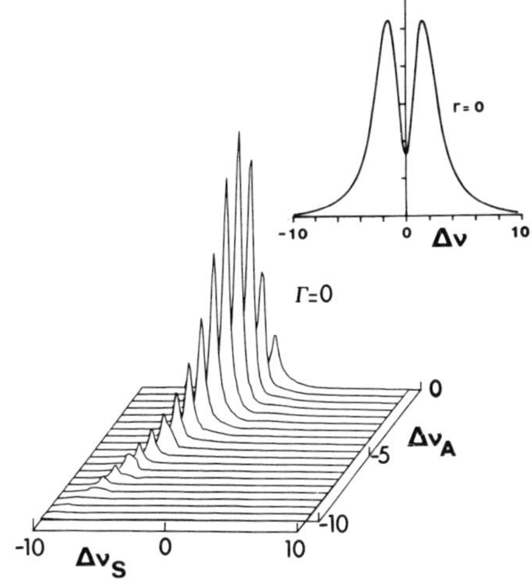

Fig. 11. Quasi-static limit of the results shown in fig. 10.

corresponding to collisionally populated levels or the absorbing profiles in the general case. No integration of transfer equation has been performed until now with profiles like swhon in fig. 10 and 11.

Experimental verification of these results could be done with line-plasmas produced by using cylindrical optics in focusing a laser-beam on a suitable target (see sec. 5). High resolution device must be used for measuring the experimental profiles.

4. COMPUTATIONAL MODELLING OF RADIATIVE TRANSFER

Between the spectral profiles discussed in the previous sections and the observable line shapes takes place the radiation transfer through the medium which has finite dimensions. The physics of radiation transfer involves plasma characteristics such as:

- plasma geometry (planar or spherical),
- plasma velocity (planar or spherical expansion),
- plasma inhomogeneities (density and temperature gradients),
- electron population distribution (possibly coupled with radiation),
- optically thin profiles,
- frequency redistribution.

A large part of the works performed so far in relation with dense laboratory plasmas had in view either line intensity calculation in more or less complex spectra or an estimation of the ambient radiation field which influences the dynamics of the plasma[40-43]. It is generally necessary to connect the radiation transfer computation to another code including hydrodynamics and ionization. The details of line shape are of little importance in these works where the largest attention is payed to the plasma geometry as well as to population-radiation coupling. Here we restrict ourselves to the calculations which are directly related to the experimental line shape interpretation. In this case the following assumptions are currently made:

- the level electron populations are fixed by collisions and radiative decay,
- the plasma geometry can be considered as planar,
- the frequency redistribution is complete,
- the plasma exhibits strong inhomogeneities and is fast expanding.

Owing to the first of these assumptions, it is not necessary to proceed to angular integration of transfer equations since the source function does not depend on the radiation. Then the choice of plasma geometry has

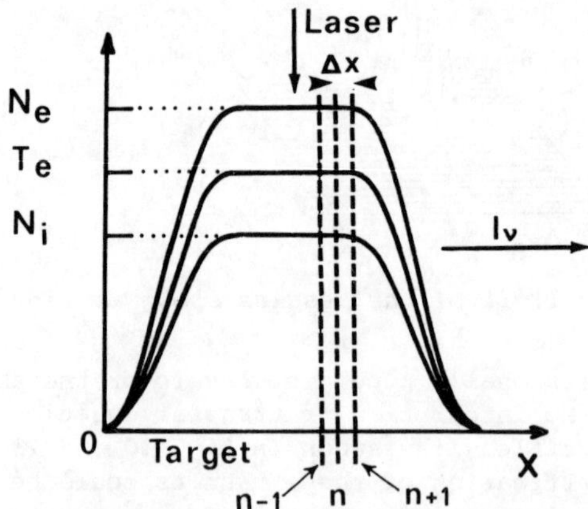

Fig. 12. Sketch of plasma modelling for radiation transfer calculation perpendiculary to the laser axis.

less importance and a planar geometry can be used except perhaps in the weakly absorbing far wings. The third assumption, which has been discussed in the previous section, allows to introduce one and the same profile for emission and for absorption in the transfer equation. Then the main features of emerging line profiles will be produced by the plasma inhomogeneities and velocity, on the one hand, and by the optically thin profiles, on the other hand.

A simple way to calculate the line profiles consists in dividing the plasma in layers of thickness Δx, small enough for the variation of the plasma parameters being negligible along Δx (fig. 13). Considering the layers n and n-1, between 1 and n_{max}, we can write the well known solution of transfer equation in homogeneous medium[44]:

$$I_n = \frac{j(\nu)}{\alpha(\nu)} \left(1 - e^{-\alpha(\nu)\Delta x}\right) + I_{n-1} e^{-\alpha(\nu)\Delta x} \qquad (29)$$

and the outgoing radiation will be:

$$I(\nu) = I_{n_{max}} \qquad (30)$$

Building the emission and absorption coefficients $j(\nu)$ and $\alpha(\nu)$ from experimental and theoretical data, then calculating (30) for comparison with observed spectra, provides very large possibilities of testing electron

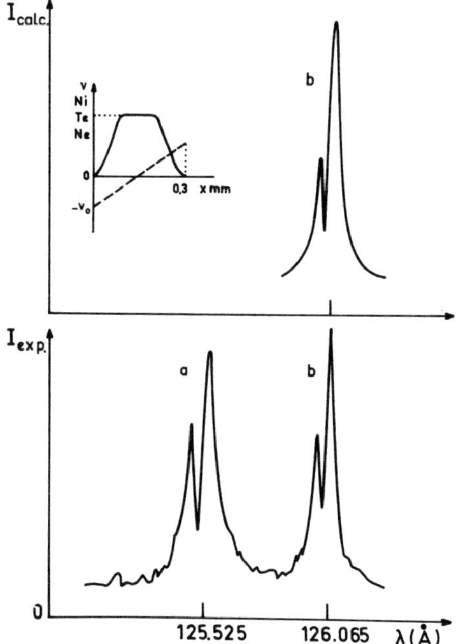

Fig. 13. For the experimental soft X-ray spectrum of Al^{4+} ions displayed in the fig. the radiation transfer calculation shows that the line splitting is a consequence of reabsorption in expanding cooling external shells of the plasma (ref. 5).

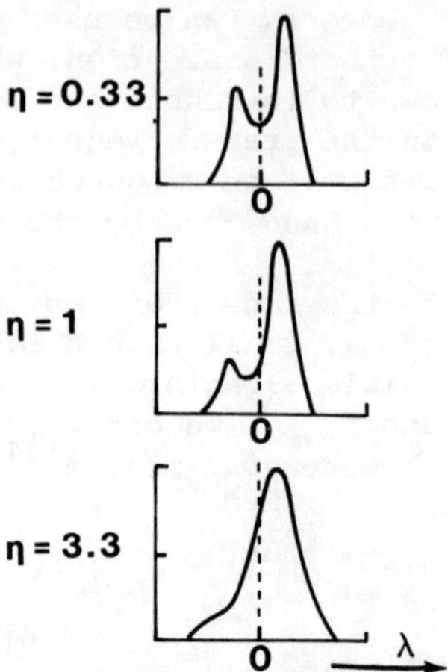

Fig. 14. Effect of the ratio η = (motional Doppler broadening)/(thermal Doppler broadening) on line shape. As η increases, the asymetry increases and the self-reversed profile becomes an asymetric shifted profile (ref.45)

level populations, plasma topography, optically thin line widths and ion abundances. Basically, these coefficients are written:

$$j(\nu) = \frac{h\nu}{4\pi} A_{21} N_2 \Phi(\nu) + \ldots + \ldots \tag{31}$$

$$\alpha(\nu) = \frac{h\nu}{c} B_{12} (N_1 - \frac{g_1}{g_2} N_2) \Phi(\nu) + \ldots + \ldots \tag{32}$$

where the first term represents the contribution of a single discrete transition, other contributions from bound-bound, free-bound and free-free processes being added up if necessary.

As an example, fig. 13 shows a 2p-3s line profile of Al^{4+}, calculated for interpreting the splitting observed at high density (see also fig. 3). A Lorentzian optically thin profile has been assumed, such as:

$$\Phi(\nu) = \frac{1}{2\pi} \frac{\Delta\nu}{(\nu - \nu_o)^2 + (\Delta\nu/2)^{1/2}} \tag{33}$$

where the width $\Delta\nu$ varies as $N_e . T_e^{-1/2}$ and the central frequency ν_o is Doppler-shifted proportionally to $(1+V_o(x)/c)$, $V_o(x)$ being the x-dependent expansion velocity component in the direction of observation; N_e, T_e, N_2 are functions of x (inset in fig. 13) and N_1/N_2 is given by Boltzman's law. From this calculation it is concluded that the splitting occurs as a consequence of the reabsorption in the external cooling expanding plasma shells. Thus it is no need to call forth some peculiar Stark effect in the understanding of the experimental profile. The effect of expansion and thermal velocities on line shape has been studied in detail by Irons[45]. This work shows, for instance, the features of line shape to be closely related to the ratio of both classes of velocity. This can be seen in fig. 14 where the profile of the line looks first like a "blue satellite" structure, then turns into an "asymetric red-shifted" line when the relative value of expansion velocity increases. A similar study has been made for the role of the optically thin line width[5].

5. STIMULATED EMISSION

In (31) and (32) both spontaneous and stimulated emission are included *via* the upper level population and Einstein's coefficients A_{21} and $B_{21} = (g_1/g_2)B_{12}$. It is easy to see that stimulated emission provides a significant reduction of the absorption coefficient[46], even when the levels are in Boltzmann equilibrium, for transitions such as:

$$h\nu < 2.5 x k T_e$$

i.e. in the X-UV wavelength range. Generally speaking, in this spectral range, to decide on neglicting, or not, stimulated emission needs to solve the rate equations giving the populations of the excited levels in the specific conditions under consideration. The important fact emphasizer here is that, in laser produced plasmas, the recombination which occurs to the end of pulse may drive an increasing of high excited level populations, leading to population inversions with respect to lower levels of the same ion. This has been evidenced for hydrogenic ions[47-48] and Lithium-like ions[8,49].The necessary development of soft X-ray gain diag-

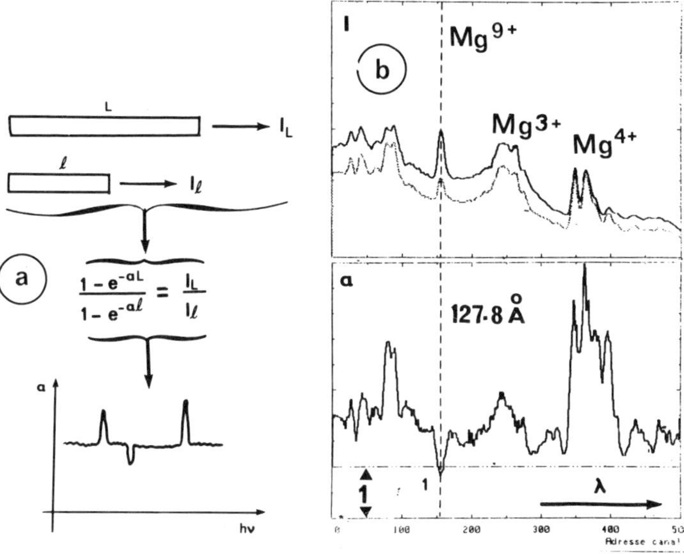

Fig. 15. a) Method of measurement of absorption coefficients,α(ν), using two plasma columns of different lengths; as a rule, a discrete transition appear as an absorption peak standing on the continupus spectrum; a reversed peak is the signature of a population inversion (negative absortion or gain); b) result achieved for the 3d-5f transition of Lithium-like Magnesium (L≃1cm, pulse duration: 25 ns)

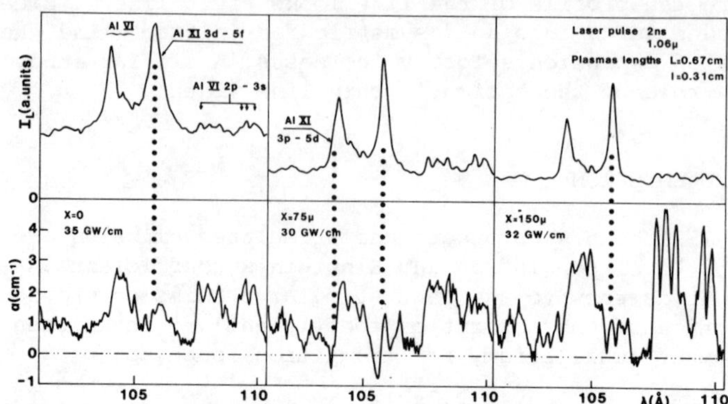

Fig. 16. Change of absorption coefficient for the 3d-5f transition of Lithium-like aluminum (λ=105.6 Å) with the distance to the target; at X=75 μ, a reversed peak shows the stimulated emission to induce a gain; a similar phenomenom occurs for the 3p-5d transition (λ=103.5 Å) which is unresolved, in the emission spectrum, among bundled lines belonging to other ions.

nostics, in view of X-ray laser achievement, is one of the best stimulants for the studies on plasma radiation transfer and will likely open new questions about line shape theory in the future. Considering again the solution of the transfer equation in an homogenous plasma of length X:

$$I(\nu) = S(1 - e^{-\alpha(\nu)X}) \tag{34}$$

where S is the source function:

$$S = \frac{j(\nu)}{\alpha(\nu)}$$

the absorption coefficient (32) can be deduced from measuring the intensity dependence versus the plasma length. With two plasma columns of lengths L and l , whose the emission intensities are $I(\nu)_L$ and $I(\nu)_l$, we can write the equation:

$$\frac{1 - e^{-\alpha(\nu)L}}{1 - e^{-\alpha(\nu)l}} = \frac{I(\nu)_L}{I(\nu)_l} \tag{35}$$

This provides the principle of methods suitable for: i) absorption coefficient measurement ($\alpha(\nu)$) and related population determination, ii) gain coefficient measurement ($G=-\alpha(\nu)$) when population inversion does occur, iii) experimental determination of optically thin profile from optically thick line shape.

Details about the way of making use of such a method may be found elsewhere[49], as well as a description of mechanisms generating population inversions. A sketch of the experimental procedure is given in fig. 15 with an example showing an absorption spectrum of magnesium, including a peak of stimulated emission at 127.8 Å. The upper part of fig.

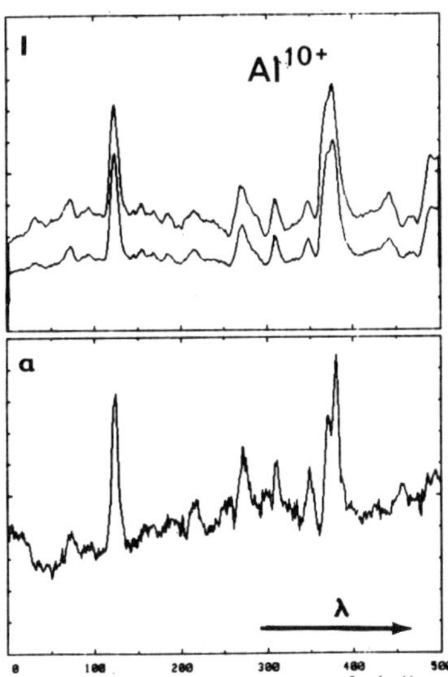

Fig. 17. Optically thick lines of 2s-3p (48.3 Å) and 2p-3d (52.4 Å) transitions of Al^{10+}. In the emission spectrum (upper frame), the lines are broadened by reabsorption (L=8 mm, l=4 mm); the absorption spectrum (lower frame) restores the optically thin profiles; it shows, in particular, the two components $2p\text{-}3d_{5/2,3/2}$ which are mixed by reabsorption in the above spectra.

15 b) displayes the emission intensities for the two lengths of plasma columns. Emission and absorption 3-5 spectra are shown for Lithium-like aluminum in fig. 17. Near the target (X=0), the plasma density is maximum and all the lines produce positive absorption peaks; farther (X=75μ), stimulated emission is seen to reverse the absorption peaks at the wavelengths of the 3d-5f and 3p-5d transitions; still farther (X=150μ), recombination is nearly terminating for Al^{10+} ions leading to the end of population inversions; at the same time, the absorption lines of ions of lower Z (Z=4,5) are rising highly about λ=104-105 Å and λ=107-110 Å

As regards the determination of the optically thin profile from an optically thick line shape, its feasability by this method , in the case of separated discrete transitions, results directly from the form of $\alpha(\nu)$ in (32). The profile will be so much the more accurate that the instrumental resolution will be large. A preliminary result is given, as an example, in fig. 17 which shows the width of absorption peaks to be narrower than the width of emission lines, so that the unresolved components of the 2p-3d transition are separated in the absorption spectrum.

REFERENCES

1. H.G. Ahlstrom, in *Laser Plasma Interaction*, Les Houches session XXXIV; ed. by R. Balian and J.Cl. Adam, North Holland, 1982.

2. G. Jamelot, Thèse de 3° Cycle, Orsay (F), 1971.

3. A. Carillon, P. Jaeglé, G. Jamelot, A. Sureau, in *Inner Shell Ionization Phenomena and Future Applications*, ed. by R.W. Fink, S.T.Manson, J.M. Palms, P.V. Rao, US AEC Conference, 72-0404, p. 2373, 1973.

4. G. Jamelot, P. Jaeglé, A. Carillon, *Annales de Physique Fr.*, 7, 399, 1982.

5. P. Jaeglé, G. Jamelot, A. Carillon, A. Sureau, *Journal de Physique*, 39, C4-75, 1978.

6. V.A. Boiko, O.N. Krokhin, S.A. Pikuz, A.Ya. Faenov, *JETP Lett.*, 20, 50-1, 1974.

7. P. Jaeglé, A. Carillon, G. Jamelot, Cl. Wehenkel, *Journal de Physique Lettres*, 40, L-551, 1979.

8. P. Jaeglé, G. Jamelot, A. Carillon, A. Klisnick, A. Sureau, H. Guennou, in *Laser Techniques in the Extreme Ultraviolet* (OSA, Boulder,Co, 1984), ed. by S.E. Harris and T.B. Lucatorto, American Institute of Physics, New York, p. 468, 1984.

9. H.R. Griem, *Spectral Lines Broadening by Plasmas*, Academic Press, 1974.

10. I.I. Sobelman, L.A. Vainstein, E.A. Yukov, *Excitation of Atoms and Broadening of Spectral Lines*, Springer Verlag, 1981.

11. M. Caby, Thèse d'Etat, Paris 1978.

12. R.J. Tighe, C.F. Hooper, *Phys. Rev. A*, 14, 1514, 1976.

13. R.W. Lee, *J. Phys. B*, 12, 1129, 1145, 1165, 1979.

14. H.R. Griem, M. Blaha, P.C. Kepple, *Phys. Rev. A*, 19, 2421, 1979.

14*. Nguyen Hoe, J. Grumberg, M. Caby, E. Leboucher, G. Coulaud, *Phys. Rev. A*, 24, 438, 1981.

15. H.R. Griem, G.T. Tsakiris, *Phys. Rev. A*, 25, 1199, 1982.

16. R. Cauble, H.R. Griem, *Phys. Rev. A*, 27, 3187, 1983.

17. Nguyen Hoe, M. König, G. Coulaud, *Proceedings of Seventh International Conference on Spectral Line Shape*, (Aussois, F.) 1984, De Gruyter, in press

18. R.W. Lee, J.D. Kilkenny, R.L. Kaufman, D.L. Matthews, *J. Quant. Spectrosc. Radiat. Transfer*, 31, 83, 1984.

19. B. d'Etat, Inv. Lect., *Proceedings of the Seventh International Conference on Spectral Line Shape*, (Aussois, F.) 1984, De Gruyter, in press.

20. Nguyen Hoe, B. d'Etat, J. Grumberg, M. Caby, E. Leboucher, G. Coulaud, *Phys. Rev. A*, 25, 891, 1982.

21. B. d'Etat, E. Leboucher, Nguyen Hoe, A. Carillon, M. Lamoureux, G. Jamelot, P. Jaegle, *Annual Report 1983*, GRECO "Interaction Laser-Matière", p. 166.

22. S. Skupsky, *Phys. Rev. A*, 21, 1316, 1980.

23. T.L. Pittman, P. Voigt, D.E. Kelleher, *Phys. Rev. Letters*, 47, 723, 1980.

24. R.W. Lee, *J. Quant. Spectrosc. Radiat. Transfer*, 27, 249, 1982.

25. R. Cauble, *J. Quant. Spectrosc. Radiat. Transfer*, 28, 41, 1982.

26. H.R. Griem, *Phys. Rev. A*, 27, 2566, 1983.

27. J. Cooper, D.E. Kelleher, R.W. Lee, *Proceedings of the Second International Conference Workshop on the Radiative Properties of Hot Dense Matter*, (Sarasota, Fl.) 1983, *J. Quant. Spectrosc. Radiat. Transfer* in press.

28. D.E. Kelleher, J. Cooper, *Proceedings of the Seventh Inernational Conference on Spectral Line Shape*, (Aussois, F.) 1984, De Gruyter, in press.

29. D.B. Boercker, C.A. Iglesias, *Phys. Rev. A*, 30, 2771, 1984.

30. see ref. 5.
31. S. Hashimoto, N. Yamaguchi, *Phys. Letters*, 95A, 299, 1983.
32. J.C. Adcock, H.R. Griem, *Phys. Rev. Letters*, 50, 1369, 1983.
33. B.F. Roznai, *J. Quant. Spectrosc. Radiat. Transfer*, 15, 695, 1975.
34. V.V. Ivanov, *Transfer of Radiation in Spectral Lines*, NBS Special Publication 1973, p. 385.
35. M.M. Basko, *Sov. Phys. JETP*, 4, 48, 1978.
36. M. Machacek, *Bull. Astron. Inst. Czecoslv.*, 29, 268, 1978; 30 , 23, 1978.
37. K. Burnett, J. Cooper, R.J. Ballagh, E.W. Smith, *Phys. Rev. A*, 22, 2005, 1980.
38. K. Burnett, J. Cooper, *Phys. Rev. A.*, 22, 2027, 1980; 2044, 1980.
39. L. Klein, B. Talin, V.P. Kaftandjian, R. Stamm, Inv. Lect., *Proceedings of the Seventh International Conference on Spectral Line Shape*, De Gruyter, in press, 1984.
40. J.P. Apruzese, J. Davis, *NRL Memorendum Report 3277*, NRL Washinggton DC, 1976.
41. D. Duston, J. Davis, *Phys . Rev. A*, 21, 1664, 1980.
42. E. Berthier, F. Garaude, B. Degove, *CEL Annual Report 1980*, 94190 Villeneuve St. Georges (F).
43. J.P. Apruzese, P.C. Kepple, K.G. Whitney, J. Davis, D. Duston, *Phys. Rev. A*, 24, 1001, 1981.
44. S. Chandrasekar, *Radiative Transfer*, Dover Publication, New York, 1960.
45. F.E. Irons, *Aust. J. Phys.*, 33, 25, 1980; *J. Phys. B*, 8, 3044, 1975.
46. A. Carillon, P. Jaeglé, G. Jamelot, A. Sureau, P. Dhez, M. Cukier, *Phys. Letters*, 36A, 167, 1971.
47. G.J.Pert, L.D. Shorrock, G.J. Tallents, R. Corbett, M.J. Lamb, C.L.S. Lewis, E. Mahoney, R.B. Eason, C. Hooker, M.H. Key, *Laser Techniques in the Extreme Ultraviolet*, (OSA, Boulder, Co.,1984), ed. by S.E. Harris and T.B. Lucatorto, American Institute of Physics, New York, p. 480, 1984.
48. S. Suckewer, C. Keane, H. Milchberg, C.H. Skinner, D. Voorhees, *Laser Techniques in the Extreme Ultraviolet*, (OSA, Boulder, Co, 1984), ed. by S.E. Harris and T.B. Lucatorto, American Institute of Physics, Nex York, p. 55, 1984.
49. A. Klisnick, P. Jaeglé, G. Jamelot, A. Carillon, Inv. Lect., *Proceedings of the Seventh International Conference on Spectral Line Shape*, De Gruyter, in press, 1984

OBSERVATIONAL PROBLEMS IN SPECTRAL LINE FORMATION

M. Hack
Astronomical Observatory of Trieste
Via Tiepolo 11
I 34131 Trieste
Italy

ABSTRACT. The main problem for the interpretation of stellar spectral lines is to explain superionization and stellar winds. Results giving information on these two phenomena in solar-type stars, hot dwarfs and supergiants, A-type dwarfs and supergiants are reviewed.

1. INTRODUCTION

The main problem for students of stellar atmospheres today is to explain the phenomena of superionization and mass loss. This kind of phenomena is observed in the ultraviolet spectra of almost all the spectral types and luminosity classes, but is quantitatively enhanced in interacting binaries, where we have direct evidence of the existence of shock waves produced by the interaction of winds from the two stars or from one star's mass-flow and the accretion disk around the companion.

Hence, on the basis of purely observational arguments, we infer that the origin of superionization and mass-flow in single stars is also due to non-thermal (mechanical and magnetic) energy which is found in the reservoir of mechanical and magnetic energy due to rotation, convection, oscillations, etc., originating in the photospheric and subphotospheric layers. It is therefore important to devote more attention to those recent sophisticated methods used for studying the photospheric spectral lines which can give more or less direct information on motions and magnetic fields in the photospheric energy reservoir.

Moreover, simultaneous observations of variability of photospheric lines (optical range), chromospheric lines (ultraviolet range) and coronal emission (X-ray range) and the search for possible

J. E. Beckman and L. Crivellari (eds.), Progress in Stellar Spectral Line Formation Theory, 265–277.

correlations, can provide insight into the hydrodynamic mechanisms causing the superionization and the mass-flow.

Hence it is now obvious that the two classical parameters- the radiative energy flux (or effective temperature T_{eff}) and the gravity - which are adequate to describe the photospheric spectrum, are completely inadequate to describe the whole atmospheric complex, i.e. a model able to reproduce the chromosphere-corona-wind complex. We must add non-radiative flux and mass-flux. The observations indicate that these additional parameters are only loosely related to the T_{eff} and g. They are determined by the sub-atmospheric structure (e.g. rotational currents, convective currents, macro-microturbulence, oscillations, magnetic fields- all quantities loosely related to T_{eff} and g). Hence to build realistic theories, we must start looking at what the observations tell us about the photospheric micro-structure.

We have two definitions of the outer layers: 1) The outer layers start at a depth in the photosphere where non-radiative energy dissipation occurs. The chromosphere-transition region-corona complex is optically thin. Non-thermal energy is dissipated there and reradiated in a non-efficient way. The result is an electronic temperature T_e of the medium that is much higher than that expected in radiative equilibrium. This phenomenon has been called superionization. 2) The outer layers start at a depth in the photosphere where mass-flux perturbs the state of hydrostatic equilibrium. The result is an expanding circumstellar envelope and velocity fields. We need to unify this dual description of these two phenomena: superionization (parameter T_e) and mass-flow (parameter velocity law).

In the following I will describe:

1) Results from high resolution studies of photospheric lines that give insight into phenomena occuring in photospheric and subphotospheric layers.

2) Observational problems posed by spectral line profiles in hot stars (O, WR, B, Be and hot supergiants).

3) The outer layers of main sequence and supergiant A-type stars.

2. HIGH RESOLUTION STUDIES OF PHOTOSPHERIC LINES

The slight asymmetries and wavelength shifts of photospheric stellar lines that are measurable in high resolution spectra (and which are not detectable in the usual photographic spectra having a resolution lower than 0.05 A) contain information on the structure and dynamics of gas flow in stellar atmospheres and can be used to study atmospheric inhomogeneities, as for instance granulation, chaotic motions, and oscillations, which, all together, produce the effect called turbulence. Rotational broadening can be separated from turbulence by Fourier analysis, if the spectral resolution is sufficiently high, as indicated by several works by Gray, and Gray and Smith (see for instance the review by Gray, 1982).

An example of the atmospheric models necessary for studying the granulation in different stellar atmospheres is given by the extended paper by Dravins (1982). He shows that it is possible to measure stellar convection parameters from the wavelength shifts and the asymmetries of the photospheric lines. In fact, he compares the shifts and asymmetries observable in the solar spectrum at the center of the disk and at different distances, from the center to the limb, as well as in the light averaged over the whole disk. In the latter case the effects are smaller but still measurable. However, it is necessary to be able to measure shifts on the order of a few 100 m/s, which is within the possibilities of modern methods. However, to measure reliable line asymmetries we need not only very high resolution and very low noise spectrograms, but also we need to measure unblended lines. Hence we have to select only stars with very low rotational velocities. The new class of large telescopes now being planned or built may become very important for this kind of problem. It would be important to measure variations in line asymmetries in order to look for solar type cycles, and to extend this kind of measurement to different spectral types, luminosity classes and stellar populations.

Actually, convection is an important but not fully understood parameter in the computation of stellar models and the determination of stellar ages, because it influences both the energy transport through the atmosphere and the replenishment of nuclear fuel in the core. The motions in stellar convection zones probably supply the energy for heating stellar atmospheres and coronae, for driving stellar winds, for generating magnetic fields and for many other non-thermal phenomena.

The majority of these studies concern solar type stars, where the presence of convective zones is well-known and ascertained, and where several slowly rotating stars are found. However, there is at least one case showing that similar effects can be observed also in early type stars and become observable in the few ones which are seen pole-on. This case is that of Tau Sco (BOV, v sin i = 5 km/s). Smith and Karp (1978, 1979) have studied the photospheric line asymmetries in its visual and ultraviolet spectrum. Almost all the lines show an asymmetry, in the sense that they have a slightly more extended shortward absorption wing than the longward wing. If this fact is interpreted as evidence of stellar wind, velocities of up to 1000 km/s result, and hence an unreasonably large mass flow (10^{-5} $M_\odot$/yr) for a main sequence star. A better explanation seems that proposed by Smith and Karp, i.e. that the line asymmetries are due to convective motions due to the He II ionization zone just below the photosphere.

2.1 Main results obtained from the study of the profiles of the photospheric lines in solar-type stars

The method of measuring the asymmetries by plotting the position of the bisector (in m/s) vs the flux in the line relative to the continuum F/F_c permits us to derive the granulation velocities for different spectral types and luminosity classes; the Fourier analysis permits us to derive the dependence of macroturbulence upon the spectral type and luminosity class. There is evidence that the granulation velocity presents a minimum around G9 and is lower in dwarfs than in giants (Gray, 1982). Macroturbulence (which includes granulation, together with other kind of motions, like chaotic motions and oscillations) decreases toward later types, and is systematically higher in giants than in dwarfs (Gray, 1982). Comparison of the macroturbulence values derived from strong and weak lines for solar type giants suggests that macroturbulence decreases with height, because weak lines yield higher values than strong lines (Gray, 1982).

Several interesting correlations have been found between photospheric motion and photospheric flux with chromospheric activity (indicated by the intensity of the chromospheric emission lines): a) macroturbulence and the flux in the k line of the Mg II chromospheric component are positively correlated (Gray, 1982). b) the flux in the C IV emission in solar type giants is dependent on the color index B-V: a maximum flux in the C IV emission is reached at G0 III (where also the granulation velocity is maximum (Simon, 1984). c) this same flux at C IV is positively correlated with the projected rotation velocity (Simon, 1984). d) the flux in the chromospheric components of H and K lines of Ca II and in the h and k chromospheric components of the Mg II lines in late-type dwarfs decreases with increasing rotation period (determined by rotational modulation of stellar luminosity, and therefore indepedent of the star's aspect (Noyes et al., 1984; Hartmann et al., 1984). e) a dependence of the chromospheric activity on age appears evident, old stars having less active chromospheres than younger ones (Noyes et al., 1984). f) the same dependence on spectral type is found for the mean rotational velocity, the X-ray flux normalized to the bolometric flux - F_x/F_{bol} - and the C IV emission flux normalized to the bolometric flux - F_{CIV}/F_{bol} (Gray, 1982). All these data indicate strict correlations between chromospheric and coronal activity (as measured by the Mg II, C IV, X-ray fluxes) and the mechanical plus magnetic energy (measured by rotation plus convection), which, in turn, seem to depend on age.

2.2 Attempts to measure magnetic fields in solar type stars

Since there is some recent evidence that magnetic fields suppress convection (Livingston, 1983) it is possible, at least in principle, to detect the effect of magnetic fields in stars, by studying the variation of granulation velocity during the period of rotation and during longer cycles of activity. In fact, there is much evidence for the presence of spots in stars: photometric variations, variations in H and K emission intensity, both during the rotational period and during long solar-type cycles. More direct measurements of the field are also possible. Since the field affects the shape of the line profiles, because of the Zeeman splitting, it is possible, at least in principle, to measure the magnetic field by comparing the profiles and the changes of magnetically sensitive lines with slightly or not at all sensitive lines, by the Fourier deconvolution method. Such an attempt has been made by Robinson et al. (1980) for ξ Boo A and by Marcy (1981) for the same star. While the first investigators detected a field of about 2650 gauss covering 45% (during one night) and 20% (the next night) of the stellar surface, the latter obtained no convincing evidence for the presence of a magnetic field. It is possible that these contradictory results indicate that the magnetic field is rapidly variable.

3. O-AND EARLY B-TYPE STARS

The ultraviolet spectra show the typical profiles indicating mass-flow and velocity gradients. Moreover there are superionization effects as indicated by the presence of the resonance lines of O VI and NV (Garmany, 1982). However, the velocity law is largely unknown. For instance, the strong resonance lines often present extended absorption with a broad minimum, while a monotonic velocity law predicts profiles with a sharp minimum (Lucy, 1982). Actually we don't know if the velocity law is a steep or a shallow function of the height z over the photosphere, if it is monotonic or non-monotonic, isotropic or non-isotropic, if it is a function of the spectral type and evolutionary stage, or if individual different laws are observed for stars of the same type. A well-recognized phenomenon is stellar individuality: objects with identical photospheric spectra have different velocity fields and different degrees of superionization.

A characteristic phenomenon is the presence, in the broad absorption profiles of the strong resonance lines, of several narrow absorption components, generally displaced at about 75% of the edge velocity (Abbott et al., 1982). Several explanations have been proposed. Lamers et al. (1982) for instance suggest that the star experiences discrete expulsions of matter (puffs). Underhill (1983) and Underhill and Fahey (1984) provide arguments for explaining the narrow components with "magnetic clouds". They suggest modelling the envelopes of O-B stars in terms of the configurations of closed and open magnetic loops. Bipolar magnetic regions (with fields lower than a few hundred gauss, and therefore not directly measurable) could explain the observed spectral features and their longevity (years). Ejection from local spots above the photosphere will produce several narrow absorptions. Since the only agents acting in the outer parts of the atmosphere and having local properties are the magnetic flux lines from bipolar magnetic regions, a bipolar field is inferred. Gry, Lamers and Vidal-Madjar (1984), using the data collected from Copernicus, found that transient components are present in the Lyman lines of hot (O9-B1) stars. These components are always observed at negative velocities (-70, -220 km/s) in 5 of 8 stars examined, on time scales of 2 to 11h (the time required by Copernicus for scanning that part of the spectrum).

They criticize the suggestion advanced by Underhill and Fahey for explaining these components. Actually they object that it is unlikely that hot stars have magnetic fields concentrated in tubes like the sun. In fact, in an atmosphere dominated by radiative tranfer rather than by convective transfer, horizontal temperature gradients are damped by radiation transfer, hence the formation of strong magnetic flux tubes or prominences in hot stars is improbable (Zwaan, 1977, 1981). They suggest, on the contrary, that the transient ejection of gas (puffs) occurs, in addition to a constant wind. This mechanism could be the same as that operating during the oscillations of Beta Cephei variables. After all, the Beta Cephei variables fall almost in the same region of the HR diagram.

4. WR STARS

WR stars are characterized by a mass loss one to two orders

of magnitude higher than that of O-type stars of the same luminosity. Mass-loss rate does not show any apparent dependence on the type (WC or WN or subtype), or on whether the star is single or a member of a binary system (Willis, 1982). A non-radiative mechanism is necessary to explain such a higher mass loss rate. Magnetic fields may play an important role; fields larger than about 10^4 gauss are necessary to drive the whole mass loss. It is possible to detect such fields by the Borra and Landstreet (1979) method, which has been successfully applied to the He-rich hot stars, which are also fast rotators.

It is difficult to model WR stars: what we observe is a dense wind rather than a true photosphere. It is difficult to determine the effective temperature, because a significant fraction of the observed continuum arises in the winds themselves. Estimates of T_e utilizing long wave-length ranges-from UV to IR- give values included between 30000 and 40000 K, the earlier types having the higher T_e, as suggested also by their line spectra. The difficulty in modeling is companied by the uncertainty in chemical composition. Differences in composition between WC, WN and the newly discovered WO are now believed to be real and due to different evolutionary stages. (WN are probably in the later stages of H-burning, with CNO producing as result the enhancement of N over C; WC are in the phase of core helium-burning, the 3 α processes giving enhancement of C; W O are in a later stage, $3\,\alpha + \alpha$ giving an excess of O). Extensive atmospheric stripping is due both to the strong mass loss rate as well as to the mass exchange in binary systems (a large number of WR are close binaries).

4.1 Line profiles

It is well-known that WR stars show a variety of profiles in a same star:

1) round-topped with steep sides
2) round- topped with extended wings
3) flat-topped

Castor (1970) was able to reproduce theoretically the three profiles, by using a velocity field $v(r) = \sqrt{(1-r_c/r)} \cdot v_\infty$ where r_c is the radius of the region giving the continuum (pseudo-photosphere), and using different mixtures of the following 4 parameters:

1) optical depth $\tau_o(r) = (\pi e^2/mc)(gf)_{1,u}\ r\ [\, c/_{\nu o} v(r)\,] \left[\frac{N_1}{g_1} - \frac{N_u}{g_u}\right]$

where 1 and u stand for the low and upper levels

2) the ratio of collisional to total de-excitation of the upper level of the transition ϵ
3) the Planck function of the local electron temperature B_ν
4) the emergent intensity of the continuum pseudo – photosphere radiation field Ic

It appears that flat-topped profiles can be obtained only when $\tau_o < 1$ in most of the envelope, i.e. when τ_c is small in region where $v_{exp} << v_{max}$. However, Castor points out that the solutions are not unique. What can be derived from the observations about the velocity fields?

The main results indicate that the edge absorption velocity v_a increases with increasing I.P.

The central absorption velocity v_o increases with decreasing E.P. (i.e. resonance lines have higher v_o).

The terminal velocity v_∞ increases going from WN8 to WN 3 and from WC9 to WC4.

The emission line-width increases from low values at WC9 toward higher excitation classes.

These relations can be understood in terms of an accelerating wind whose density is decreasing outward.

5. B-Be TYPE STARS

5.1 B V-III versus B Ia, Ib

Let us compare dwarf and giant- B stars with B-supergiants. Superionization and mass-loss are rarely observed in B V and B III stars, while they are always present in B I stars. The spectra of B supergiants almost always present a Balmer progression, variable in time, and H_α has a radial velocity that is more negative than that of the other Balmer lines, indicating that the atmosphere is expanding. These results were later confirmed by ultraviolet observations, because the profiles of the C IV, N V, O VI resonance lines indicate outflow at velocities higher than escape velocity.

It is well known that the photospheric density decreases, going from Ib to Ia and Ia+, as indicated by the quantum number of the last observable Balmer line. In the outer envelope, on the contrary, the density or the extension or both, increase from Ib to Ia and Ia+, as indicated by the intensity of the Si IV, C IV, N V and O VI lines, and by the strength of the Hα emission. Tau Sco is a very notable exception among B dwarfs. It shows clear evidence of superionization and velocity gradients, with $v_{edge} = -800$ km/s .

As we have seen in a previous section, it shows asymmetries also in the photospheric lines, both in the visual and in the ultraviolet range of the spectrum. Such asymmetries have not been observed in other sharp-lined B dwarfs. However, the majority of early-B stars with sharp line spectra belong to the class of Beta Cephei variables. They show evidence of non-radial pulsations in their atmospheres, producing a distortion of the profiles, which can mask the asymmetries if they are present. Actually, these non-radial oscillations could play the same role that convection plays in late-type stars in originating superionization and mass flow.

5.2 B stars versus Be stars

Let us now compare normal B stars and Be stars. Abbott et al. (1982), from a study of 53 O6.5 V to B2.5 IV stars, show that N V is present up to BOV, B1 III and B2 I. Slettebak and Carpenter (1983), from a study of some Be stars, observe that superionization extends up to B8e-B9e. Hence, although Be stars and normal B stars have almost indistinguishable spectra in the ultraviolet, there are difference in the degree of superionization effects. The same authors observe that Be stars with high v sin i show a whole range of C IV line strengths (generally in absorption), while Be with low v sin i show only weak C IV lines. Hence C IV lines are only loosely correlated with v sin i, while the visual shell lines are much more strictly correlated with v sin i. This fact indicated that the hot envelope (ultraviolet) is not strictly confined to the equatorial plane, while the cool envelope (visual range) is strictly confined to the equatorial plane.

Both Be and B supergiants are characterized by the presence of extended envelopes and mass loss. Hence we can ask what the difference is between Be and B I. This is evident in the visual range; hot supergiants show emission components at Hα, possibly at Hβ, and radial velocity Balmer progression. Be, on the contrary, show the whole Balmer series in emission, and often also several Fe II emission lines, besides the radial velocity Balmer progression. Be stars also show a larger spectral variability than B supergiants. Actually Be represent a transient phase of normal B stars, as indicated by several members of this class (Doazan, 1982). The cases of some Be stars (Zeta Tau and 48 Lib are typical examples) which show phases of accelerated and decelerated expansion, with quasi-periodic variations, must be noted. Zeta Tau, for instance, shows a periodicity of about 7 years, but if we go back to the years 1920-50 we find no trace of it.

6. A- TYPE STARS

We have found strong evidence for superionization in the ultraviolet spectra of OB stars, where absorption lines of ions like O VI and N V have been observed. We have strong evidence of superionization from the solar coronal spectrum and from the emission lines observed in the far ultraviolet of solar-type stars. The only class where no superionization effects have been detected is that of the A-type stars. In fact, UV observations have given no or very little evidence for the presence of a chromosphere in A-type stars, although there is evidence, in some cases, of a corona, (as suggested by X-ray observations). A-type supergiants show little or no evidence of chromospheres, but give evidence of variable winds.

6.1 A-type dwarfs

Let us consider first A-type dwarfs. Several attempts to observe emission components in the Mg II resonance doublet, as a chromospheric signature, have been made, but without success. Attempts to detect lines of Si IV or C IV have had no success either. But there are some noticeable exceptions. Some scanty evidence for a Mg II emission components in Alpha Aql (A7 V) is given by Kondo et al. (1977) and by Blanco et al. (1980). They observed slight bumps on either side of the sharp chromospheric absorption cores, which they interpret as emission components. Moreover, the flux at the center is not zero, suggesting the contribution of a chromospheric emission. Molaro et al. (1983, 1984)in the course of a systematic study of very fast rotators of type A, have found that HD 119921, v sin i = 435 km/s, shows C IV and Si IV absorption lines at v_o = -70 km/s; the stronger components of the doublet for both ions give v_{edge} = -550 km/s. No N V lines have been observed. This is the only A type star showing superionized lines. No correlation with the high rotational velocity exists, because four other stars of the same spectral type and same rotational velocity do not show superionization. This star is at 80 pcs; it is therefore improbable that the Si IV and C IV are of interstellar origin. Their profiles, moreover, clearly show the typical velocity-gradient shape.

Another A-type star with peculiar behavior is HR 5999. Its spectral type is A8 III; the far UV continuum fits a Kurucz model T_e = 8000 K, log g = 4, and presents several emissions (among them N V, C IV and possibly 1218,4 O V). Mg II presents emission wings. The anomalous and variable interstellar reddening suggests that it is embedded in a dusty medium; it could belong to the Herbig nebular variables (Tjin A Djie et al. (1982). Several other stars of this class have been studied by Felenbok et al. (1983). All of them have Mg II profiles with emission wings and a typical shape indicating the presence of velocity gradients and mass loss.

6.2 A-type supergiants

Emission components have been observed by Evans and Jordan (1975) at the center of the Mg II lines in the spectrum of Alpha Car (A9 Ib). The profile is truncated and the flux at the center is different from zero. However, other observations made by Praderie et al. (1980) do not confirm this result; the flux at the center is zero and the shape of the profile is not distorted. A similar effect has been observed by Hack and Selvelli (1979) for Epsilon Aur (A9 Ia). The profile of Mg II lines observed in 1978 is truncated and the flux at the center is different from zero, clearly suggesting the presence of emission at the center of the absorptions. However, other observations made in 1982 before the beginning of the eclipse do not show the central emission component and the flux at the center is zero. These results suggest that the central emission is variable in time. On the contrary, before, during and after the eclipse it is possible to observe an emission wing on the longward side of the central absorption core. Since the strength of the emission is not affected by the eclipse, it must be formed in an extended envelope surrounding the whole system.

Evidence of velocity gradients in the atmosphere of A-type supergiants is given by the Balmer progression. Usually the lower members of the series have more negative velocities than the upper members, indicating an expanding atmosphere. One exception is known: HD 21389 (Aydin 1972), which, at least at one epoch, shows the reverse behavior.

In the case of Epsilon Aur there is very little evidence of velocity gradients and mass loss; probably the largest quantity of mass loss occurred in earlier stages of its evolution, near the main sequence.

The ultraviolet spectra of Alpha Cyg (A2 Ia), Eta Leo (A0 Ib), HR 2874 (A5 Ib) and HR 1040 (A0 Ib) do not show evidence of P Cyg contours. However, the resonance lines of the abundant elements (C II, Si II, Fe II, Mg II) are shortward shifted by 240 km/s in Alpha Cyg, by 150 km/s in HR 1040 and HR 2874, while no shift is observable in Eta Leo. The Mg II profiles of Alpha Cygni are very broad and saturated. In Beta Ori (B8 Ia) a sharp component is observable at -190 km/s, suggesting the presence of a shell.

It would be very important to make simultaneous observations of the visual and ultraviolet spectra in order to look for possible correlations between the phenomena occurring in the photosphere and those occuring in the chromosphere, and identify the hydrodynamic mechanisms causing the winds.

7. REFERENCES

Abbott, D.C., Bohlin, R.C., Savage, B.D.: 1982, Astroph. J Suppl. 48, 369.

Aydin, C.: 1972, Astron. Astroph. 19, 389.

Blanco, C., Catalano, S., Marilli, E.: 1980, 2nd IUE European Conference, ESA SP-157, p. 63

Borra', E.F., Landstreet, J.O.: 1979, Astroph. J. 228, 809.

Castor, J.I.: 1970, Mon. Not. 149, 111.

Doazan, V.: 1982, in B Stars With and Without Emission Lines, NASA SP-456, p. 279.

Dravins, D.: 1982, Ann. Rev. Astron. Astroph. 20, 61.

Evans, R.G., Jordan, C.: 1975, Mon. Not. 172, 585.

Felenbok, P., Praderie, F., Talavera, A.: 1983, Astron. Astroph. 128, 74.

Garmany, C.D.: 1982, in Observational Basis for Velocity Fields in Stellar Atmospheres, Ed. R. Stalio, Trieste-Workshop, p. 83.

Gray, D.F.: 1982, Mem. Soc. Astron. It. 53, 931.

Gry, C., Lamers, H.J.G.L.M., Vidal-Madjar, A.: 1984, Astron. Astroph. 137, 29.

Hartmann, L., Baljunas, S.L., Duncan, D.K., Noyes, R.W.: 1984, Astroph. J. 279, 778.

Hack, M., Selvelli, P.L.: 1979, Astron. Astroph. 75, 316.

Kondo, Y., Morgan, T.H., Modisette, J.L.: 1977, Astron. Soc. Pacific 89, 675.

Lamers, H.J.G.L.M., Gathier, R., Snow, T.P.: 1982, Astroph. J. 258, 186.

Livingston, W.C.: 1983, in Solar and Stellar Magnetic Fields: Origins and Coronal Effects, IAU Symp. No. 102, J. Stenflo ed.

Lucy, L.B.: 1982, Astroph. J. 255, 278.

Lucy, L.B.: 1982, Astroph. J. 255, 286.

Marcy, G.W.: 1981, Astroph. J. 245, 624.

Molaro, P., Morossi, C., Ramella, M., Franco, M.: 1983, Astron. Astroph. 127, L3.

Molaro, P., Morossi, C., Ramella, M., Franco, M.: 1984, 4th IUE European Conference, ESA-SP 218, p. 223.

Noyes, R.W., Hartmann, L.W., Baljunas, S.L., Duncan, D.K., Vaughan, A.H.: 1984, Astroph. J. 279, 763.

Praderie, F., Lamers, H.J.G.L.M., Talavera, A.: 1980, Astron. Astroph. 86, 271.

Robinson, R.D., Worden, S.P., Harvey, J.W.: 1980, Astroph. J. Lett. 236, L155.

Simon, T.: 1984, Astroph. J. 279, 738.

Slettebak, A., Carpenter, K.G.: 1983, Astroph. J. Suppl. 53, 869.

Smith, M.A., Karp, A.H.: 1978, Astroph. J. 219, 522.

Smith, M.A., Karp, A.H.: 1979, Astroph. J. 230, 156.

Snow, T.P., Morton, D.C.: 1976, Astroph. J. Suppl. 32, 429.

Tjin A Djie, H.R.E., Thè, P.S., Hack, M., Selvelli, P.L.: 1982, Astron. Astroph. 106, 98.

Underhill, A.B.: 1983, Astroph. J. 265, 933.

Underhill, A.B., Fahey, R.P.: 1984, Astroph. J. 280, 712.

Willis, A.J.: 1982, in Observational Basis for Velocity fields in Stellar Atmospheres, Ed. R. Stalio, Trieste-Workshop, p. 111.

Zwaan, C.: 1977, Mem. Soc. Astron. It. 48, 525.

Zwaan, C.: 1981, in Solar Phenomena in Stars and Stellar systems eds. R.M. Bonnet, A.K. Dupree, Reidel, Dordrecht, p. 463.

CURRENT PROBLEMS OF LINE FORMATION IN EARLY-TYPE STARS

David C. Abbott*
Department of Physics and Astronomy, University College
London, Gower Street, London, WC1E 6BT, England

ABSTRACT. Extensive observations of UV P Cygni lines in the winds of early-type stars reveal the following four major problems with line formation theory: i) the evidence for photon destruction even in resonance lines; ii) the presence of strong wind shocks and the resultant nonmonotonic flow; iii) the mystery of the UV discrete components, and; iv) line blending of P Cygni profiles and the multiple scattering of photons. Observations of optical photospheric lines with ground-based detectors of high precision are raising the following two additional problems of line formation: i) the presence of stellar pulsations, and ii) the re-determination of fundamental stellar parameters from classic photospheric analysis. A first comparison of CCD observations of an O star (ζ Pup) to model atmospheres, both with and without wind-blanketing, shows that the spectral classification can depend as strongly on mass loss as on the usual parameters of T_{eff} and log(g).

1. INTRODUCTION

Observations of line formation in early-type stars are at present limited mainly to optical and ultraviolet wavelengths, although instruments now under development offer exciting possibilities for spectroscopy at infrared, EUV, and X-ray wavelengths. The observed lines are of basically two types. The first are lines formed in the stellar wind, which are principally ultraviolet, resonance, P Cygni lines. These have been extensively modeled to derive the mass loss properties of hot stars, such as the mass loss rate $\dot{M}$ and velocity law

---

*Permanent address: Joint Institute for Laboratory Astrophysics, University of Colorado and National Bureau of Standards, Boulder, Colorado 80309 USA.

J. E. Beckman and L. Crivellari (eds.), Progress in Stellar Spectral Line Formation Theory, 279–304.

V(r) (e.g. Castor and Lamers 1979; Garmany *et al.* 1981; Gathier, Lamers and Snow 1981). The second are the absorption lines formed in the hydrostatic layers beneath the wind, which are termed "photospheric" lines. These are principally hydrogen and helium lines at optical wavelengths, as the observed ultraviolet photospheric features are metallic lines which are difficult to model and are subject to blending and poor photometric accuracy. The photospheric lines are analyzed to derive the basic stellar parameters of effective temperature T_{eff}, gravity log(g), and chemical composition (e.g. Auer and Mihalas 1972).

Presently, several hundred early-type stars have been observed in the ultraviolet, primarily by the *Copernicus* and International Ultraviolet Explorer (IUE) spacecraft. From this data set four major problems of the formation of wind lines have emerged which challenge current theory, and cloud any interpretation of the mass loss properties of these stars. These are:

- the presence of photon destruction and creation
- evidence for nonmonotonic outflow
- variable, discrete components in P Cygni profiles
- line blending and multiline radiative transfer.

There is an added strong motivation for attacking these problems, because the Space Telescope will soon provide ultraviolet observations of extragalactic OB and Wolf-Rayet stars with a precision and resolution comparable to that provided by *Copernicus* and IUE for the nearby galactic stars. Obviously, our understanding of these distant stars and their extragalactic environment will be limited by our ability to overcome the problems of interpretation of existing data in nearby stars.

In the optical, the recent development of devices of high linearity, such as CCD and Reticon, are allowing the study of photospheric line profiles at accuracies better than 1%. It is almost an axiom of astronomy that any order of magnitude increase in observational ability will automatically produce new and unexpected discoveries. Already, ground-based observations with these new detectors are finding pulsational activity in a much broader class of early-type stars than previously thought (e.g. Vogt and Penrod 1983). These observations also have the potential to tell us much more about the stellar photospheres. The theoretical challenge is to develop models whose accuracy is also of the order of 1%, so that the information contained in these observations can be tapped.

2. ULTRAVIOLET OBSERVATIONS OF LINES FORMED IN STELLAR WINDS

2.1. A Simple Model of Line Formation

In stark contrast to the very sophisticated and complex treatments of line transfer discussed throughout this workshop, observations of P Cygni profiles in hot stars are analyzed with a model whose dominant characteristic is one of simplicity (e.g. Castor 1970; Lucy 1971). Discrepancies between observation and theory can come about from over-simplification of the model, as well as from omission of some physical process. Specifically, the models assume spherically symmetric, homogeneous, steady outflow with a rapid and monotonic increase in velocity with radius to a maximum, or terminal, velocity V_∞. With these assumptions and the large values of $V_\infty/V_{thermal}$ observed for these winds, the intrinsic width of the line is negligible compared to the Doppler shift caused by the wind except at line center, and the radiative transfer is calculated in the Sobolev approximation. Further, the lines are assumed to be pure scattering, as the ratio ε of the rate of collisional de-excitation to that of radiative de-excitation is smaller than $\varepsilon \sim 10^{-5}$, except in the dense winds of the Wolf-Rayet stars (e.g. Castor and Lamers 1979). Complete frequency redistribution in the frame of the atom and angular redistribution from the escape probability are the final approximations.

Examples of P Cygni profiles calculated with this model are shown in Figure 1. The lines are all completely saturated, meaning that every photon with the appropriate co-moving frequency is scattered, and the profiles show a sequence of adopted velocity laws $V(r)/V_\infty$ which are increasingly steep. The P Cygni profile is usually pictured as a summation of an emission component, which is nearly symmetric about a peak at line center frequency ν_0, and a continuum component in

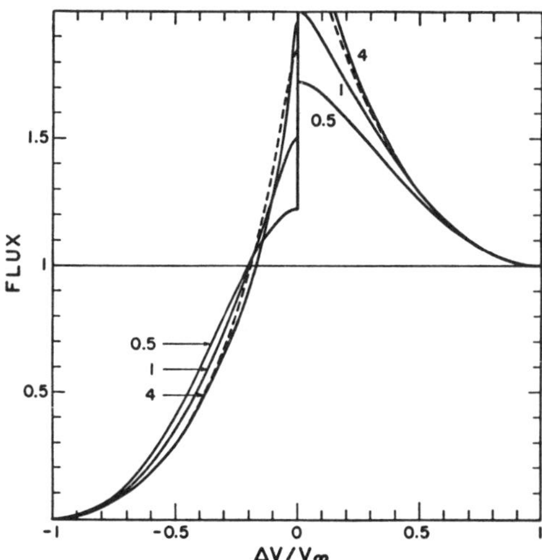

Fig. 1. Examples of P Cygni profiles computed using the standard model of line formation in the wind. Solid lines were computed with the velocity law $V/V_\infty = 0.01 + 0.99(1 - r_*/r)^\beta$ with the value of β indicated. Dashed line was with $V/V_\infty = 1 - (r_*/r)^{-1/2}$. All lines are pure scattering with infinite optical depth (from Castor and Lamers 1979).

which photons in the interval $(1 + V_\infty/c)\nu_o \leq \nu \leq \nu_o$ have been removed by scattering (e.g. Mihalas 1978, §14.2). The profiles in Figure 1, and their elaborations in Castor and Lamers (1979), are characterized by the following general features:

1) The absorption equivalent width is insensitive to the velocity law, and therefore is a good diagnostic of $\dot{M}$, if the ionization fraction of the absorbing ion can be determined independently. The violet edge of the absorption trough of a saturated profile accurately measures V_∞.

2) The ratio of the emission equivalent width to the absorption equivalent width, W_E/W_A, is a good diagnostic of the velocity law $V(r)/V_\infty$. For a pure scattering line, conservation of photons guarantees that the number of emitted photons equals the number of scattered continuum photons minus those photons reflected back onto the stellar core. A steeper velocity law reflects a larger fraction of scattered photons onto the stellar core, because of the greater solid angle subtended by the core, which results in a smaller value of W_E/W_A. For the example of Figure 1, the ratio varies from $W_E/W_A = 0.61$ to $W_E/W_A = 0.87$.

3) No profile can ever by flat-bottomed, i.e., the absorption cannot extend down to zero flux for a finite interval of frequency. This is because the lines scatter the photons equally in the forward and backward direction. (The angular redistribution function scales as $f(\mu) \propto \mu^2 dV/dr + (1-\mu^2)V/r$ for opaque lines.) Thus, at all velocities except the very maximum, some of the scattered photons must be re-emitted in the forward direction, thereby filling in the absorption trough.

2.2. Evidence for the Destruction of Photons

An example of the application of the simple theory to observations is given in Figure 2. While the model gives a reasonable fit to the absorption trough, which is the intent of the authors who were deriving $\dot{M}$, there is a glaring discrepancy between the observed and predicted emission. This absence of emission in the observed profiles is quite common. In a survey of N V wind profiles, Abbott, Bohlin and Savage (1982) found that 1/3 of the stars with optically thick absorption showed an obvious absence of emission. In almost every case, the star was of the main sequence or giant spectral type. This result is also apparent in the data of Snow and Morton (1976).

Abbott, Bohlin and Savage (1982) considered three explanations for the lack of emission. None were adequate. I summarize these below:

1) Steep Velocity Law. Since the W_e/W_a ratio is supposed to measure $V(r)/V_\infty$, the most straightforward explanation is that the stars without emission have a very steep rise in velocity with radius. To get the observed ratio in many stars, however, requires ridiculously steep laws, i.e., $V/V_\infty = (1 - r_*/r)^\beta$ with $\beta \leq 0.1$, which violate other observational constraints.

2) Underlying "Photospheric" Absorption. If the P Cygni line is superimposed on a continuum containing strong photospheric absorption

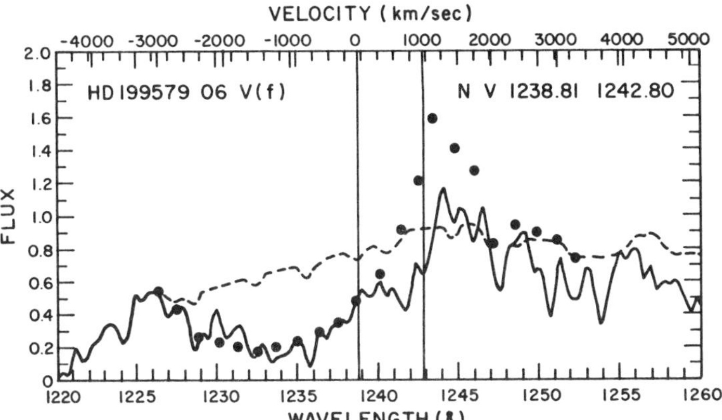

Fig. 2. An example of a fit of the standard model to the observed N V doublet at 1240 Å. Solid line -- Copernicus U2 observations. Filled circles -- best fit model. Dashed line -- example photospheric spectrum from a comparison star. Vertical lines indicate the rest wavelengths of the blue and red components (from Gathier, Lamers and Snow 1981).

features, either from the line itself or from overlapping lines of other species, then the observed emission is reduced below model expectations. There are two ways to estimate the strength of these photospheric blends. One is to calculate them from line lists and f-values (e.g. Cassinelli and Abbott 1981). The other is to estimate them empirically from comparison stars of similar T_{eff}, but without P Cygni profiles (see, for example, Figure 2). Both methods suggest that photospheric blends are unimportant to the measured W_e/W_a, especially for the N V line, which is a superionized species with no photospheric component of its own.

3) Anisotropic Mass Loss. If mass loss occurs mainly from latitudes along the line of sight, the W_e/W_a ratio can be reduced significantly. The main argument against this explanation is that it requires a special geometry between the star and observer, and hence it cannot apply to a statistically large sample. There is also no evidence that the mass loss is strongly anisotropic in early-type stars, except in the rapidly rotating Be and Oe stars. Abbott, Bohlin and Savage (1982) found no correlation between the observed W_e/W_a and V sin i.

Given the failure of all explanations so far, one is forced to the conclusion that real destruction of photons is occurring in these winds, even though $\varepsilon \sim 10^{-5}$. One way to accomplish this is to trap the photons in opaque lines. For example, while Sobolev line transfer pictures a photon as scattering once at the resonant point in the wind and then escaping, the photon may actually be scattered locally a large number of times at the resonant point. The number of scatterings of such a trapped photon is roughly $(PE)^{-1}$ where PE is the escape probability, given by (Castor 1970)

$$PE = 1/\tau_o \quad , \tag{1}$$

where τ_o is the transverse optical depth given by

$$\tau_o = \frac{\pi e^2}{mc} \lambda_o f n_i \; r/V \quad . \tag{2}$$

The number of scatterings experienced by a photon can be further increased if the outflow is nonmonotonic, as noted by Lucy (1983). Considering the example of N parallel, distinct resonant surfaces which are all opaque, a photon will undergo an average of N times the scatterings from a single surface before its ultimate emergence in the forward or backward direction. Thus, the effective destruction probability becomes

$$(\varepsilon)_{eff} = N \, \varepsilon \, \tau_o \quad . \tag{3}$$

For a star with a strong wind like ζ Pup, appropriate estimates are $N = 20$ (Lucy 1983) and $\varepsilon\tau_o \simeq 10^{-4} \, r_* V_\infty/(rV)^3$, so that $(\varepsilon)_{eff}$ exceeds 10% for $V/V_\infty \lesssim 0.2$. Thus, it is possible to get a weak photon destruction in saturated lines. It is difficult, however, for these ideas to explain the lack of emission observed in some stars with very weak winds. In addition, one must worry that the inverse process of photon creation does not balance the effect.

2.3. Nonmonotonic Outflow

The absorption trough of the P Cygni profile is also an area of contradiction between observation and the simple model. One discrepancy is the large frequency interval of zero residual intensity, or the extent of the "blackness," which is observed in every star which has saturated wind profiles. Examples are shown in Figure 3. Lucy (1982, 1983) interprets this blackness, which is in fundamental contradiction of the simple model, as evidence for nonmonotonic outflow, which is a natural consequence of the presence of strong shocks in the wind.

The best observed example of this feature is the *Copernicus* U1 observations of ζ Pup (O4f) by Morton (1976). For the N V doublet at 1240 Å, the residual flux is less than 3% of the continuum over a frequency interval corresponding to roughly 1400 km s^{-1}. By contrast, the residual intensity in the simple model has a value of 17% at the point of doublet overlap at 900 km s^{-1}. Figure 4 shows these observations and the fit by Lucy (1983) using the simple model modified to account for nonmonotonic outflow. For contrast, the monotonic line profile with the same wind parameters is shown. The convincing fit of the nonmonotonic model to the observations, and the physical basis for expecting nonmonotonic flow all argue strongly that Lucy's explanation is correct.

There are two obvious implications of the presence of nonmonotonic outflow for the interpretation of the P Cygni profiles in these stars. First, as emphasized by Lucy (1983), the extent of the blackness

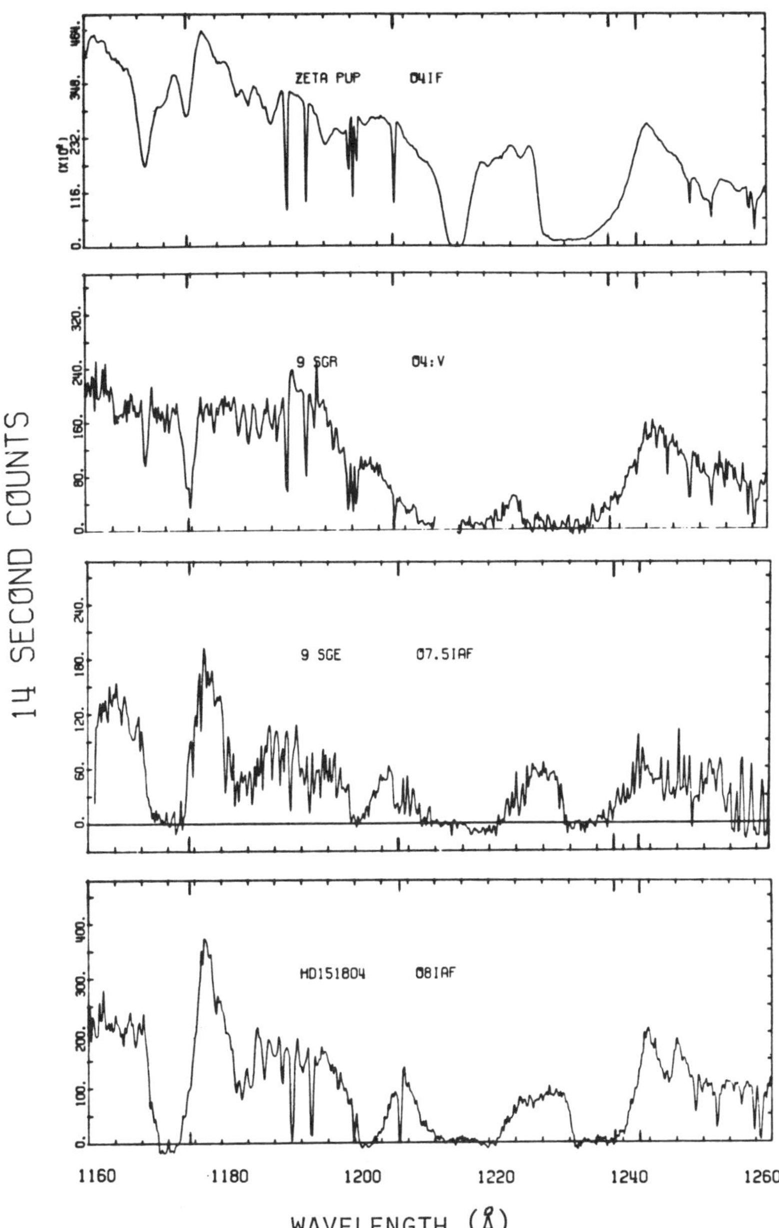

Fig. 3. Observations of the N V P Cygni profile in four example stars, which illustrates the large wavelength interval of blackness that is always present in saturated lines (from Snow and Morton 1976).

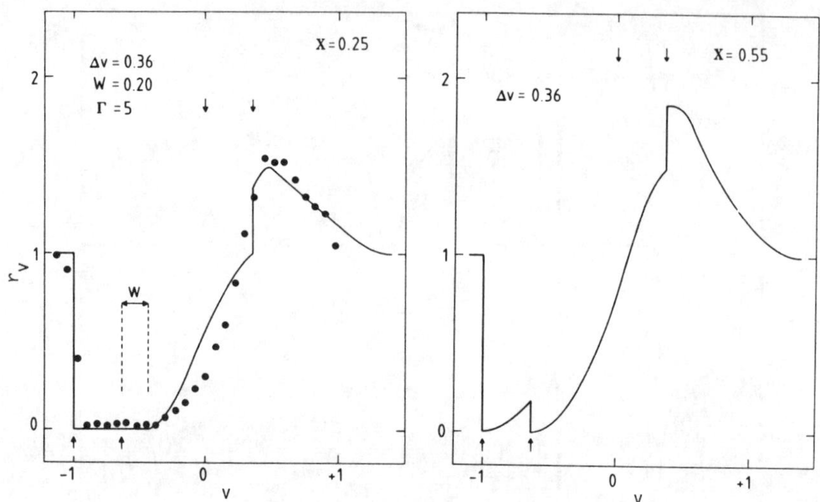

Fig. 4. Fits to the Copernicus U1 observations of N V in Zeta Puppis. Left panel -- solid line is a model with nonmonotonic flow and a shock amplitude of $W = 0.2\ V_\infty$. Filled circles are the observations. Right panel -- solid line is comparable model with monotonic flow. Arrows show the doublet separation (from Lucy 1983).

provides an unambiguous diagnostic of the characteristic amplitude of the shock velocities in the wind. For the example of ζ Pup, Lucy finds 480 ± 50 km s^{-1}. A systematic study of this diagnostic from existing UV data would be valuable to search for correlations between the characteristic shock velocities and the X-ray luminosity, wind strength, etc. of the stellar wind.

The presence of nonmonotonic flow may also affect the values of the mass loss rate derived from fits to the absorption profile. For lines which are opaque, the profile will be much deeper than its monotonic counterpart, and hence the use of the standard model will overestimate the derived $\dot{M}$. For lines which are optically thin, it is not clear what, if any, effect nonmonotonic flow will have on the mass loss rate diagnostic. In addition, as pointed out by Lucy (1983), the terminal velocity of the wind may be overestimated by an amount equal to the characteristic velocity of the shock motions.

2.4. Discrete Components

The observation of discrete components, also called UV shell lines, provides a second major discrepancy with theory in the absorption part of the P Cygni profile. Figure 5 shows an example of such a discrete component, which appeared in the profile of HD 93250 (O3 V) during a

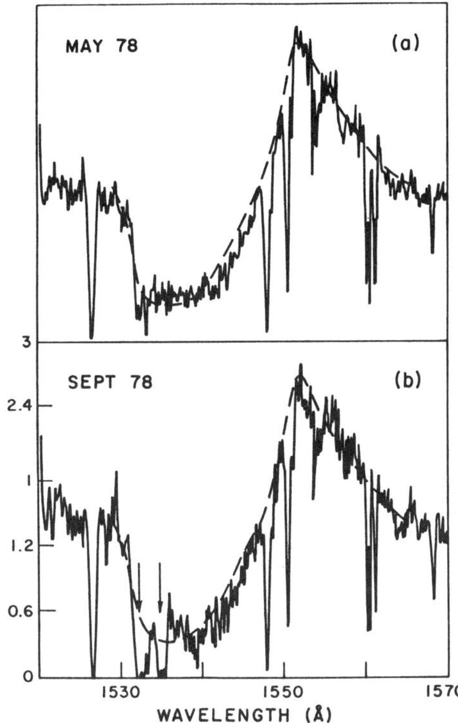

Fig. 5. Two IUE observations of the star HD 93250 (O3 V) separated in time by three months. Well-developed discrete components are indicated by the arrows in the bottom panel (from Henrichs, 1984, private comm.).

time interval of less than three months. Unlike the "blackness" of nonmonotonic flow, discrete components can only be detected in lines which are not saturated. The origin of the discrete components is unknown. Discrete components were noticed almost from the first in the satellite UV spectra of hot stars (e.g. Morton 1976, Snow 1977). They are present in roughly 50% of all UV spectra of early-type stars exhibiting mass loss. The first systematic study devoted to these components was by Lamers, Gathier and Snow (1982). The current state of our knowledge of discrete components is comprehensively reviewed by Henrichs (1984).

There is no formal definition of what constitutes a discrete component, largely because, as shown in Figure 5, one knows them when one sees them. The following characteristics provide a working definition of a discrete component (see also Henrichs 1984):

1) Velocity Width. The narrow absorption has a velocity width much larger than the interstellar line width, but much smaller than the characteristic wind velocity V_∞ or the photospheric rotational velocity V sin i. Typical widths are 100-200 km s^{-1}.

2) Line Identification. Discrete components are only found in lines which show strong wind profiles, i.e., UV resonance lines or lines arising from a metastable state. Since these are mainly doublets, the

presence of the absorption feature at the same velocity in both components is strong evidence both for their reality and for their identification with the underlying line. Discrete components at a given velocity are almost always found either in all wind line profiles, or in none.

3) Variability. The features vary in strength on time scales of days. However, it is common for some sort of feature to be present at roughly the same velocity for a time scale of years.

4) Velocity. The features have only been observed at high velocities, typically $0.5 \lesssim V \lesssim 0.9\ V_\infty$ in cases where V_∞ could be measured. In the majority of cases, features are only seen at one velocity. However, there are many well-documented cases of features at multiple velocities. A further illustration of these characteristics is provided by observations of γ Cas (B0.5 IVe) in Figure 6.

These UV discrete components are similar in many respects to the narrow absorption features, termed "shell" lines, which have been observed for many years in the optical spectra of some stars, usually of the Be-type (e.g. Hack and Struve 1970). It is perhaps not surprising then, that the favored explanation for the discrete components has evolved to the model developed for the shell stars, that is the absorption arises from a real density enhancement expanding at large radii from the star, i.e., a moving shell. Early attempts to explain the absorption as an apparent concentration of material, for example, because of ionization effects or a plateau in the velocity law, are ruled out by recent, more detailed observations (e.g. Henrichs 1984).

As there is no generally agreed upon kinematic model for these shells, the radiative transfer used to interpret the discrete components has remained basic. The main difficulty is that the velocity width of the narrow component exceeds the thermal velocity width, which implies differential motions. This leaves the usual two options. One choice assumes that the shell material is characterized by a single velocity, and that the width results from small-scale random motions which are greater than thermal, i.e., microturbulence (see Henrichs et al. 1983). The other assumes that, locally, the material has the usual thermal random motions, but there is a velocity gradient between the leading and following edges of the shell, which follows the wind V(r), i.e., macroturbulence (e.g. Lamers et al. 1982).

Using one of the above approaches, the narrow absorption components have been fit by Gaussians to extract the shell velocity, velocity width, optical depth, and column density. Several features of the components are apparent from these data:

1) Memory. When a large number of observations are available, some, but not all, stars exhibit a correlation in which the larger column densities have smaller shell velocities (e.g. Henrichs 1984). Notable examples are γ Cas (B0.5 IVe), ω Ori (B2 IIIe), and ξ Per (O7.5 IIIe). Since the various components and their shell velocities are thought to be independent events, often separated by long times, this correlation is called memory.

2) Mass Loss Rate. The column density of the discrete components often rivals that of the underlying P Cygni profile, especially in the Be-type stars, where there may not even be an underlying component.

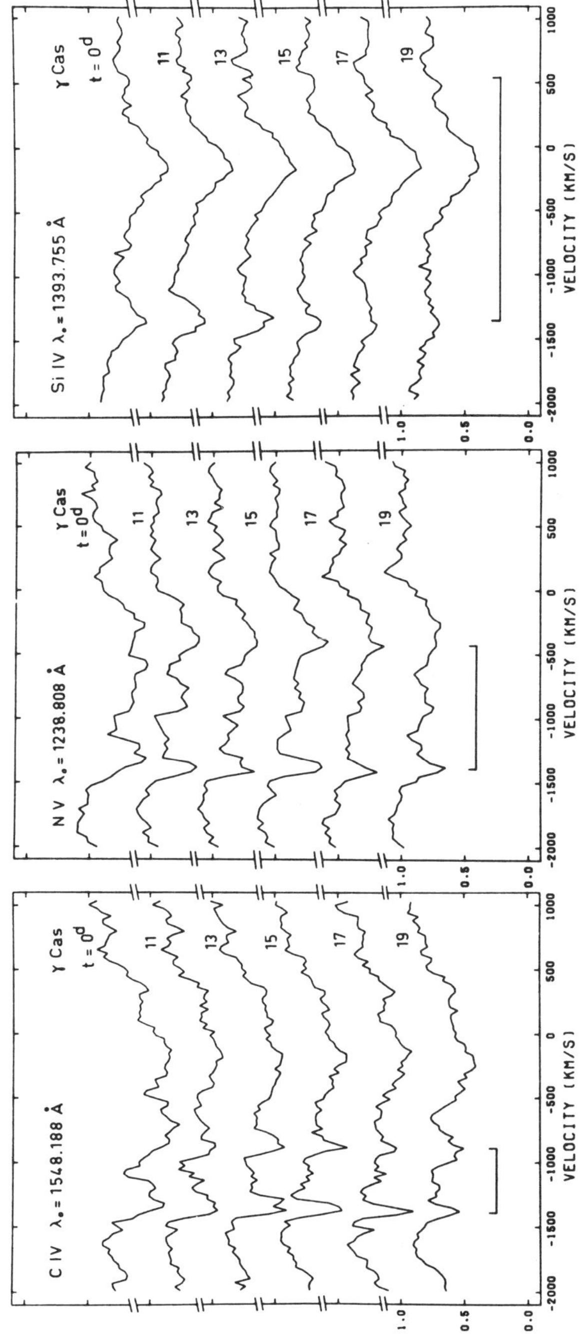

Fig. 6. A time sequence of IUE observations of γ Cas (B0.5 IVe) showing a rare case in which the discrete components are observed to decay in strength. Note that the components are present in both doublet components of three different ions, and that the velocity does not change with time (from Henrichs *et al.* 1983).

This creates obvious uncertainty in any mass loss rates derived for these stars, both in the proper treatment of the radiative transfer, and in the question of time variability of the mass loss.

3) Covering Factor. The violet-shifted absorption of the P Cygni profile is caused by material moving outward radially within the line-of-sight from the observer to the stellar continuum source. The minimum fraction of this solid angle occupied by shell material is $(1 - F_\nu/F_c)$, where F_ν is the residual flux of the discrete component and F_c is that of the neighboring continuum. This covering factor of the shell is typically 50% or more.

The discrete components cannot be explained by a model with a single shell expanding from the star. Two arguments show that such a shell could not be located near the star. First, as noted by Lamers et al. (1982), the column density of a shell of initial radius r_o and column density N_o will decrease because of spherical dilution at a rate of

$$N = N_o \left(\frac{r_o}{r_o + V_s t}\right)^2 , \tag{4}$$

i.e., the time scale for decay is r_o/V_s. Thus, for a typical shell velocity of $V_s = 10^8$ cm s^{-1}, the shell will disappear on a time scale of days for r_o smaller than 10^{13} cm. (A typical stellar radius is 10^{12} cm.) A second argument, not noted before to my knowledge, is relevant for stars with strong winds. It relies on the fact that the slower moving shell experiences on acceleration as it accretes mass from the background wind. If the wind has a mass loss rate $\dot{M}$ and velocity V_∞, the shell gains mass at a rate of $f_c\dot{M}(V_\infty - V_s)/(4\pi r_s^2 V_\infty)$, where f_c is the covering factor. Applying conservation of radial momentum to a shell of column mass $\mu m_H N$ over a time interval t gives a velocity increase for the shell of

$$\frac{\Delta V_s}{V_s} = 2.2\ f_c \frac{\dot{M}_{-6}}{N_{21}} \frac{(V_\infty - V_s)^2}{V_\infty V_s} \frac{t_1}{(r_s)^2_{13}} , \tag{5}$$

where the units are the typical values $10^{-6}\ M_\odot\ yr^{-1}$ for $\dot{M}_{-6}$, 10^{21} g cm^{-2} for N_{21}, 10^{13} cm for r_s, and 1 day for t_1. Taking $f_c = 1/2$ and $V_s = 0.75\ V_\infty$, equation (5) gives a 10% velocity increase in 1 day.

Since the discrete components are observed to persist at roughly the same strength and velocity over time periods of weeks, months, and even years, they cannot be a single expanding shell unless $r_s \gg 10^{13}$ cm. Such a single, very distant shell is unlikely for two reasons. First, the mass of such a shell would be very high, requiring an ejection mass loss rate for a period of t_1 days of

$$\dot{M}_{ejection} \approx 10^{-6} \frac{N_{21}(r_s)^2_{13}}{t_1} M_\odot\ yr^{-1} . \tag{6}$$

Very distant shells would require an almost nova-like eruption to provide the high mass required by the observed column density. A second reason is that, in every case where discrete components have been observed with high time resolution, they have shown significant variability in strength on a time scale of days.

The discrete components most likely represent a superposition of many density enhancements over a time scale of days or less. The mechanism producing the discrete components must provide a nearly continuous supply of mass at roughly the same velocity. There are currently two ideas for the source of the mass.

One is that the mass in the discrete components represents extra matter that was ejected from the surface of star during episodes of enhanced mass loss (e.g. Henrichs et al. 1983). This explanation requires a mechanism to cause the enhanced mass loss, which must operate semi-continuously in the majority of hot stars. One candidate is stellar pulsation, which evidently caused an episode of mass ejection in ζ Oph (Vogt and Penrod 1983), which correlated with changes in the UV discrete components (e.g. Henrichs 1984). As an extreme statement of this idea, Barker and Marlborough (1984) hypothesize that in some Be-type stars there is no underlying wind; all mass is lost in episodic ejections that form discrete components which, when superposed, create the impression of a P Cygni profile.

At the opposite extreme is the idea of Lucy (1983), who proposed that discrete components are clumps and inhomogeneities in the chaotic outer part of a wind whose mass loss rate is constant in time. The clumps are formed through the compressive action of the same strong shocks responsible for the X-ray emission and nonmonotonic flow, and they are accelerated to less than the maximum velocity because of shadowing of the continuum radiation field. The advantages of this model are that it relies on a physical process known to operate in all hot star winds, that it provides a continuous supply of clumped mass, and that it will vary on a characteristic time scale, r_*/V_∞, which is short. Calculations which show that the idea works in practice would be very valuable.

Unambiguous observational tests between these ideas are difficult. Both ideas can lead to a "memory" between column density and shell velocity. The absence of discrete components at low velocity, or those moving from low to high velocity, favors Lucy's hypothesis, which forms clumps only at high velocities. However, simulations of the effect of episodic mass loss on P Cygni profiles by Prinja and Howarth (1984) show that density enhancements may appear as broad and weak features at low velocity, taking on the appearance of sharp, distinct components only at high velocities. A possible first observation of such broad, low-velocity features was made in ξ Per (O7.5 III) by Prinja et al. (1984). Further, there is the exceptional case of P Cygni (B1 Ia^+) whose discrete components move from low to high velocity, with an apparent periodicity of ~400 days (see Figure 7). One possible observational test is to monitor the UV spectrum of a star with both saturated and unsaturated wind profiles, which also has fairly sharp, well-defined photospheric lines. Measurements of changes in the discrete components presumably should correlate either with the diagnostics of nonmonotonic outflow, such as the extent of the blackness,

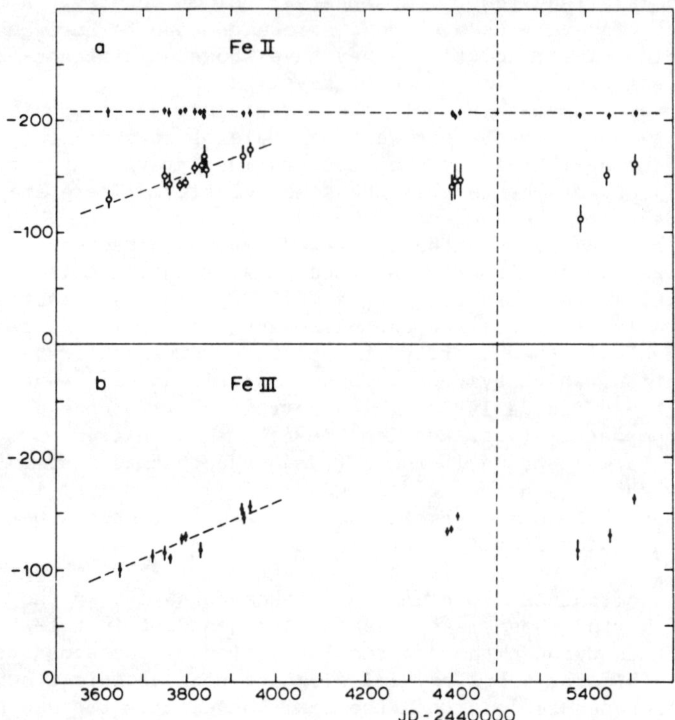

Fig. 7. Variations with time of the velocities of the discrete components in the UV spectra of P Cygni (B2 Ia$^+$). Each point represents a mean radial velocity of six individual Fe II or Fe III lines, along with the 1σ error bars. This is the only case in which a component is seen to accelerate from low to high velocity. There is a possible period of ~400 days for the recurrence of the components (from Lamers, Korevaar and Cassatella 1984).

the X-ray flux, and the curvature of the violet edge, or with diagnostics of pulsation, such as variations in the shape or radial velocity of photospheric lines.

Given the very complex, and sometimes contradictory, behavior of the discrete components observed in different stars, or at different times in the same star, it may also be that many different processes contribute to their origin. A density clump in the wind will evolve into an expanding shell in a similar manner regardless of whether it was created by a shock, by a surface eruption, or even by Roche lobe overflow (Howarth 1984). A hydrodynamic calculation of the evolution of a clump in a radiation-driven, supersonic outflow is needed.

2.5. Spectrum Synthesis

All of the above wind profiles can be analyzed in isolation from other transitions, except for the problems of blending from photospheric absorption in the background continuum. Unfortunately, in many cases of practical interest, profiles from many individual lines overlap in frequency because of the large Doppler shift introduced by the outflow of the wind. This problem is particularly acute in Wolf-Rayet stars, whose winds are so dense and so opaque that essentially every line is a wind profile. Figure 8 shows a typical example of the UV spectrum of a Wolf-Rayet star. The lines are so badly overlapped that it is impossible to distinguish an emission line from a P Cygni feature, or to even roughly define a continuum level. Even determining the identity of the lines is a major challenge. Clearly, the simple model described above is useless to interpret these spectra.

One promising approach to the problem is provided by the recent wind models of Abbott and Lucy (1985), which are based on a Monte Carlo technique which Lucy (1983) developed for the radiative transfer in P Cygni doublets. This method takes advantage of the fact that the UV wind profiles are primarily resonance lines, or transitions which mimic resonance lines. Thus, the line transfer is pure scattering, as is the continuum scattering by electrons, and the statistical equilibrium can

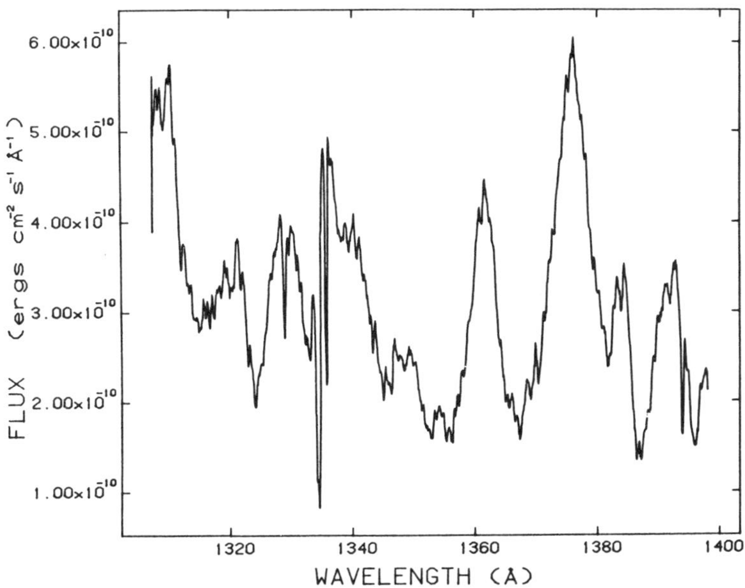

Fig. 8. An IUE spectrum of HD 50896 (WN5) showing the problems of line-blending typical in the UV of Wolf-Rayet stars. It is impossible to establish a continuum level, or even the identity of all the lines (from the IUE archives).

be realistically approximated by a treatment similar to that employed in the nebular case. The Monte Carlo method offers great advantages in terms of its programming simplicity, its ability to treat thousands of lines arbitrarily spaced in frequency, and its flexibility to be adapted to any specified geometry, velocity field, or angular redistribution function. Its weakness, which must be overcome to solve the Wolf-Rayet case, is the situation where photons are created and destroyed in the wind.

A first example of a synthetic wind spectrum with severely overlapping lines is shown in Figure 9 for a model of the EUV wavelength region in the star ζ Pup (O4f). The spectrum was made with roughly 10^5 photons simulating a wavelength resolution of ~1 Å. Many of the photons in this example experienced a frequency redshift greater than $4\ \nu_o V_\infty/c$ before emerging in the emission peaks, which are associated with gaps in frequency where no lines are present. However, at wavelengths such as 380 Å, where the lines are very densely clustered, the photons are scattered immediately and often as they emerge from the

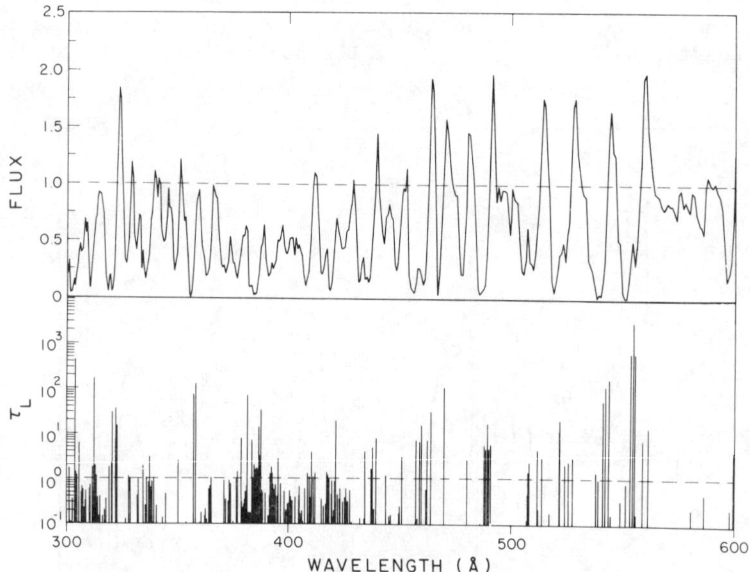

Fig. 9. The synthetic spectrum emergent from the wind of the star ζ Pup (O4f), using the model of Abbott and Lucy (1985). Top panel shows the luminosity emergent from the wind, L_ν, normalized with the luminosity emergent from the photosphere as unity. The resolution of the spectrum is $0.2\ \lambda\ V_\infty/c$. Vertical bars in the bottom panel show the optical depth τ_L of the lines included in the model at a representative velocity point of $V = 0.5\ V_\infty$. The vertical bars are at the rest wavelength of the lines.

photosphere, and the most probable outcome is reflection back onto the photosphere. This is apparent in the depressed level of the "continuum" of the flux emergent from the wind, as compared to that leaving the photosphere.

A plot of this wind albedo, defined as the ratio of the reflected to the emergent luminosity at the photosphere, is shown as a function of wavelength in Figure 10 for the same example of ζ Pup. At low frequencies, there are few lines, and the albedo is dominated by electron scattering, which is wavelength independent. In the EUV the lines dominate, and the albedo reaches a remarkable peak of ~45% at 400 Å. To appreciate the density of lines required to achieve this value, recall that the isolated line of infinite optical depth shown in Figure 1 ($\beta = 1$) reflects only 14% of the photons incident on it from the photosphere between frequencies corresponding to $V = 0$ and $V = V_\infty$.

3. OPTICAL OBSERVATIONS OF PHOTOSPHERIC LINES

The "photospheric" spectrum of a star is made up of pure absorption lines, which show no evidence for mass loss or chromospheric activity. The traditional photospheric analysis compares observations of these lines to model atmospheres, which assume hydrostatic and radiative equilibrium with plane parallel geometry, to derive the fundamental stellar properties of T_{eff}, log g, and chemical composition. The validity of this approach for hot stars was first demonstrated convincingly by Auer and Mihalas (1969, 1972), whose non-LTE models fit existing observations of photospheric lines. Their models have been

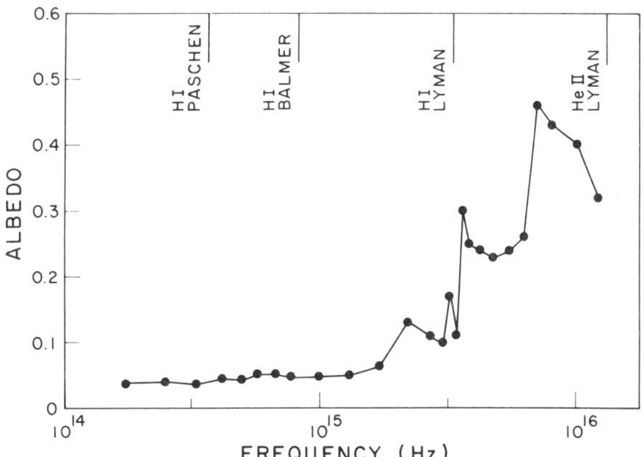

Fig. 10. The frequency dependence of the ratio of emergent to reflected radiation, or albedo, from the model of ζ Pup by Abbott and Lucy (1985). The plotted albedo is the average over frequency intervals whose central frequency is indicated by the filled circle.

extensively applied to the OB stars (e.g. Conti 1975, Auer and Mihalas 1973), and to the O subdwarfs (e.g. Kudritzki 1976). Two events since this work suggest that this procedure should be re-evaluated:

1) The discovery that all luminous early-type stars possess strong stellar winds (e.g. Conti 1978). The effects of this wind can extend deep back into the atmosphere, even to the static regions where the photospheric lines are formed, mainly through the line-blanketing effects of the backscattered radiation field, such as that shown in Figure 10 (Hummer 1982; Abbott and Hummer 1984). This phenomenon is referred to as "wind-blanketing."

2) The advent of devices, such as CCD and Reticon, whose photometric precision is an order of magnitude better than the high resolution, photographic spectra of a decade ago. Clearly the reduction of noise to a level of less than 1% provides a much more vigorous test of the models, and will inevitably lead to the discovery of new phenomenon.

3.1. Observations

An example of the improvement in data quality is provided by Figure 11, which shows the He I 4471 Å and He II 4542 Å lines in ζ Pup. All

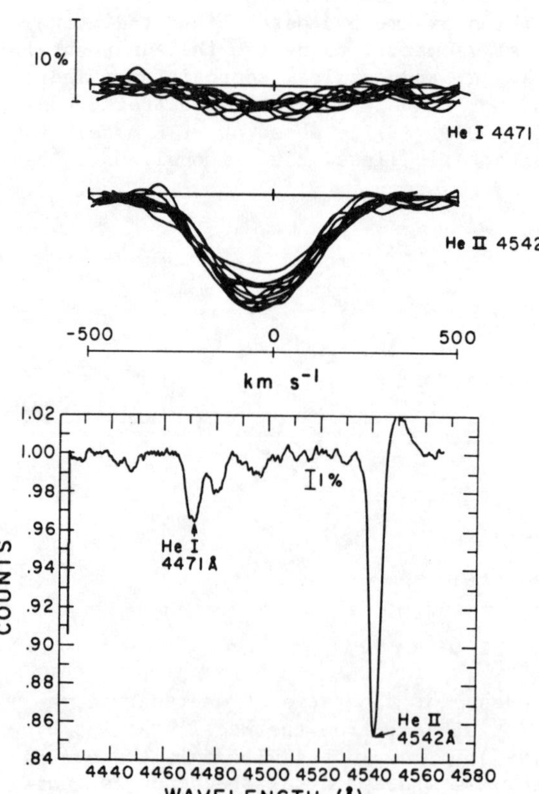

Fig. 11. Comparison of observations of photospheric lines in ζ Pup made with IIa-O plates (top panel) and an RCA CCD detector (bottom panel), but with the same coude spectrograph. The signal-to-noise is about a factor of 10 better with the electronic detector, which reveals the presence of numerous blends and asymmetries. (Top panel from Conti and Frost 1977. Bottom panel from Hummer, Abbott and Bohannan 1984.)

observations used the same coude spectrograph, but the data in the top panel are the digitized tracings of a IIa-O plate, filtered to remove plate noise, while the data in the bottom panel are the output from a CCD detector with the flat-fielding correction. Both were normalized by fitting the continuum "by eye." Note the heavy blending of the He I 4471 line, the primary temperature diagnostic, which is clearly present in the CCD spectrum. A further example of spectra obtained with electronic spectra are the observations of 17 B stars made by Heasley, Wolff and Timothy (1982). These are all stars for which the effects of mass loss are negligible, however, they found striking discrepancies between theory and observation in several He I lines.

As an example of new phenomenon already discovered with these detectors, Figure 12 shows observations of He I 6678 Å in ζ Oph (O9 Ve). At the 1% level there is structure in the line profiles which is real, and which is strongly linked to nonradial oscillations (Vogt and Penrod 1983). The discovery of nonradial pulsation in a rapidly rotating O star has fueled a growing realization that pulsational activity extends far beyond the traditional Beta Cephei region, and is probably of measurable importance in most massive stars (e.g. Smith and Penrod 1984, Vreux 1984). The consequences of this pulsational activity on the photospheric analysis is unexplored, but surely merits investigation.

3.2. Wind-Blanketed Atmospheres

A first effort to study the quantitative effects of the wind-photosphere interaction on the photospheric spectrum was made by Abbott and Hummer (1984), using a core/halo model. The "core" is formed by the hydrostatic layers of the star, where the continua and absorption lines are formed. It is represented by the usual Auer-Mihalas model atmosphere. The "halo" is formed by the stellar wind. The radiation reflected back onto the core by the halo, i.e., the wavelength-dependent albedo, is calculated in the manner of Abbott and Lucy (1985). This becomes the upper boundary condition to the model atmosphere, just as the radiation emergent from the model atmosphere is the lower boundary condition to the wind calculation. Although the boundary between core and halo is rather sharp (e.g. Castor, Abbott and Klein 1975), it cannot be defined precisely. Fortunately, no observable results are affected by the exact choice. Details of the calculation are given by Abbott and Hummer (1984).

The major effect of wind-blanketing is to raise the surface temperature of the atmosphere, from the upper boundary extending down to the layers of continuum formation. Since this includes the region of line formation, wind-blanketing causes significant changes in all lines which are temperature sensitive. Figure 13 shows the dramatic weakening of the line He I 4471 Å as the mass loss rate of the wind goes up. While all four models in Figure 13 have T_{eff} = 42,000°K and log g = 3.5, the spectroscopic spectral type dictated by the ratio He I 4471/He II 4542 varies from O3 to O5.5. For luminous, early-type stars the spectral classification is truly three-dimensional, depending on T_{eff}, log g, and the strength of the wind, as parameterized here by its mass loss rate.

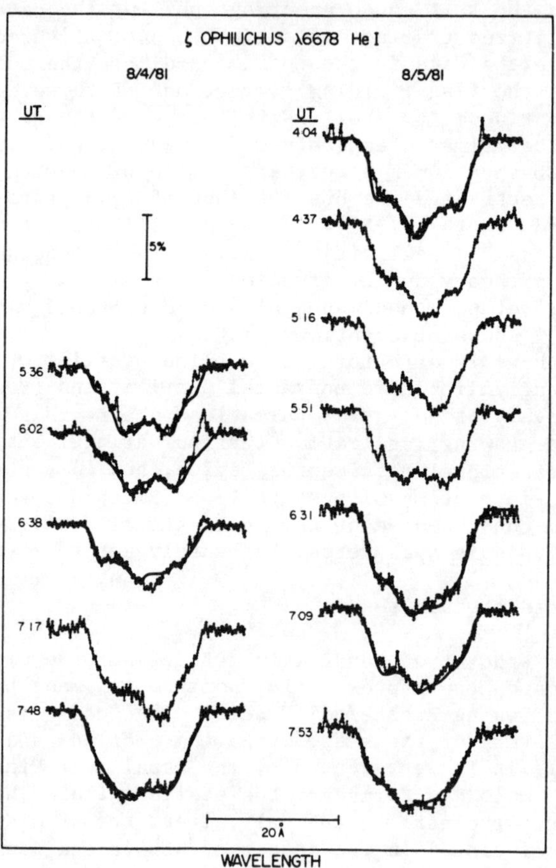

Fig. 12. Observations of the time-variable bumps and wiggles in the He I λ6678 profile of ζ Oph (O9.5Ve), which are interpreted as nonradial oscillations of high-order. The strong solid lines are model fits to the observations with $\ell = 8$ and $m = -8$ (from Vogt and Penrod 1983).

Wind blanketing essentially scales up the temperature of the surface layers uniformly, while preserving the overall shape of the run of temperature with depth. Thus, the spectrum of a wind-blanketed model is virtually identical to that of an unblanketed model at a higher T_{eff} (see also Figure 14 below). In each case, the spectral lines, especially those of He I, give a true measurement of the stellar surface temperature. The theoretical challenge is to correctly relate the surface temperature to the effective temperature. For the examples of Figure 13, the neglect of mass loss leads to errors in the derived T_{eff} as large as 10,000°K.

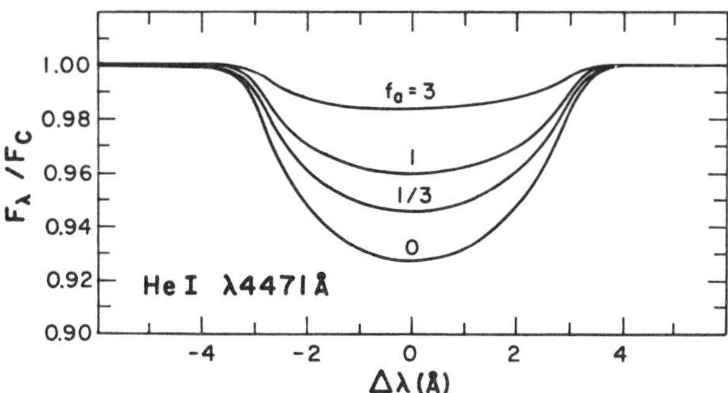

Fig. 13. Examples of the effect of wind-blanketing on the emergent line profile of He I λ4471 Å. The models all have T_{eff} = 42,000°K, log g = 3.5, [He] = 0.14, and a rotational broadening of 210 km s^{-1}. The mass loss rate of the models is parameterized by its ratio, f_A, to that observed in ζ Pup ($\dot{M} = 5 \times 10^{-6}$ $M_\odot$ yr^{-1}), where f_A = 0 is an unblanketed model (from Abbott and Hummer 1984).

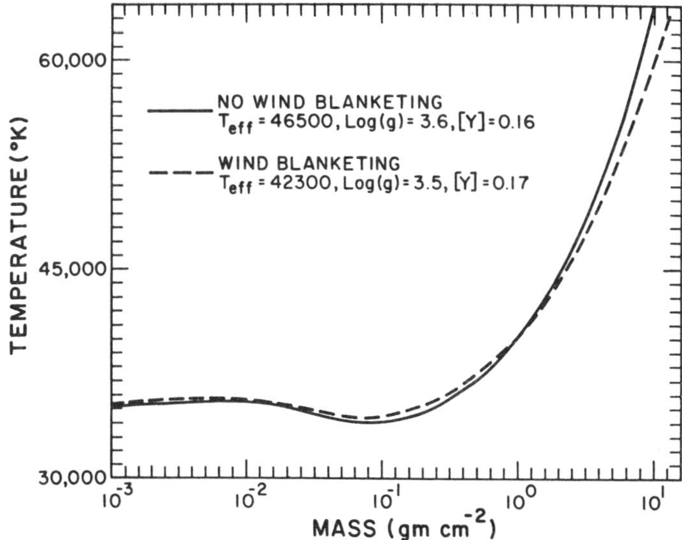

Fig. 14. The run of temperature with depth for two atmosphere models, with and without wind-blanketing, which both fit the observed photospheric spectrum of ζ Pup. The atmospheric structures are so similar that the line profiles are virtually identical. The error in the derived T_{eff} using unblanketed models is roughly 4000°K (from Hummer, Abbott and Bohannan 1984).

3.3. A Comparison of Theory and Observation for the Example of Zeta Puppis

As a first application of these model improvements, Hummer, Abbott and Bohannan (1984, and in preparation) compare the optical CCD observations of ζ Pup for 10 lines of H and He to model atmospheres, with and without wind blanketing. The comparison with additional lines, beyond the primary ones of Hγ, He I 4471, and He II 4542, reduces errors due to blends and continuum fitting, and more importantly, provides a check on the internal consistency of the models. The temperature structure of the best fit wind-blanketed and unblanketed models are shown in Figure 14. They are almost indistinguishable, except for a slightly steeper temperature gradient in the unblanketed model, which gives slightly deeper line cores and slightly steeper wings in the profiles. For ζ Pup, the effect of wind blanketing lowers the derived value of T_{eff} by about 4000°K.

Both models fit the observations to an average accuracy of about 1%. Table 1 summarizes the comparison for the individual lines. Also given are the photographic observations of Kudritzki, Simon and Hamann (1983), which were made by superimposing four individual exposures on IIIaJ plates. Examples of the actual profile fits are shown in Figure 15 for the wind-blanketed model with a rotational broadening of 210 km s^{-1}. Interestingly, the parameters of the best fit, wind-blanketed model are essentially the same as those derived by Kudritzki _et al._ from their best unblanketed model as fitted to their photographic data. In quite a coincidence, the differences between the CCD and photographic observations exactly canceled the 4000°K difference from wind blanketing.

Three conclusions are evident from this preliminary comparison:

1) The overall agreement at the 1% level between observations and model atmospheres, which are based on calculations from first principles, strongly supports the validity of a photospheric analysis for deriving fundamental stellar parameters.

2) The presence of wind blanketing cannot be directly verified from observations of optical photospheric lines, because the profiles are too similar to those of unblanketed models scaled to a higher temperature.

3) While the discrepancies in the fits are small, they exceed the observational uncertainty. These real differences imply either that the models must improve their accuracy, or that other physical processes need to be included, such as pulsation or line-blanketing in the photosphere.

3.4. Conclusions

Observations of early-type stars with electronic detectors with high signal-to-noise and high resolution are changing our ideas about the pulsational properties of these stars, and about the calibration of spectral type with T_{eff}, log g, and now $\dot{M}$. From a case study of the prototype star ζ Pup, I conclude that a "photospheric" analysis to derive T_{eff}, log g, and abundance is still legitimate, subject to the following three conditions:

Table 1. Comparison of Theory and Observations for Zeta Puppis

Line	Transition	Equivalent Width (Å)				Profiles: <Model-Obs>[a]		Standard Deviation	
		CCD	Photog[b]	Unblan.	Blan.	Unblan.	Blan.	Unblan.	Blan.
Hγ + He II	(4-10)	1.75	1.94	1.79	1.71	0.2%	0.8%	0.7%	1.1%
Hε + He II	(4-14)	1.30[c]	--	1.54	1.46	-1.0	-0.5	1.5	0.9
He I 5876	$2p^3P^o-3d^3D$	0.58	--	0.47	0.46	-0.3	-0.3	1.0	1.0
He I 4471	$2p^3P^o-4d^3D$	0.20	0.25	0.20	0.21	0.2	0.1	0.8	0.9
He II 5412	4-7	1.05	--	1.06	0.98	0.4	1.0	1.7	2.1
He II 4542	4-9	0.76	0.90	0.80	0.75	-0.1	0.4	1.1	1.6
He II 4200	4-11	0.50	0.73	0.68	0.62	-1.0	-0.5	1.3	0.7
He II 4026	4-13	0.45	0.65	0.66	0.49	-1.6	-0.2	1.8	0.7
He II 3924	4-15	0.37	0.40	0.42	0.37	-0.1	0.3	0.4	0.4
He II 6683	5-13	0.61	--	0.75	--	-0.4	--	0.6	--

[a]Value of $(F_\nu/F_c)_{model} - (F_\nu/F_c)_{observed}$ averaged across the line profile.

[b]Kudritzki *et al.* (1983).

[c]Provisional.

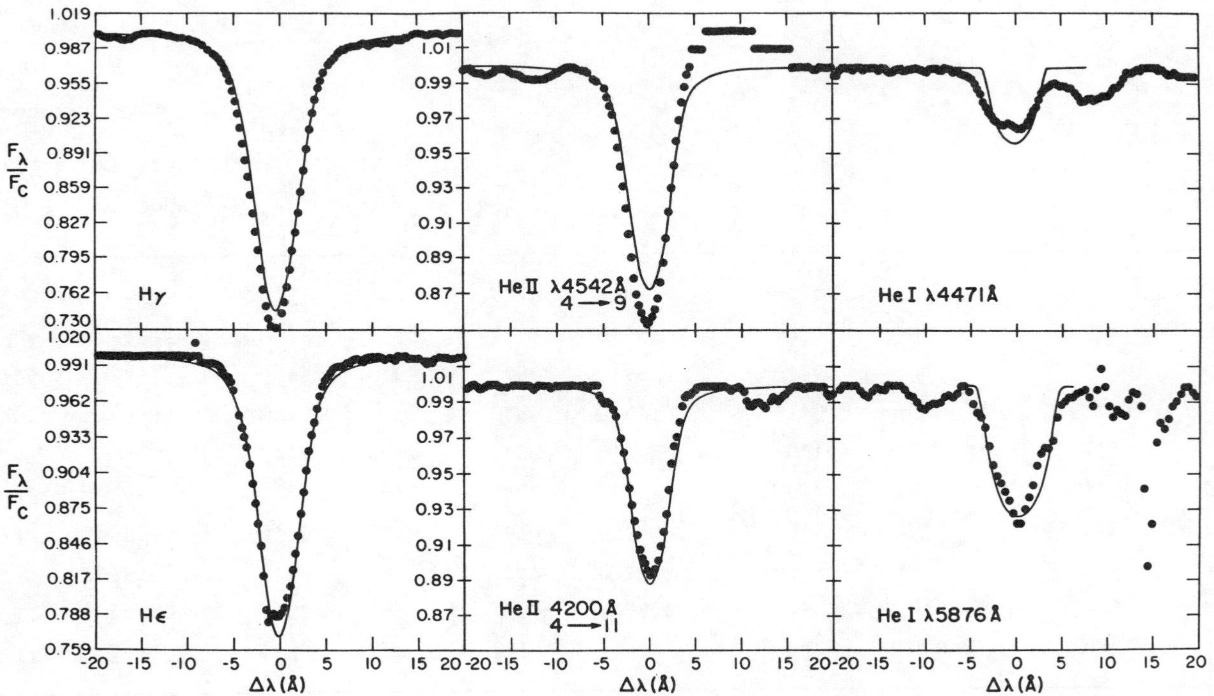

Fig. 15. Comparison of line profiles from the wind-blanketed model atmosphere of Figure 14 (solid line) with CCD observations of ζ Pup (filled circles). The overall accuracy of the fits is ~1%, although real discrepancies exist. Fits of comparable accuracy are also possible with unblanketed atmospheres with higher T_{eff}. Top and bottom panels show that discrepancies between fits to similar lines are often of the opposite sense, so it is valuable to have redundant lines for each stellar parameter being diagnosed (from Hummer, Abbott and Bohannan 1984).

1) The observations must be of sufficient precision and resolution to allow the determination of T_{eff} and log g independently from fits to line profiles.

2) The effects of mass loss must be accounted for in stars with strong winds. (This probably applies to any star with measurable emission at Hα or He II 4686.)

3) There is no other physical process which significantly alters the relationship between surface temperature and effective temperature.

ACKNOWLEDGMENTS

I thank Prof. M. J. Seaton for his hospitality at UCL, during which these remarks were prepared, and Drs. Garmany, Henrichs, and Lamers for useful discussions and their help in gathering together the illustrations. This research was supported by a fellowship from the SERC of the United Kingdom and by National Science Foundation grant AST-821835 through the University of Colorado.

REFERENCES

Abbott, D. C., Bohlin, R. C., and Savage, B. D. 1982, Ap. J. Suppl., **48**, 369.

Abbott, D. C. and Hummer, D. G. 1984, submitted to Ap. J.

Abbott, D. C. and Lucy, L. B. 1985, Ap. J., in press.

Auer, L. H. and Mihalas, D. 1969, Ap. J., **158**, 641.

_____. 1972, Ap. J. Suppl., **25**, 193.

_____. 1973, Ap. J. Suppl., **25**, 433.

Barker, P. K. and Marlborough, J. M. 1984, preprint.

Cassinelli, J. P. and Abbott, D. C. 1981, in The Universe at Ultraviolet Wavelengths, ed. R. D. Chapman (NASA CP-2171), p. 127.

Castor, J. I. 1970, M.N.R.A.S., **149**, 111.

Castor, J. I., Abbott, D. C., and Klein, R. I. 1975, Ap. J., **195**, 157.

Castor, J. I. and Lamers, H. J. G. L. M. 1979, Ap. J. Suppl., **39**, 481.

Conti, P. S. 1975, in H II Regions and Related Objects, eds. T. Wilson and D. Downes (Springer-Verlag, Heidelberg), p. 207.

_____. 1978, Ann. Rev. Astr. Ap., **16**, 371.

Conti, P. S. and Frost, S. A. 1977, Ap. J., **212**, 728.

Garmany, C. D., Olson, G. L., Conti, P. S., and Van Steenberg, M. 1981, Ap. J., **250**, 660.

Gathier, R., Lamers, H. J. G. L. M., and Snow, T. P. 1981, Ap. J., **247**, 173.

Hack, M. and Struve, O. 1970, Stellar Spectroscopy: Peculiar Stars (Trieste: Obs. Astr. di Trieste).

Heasley, J. N., Wolff, S. C., and Timothy, J. G. 1982, Ap. J., **262**, 663.

Henrichs, H. 1984, to appear in the 4th European IUE Conference, Rome, May 1984.

Henrichs, H. F., Hammerschlag-Hensberge, G., Howarth, I. D., and Barr, P. 1983, Ap. J., **268**, 807.

Howarth, I. 1984, M.N.R.A.S., **206**, 625.

Hummer, D. G. 1982, Ap. J., **257**, 724.

Hummer, D. G., Abbott, D. C., and Bohannan, B. 1984, B.A.A.S., **16**, 509.

Kudritzki, R. P. 1976, Astr. Ap., **52**, 11.

Kudritzki, R. P., Simon, K. P., and Hamann, W. R. 1983, Astr. Ap., **118**, 245.

Lamers, H. J. G. L. M., Gathier, R., and Snow, T. P. 1982, Ap. J., **258**, 186.

Lamers, H. J. G. L. M., Korevaar, P., and Cassatella, A. 1984, preprint.

Lucy, L. B. 1971, Ap. J., **163**, 95.

______. 1982, Ap. J., **255**, 278.

______. 1983, Ap. J., **274**, 372.

Mihalas, D. 1978, Stellar Atmospheres, 2nd Ed. (San Francisco: Freeman).

Morton, D. C. 1976, Ap. J., **203**, 386.

Prinja, R. K., Henrichs, H. F., Howarth, I. D., and van der Klis, M. 1984, in Proceedings 4th European IUE Conference, ESA SP-218, p. 319.

Prinja, R. K. and Howarth, I. D. 1984, submitted to Astr. Ap.

Smith, M. A. and Penrod, G. D. 1984, to appear in the Third Trieste Workshop on "Relations Between Chromospheres-Coronae and Mass Loss in Stars," Sac Peak, NM, 18-25 Aug. 1984, eds. J. Zirker and R. Stalio.

Snow, T. P. 1977, Ap. J., **217**, 760.

Snow, T. P. and Morton, D. C. 1976, Ap. J. Suppl., **32**, 429.

Vogt, S. and Penrod, G. D. 1983, Ap. J., **275**, 661.

Vreux, J. M. 1984, submitted to Astr. Ap.

STELLAR SURFACE INHOMOGENEITIES AND THE INTERPRETATION OF STELLAR SPECTRA

Mark S. Giampapa
National Solar Observatory
National Optical Astronomy Observatories*
P. O. Box 26732
Tucson, Arizona 85726 USA

ABSTRACT. I discuss manifestations of stellar surface thermal inhomogeneities in stellar spectra. I illustrate by example the effects of multi-components in stellar atmospheres on the interpretation of line diagnostics and single-component models as well as on the treatment of line transfer problems. The examples offered involve metal abundance determinations, chromospheric line diagnostics, the realistic representation of pre-main sequence atmospheres and stellar magnetic fields.

1. INTRODUCTION

The interpretation of stellar spectra through the application of the techniques embodied in radiative transfer theory has traditionally relied upon the assumptions of lateral homogeneity and a plane parallel atmosphere. A further refinement occurred with the treatment of spherically expanding atmospheres although the assumption of homogeneity (i.e., spherical symmetry) remained. However, solar investigations and related stellar studies have clearly shown that inhomogeneities are a fundamental property of stellar atmospheres. The spatial structuring of the outer atmospheres of the stars is, in turn, defined by emergent and evolving magnetic field configurations. As a result, one dimensional semi-empirical model atmospheres can, at best, offer a physical picture of the surface-averaged properties of a star at the time of observations. Furthermore, stellar surface inhomogeneities can introduce ambiguities into the interpretation of stellar spectra that can belie fundamental stellar properties, such as metal abundances, if approached in a naive manner. Conversely, the subtle signatures of surface features in spectra offer the opportunity to directly study stellar surface structure. The manifestations of stellar surface inhomogeneities and their consequences for the

*Operated by the Association of Universities for Research in Astronomy, Inc., under contract with the National Science Foundation.

J. E. Beckman and L. Crivellari (eds.), Progress in Stellar Spectral Line Formation Theory, 305–318.

development of model atmospheres are discussed in the following. In particular, I will be concerned with the effects of inhomogeneities on abundance determinations, the interpretation of chromospheric line diagnostics, and single-component versus mutli-component model approaches and their associated implications for the treatment of line formation problems.

2. ABUNDANCE DETERMINATIONS

An important input to any model atmospheres computation is the metals abundance. For example, the development of accurate models of the temperature minimum region in stellar chromospheres requires a knowledge of the metallicity since the dominant contribution to the local electron density in this region arises from ionization of neutral metal species.

The potential effects of chromospheric activity on the apparent metallicity as deduced from multi-color photometry was investigated by Giampapa, Worden and Gilliam (1979). These investigators noted that differential Strömgren uvbyß photometry of solar active (plage) and quiet comparison regions yielded an apparent difference in metallicity between these regions with the solar plage appearing metal deficient with respect to comparison quiet regions. This effect was attributed to the filling in of the line cores of neutral metal species as a result of the presence of enhanced chromospheric heating in plage rather than to any real difference in metallicity between plage and the quiet Sun.

In a related investigation, Giampapa (1984a) studied the Li I λ6707 resonance line as it appears in the quiet Sun, solar plage and sunspots. This spectral feature is utilized to infer the abundance of the element lithium which is, in turn, a crucial diagnostic for investigations of stellar interiors, stellar ages and nucleosynthesis in the cosmos. Giampapa (1984a) observed that the equivalent width of the λ6707 was reduced (relative to the quiet Sun) in plage. The relative weakening of the lithium resonance doublet can be attributed to the increased ionization of neutral lithium in the presence of the enhanced nonradiative heating that characterize solar plages. This conclusion is corraborated by schematic model calculations of the depth dependence of the lithium ionization fraction in models of active network elements given by Vernazza, Avrett and Loeser (1981). Further-more, the Li I λ6707 resonance line is strongly enhanced in sunspots due to their relatively cool temperatures.

The aforementioned investigation of the Li I λ6707 line formation problem clearly implies that the nature and filling factor of stellar surface features analogous to solar plages and sunspots can substantially alter the equivalent width of this feature in spatially unresolved stellar spectra. In general, the observed relative change in a line equivalent width is given by (Giampapa 1984a)

$$\frac{\Delta W}{Wq} = f_p \frac{\Delta W \text{ plage}}{Wq} + \alpha f_s \frac{\Delta W \text{ spot}}{Wq},$$

where f corresponds to the active (plage) region filling factor, α is a correction factor that accounts for the fainter continuum of a stellar spot, and Wq is the line equivalent width in the quiet stellar photosphere. In summary, the presence of stellar surface features can alter the strengths of line diagnostics and thus belie the metal abundances that are ultimately derived.

3. CHROMOSPHERIC LINE DIAGNOSTICS

3.1 The Ca II Resonance Lines

The resonance lines of Ca II, known as the H and K lines, are notable as diagnostics of stellar chromospheric properties as well as for their accessibility to earth-based observation. The occurrence of chromospheric H and K core emission in spectral types later than F5 V (Wilson 1973) provides a valuable tool with which to study analogues of the solar chromosphere and the chromospheres of cool stars.

During approximately the past 10 years extensive observational and theoretical effort has been devoted to the development of semi-empirical stellar model chromospheres based on observed, high spectral resolution, flux-calibrated Ca II resonance line profiles (see the review by Linsky 1980). Of course, the stellar disk is unresolved and we therefore observe an integrated Ca II line profile. The semi-empirical models based on these profiles consequently reflect properties of the stellar chromosphere that have been spatially averaged in some unknown fashion. Hence we must ask the following questions: What do single-component, semi-empirical models based on the Ca II lines physically represent? What kind of regions are the dominant contributors to the observed profiles?

I present in Figure 1 the core of the integrated *solar* Ca II line as obtained by W. C. Livingston during the course of his solar Ca II monitoring program at the National Solar Observatory in Tucson, Arizona. The emission cores are weak compared to other active chromosphere, solar-type stars (e.g. Linsky *et al.* 1979). A single-component, semi-empirical model chromosphere designed to best fit this feature would be characterized by relatively low densities at chromospheric temperatures. However, I display in Figure 2(a)-(c) the Ca II K line profiles observed in solar plage during the course of my (previously cited) investigation of the appearance of the Li I λ6707 lines in various solar active and quiet regions. These profiles are strikingly different from each other (and the profile in Fig. 1) both in shape and intensity. Yet they all contribute to the observed, spatially-averaged solar K line profile given in Fig. 1. Therefore, what does a model based on the integrated Ca II K line really represent? Clearly, surface-averaged, single-component, semi-empirical

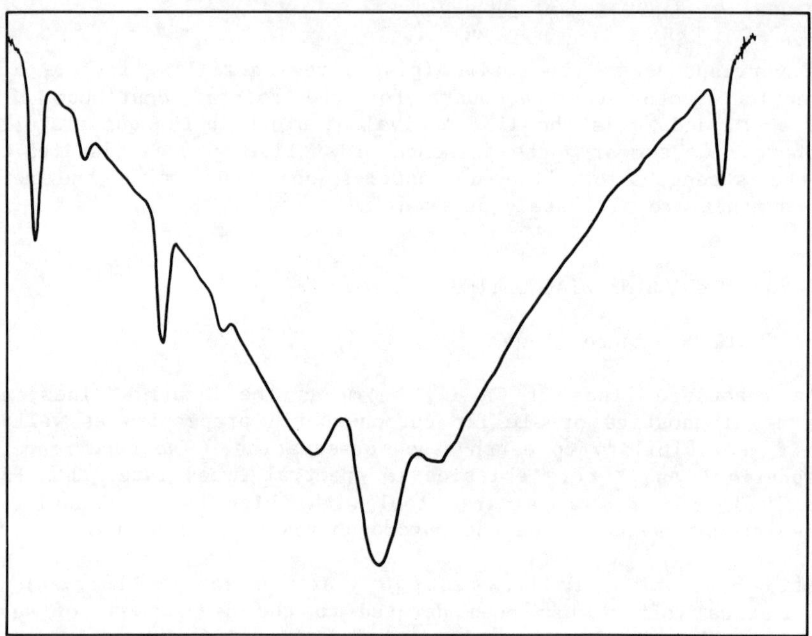

Figure 1. The core of the solar Ca II line as seen in integrated sunlight. Courtesy of W. C. Livingston, National Solar Observatory.

models, by their theoretical and observational nature, do not reflect the <u>range</u> of possible atmospheric structures that may be present on stellar surfaces. The integrated solar Ca II K line itself likely arises predominately from network and cell elements (Skumanich <u>et al.</u> 1984).

3.2 The Hα Line

Chromospheric models have traditionally been based on observed Ca II resonance line emission cores and wings. However, other important chromospheric line diagnostics are present in the optical but they are not obvious because they appear in <u>absorption</u> rather than emission. One such feature is the Hα line.

3.2.1. The Hα line in solar-type stars. The Hα line appears as a strong absorption feature in the spectra of solar-type stars. This line has been regarded as a purely photospheric diagnostic. However, a recent study by Zarro (1983) disclosed the presence of residual Hα

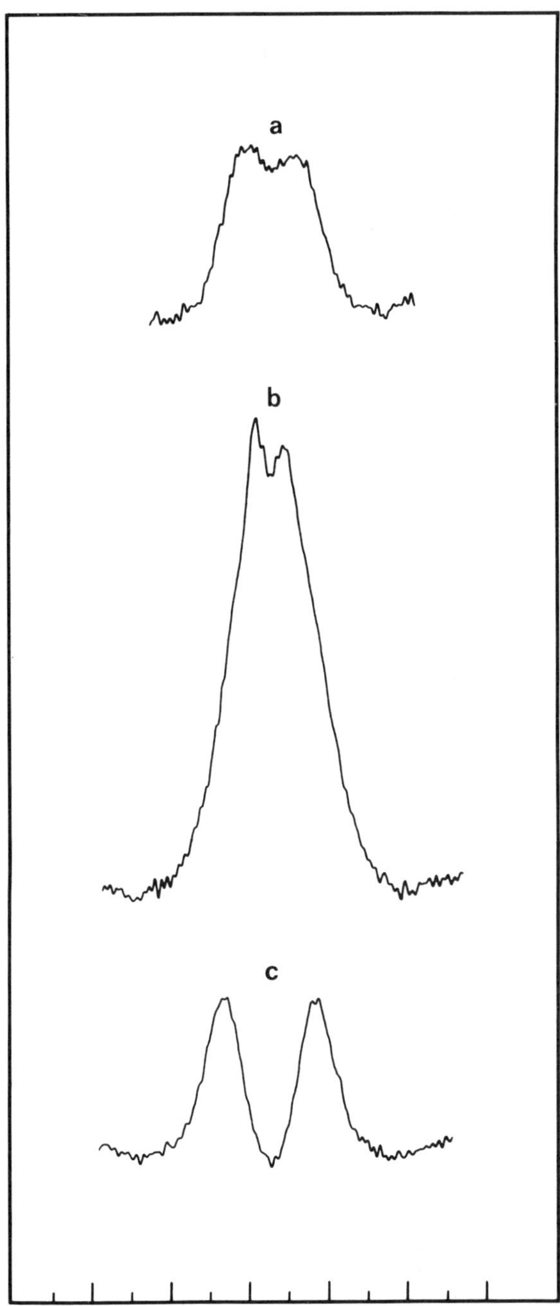

Figure 2. The core of the Ca II K line as observed in three randomly selected solar plages.

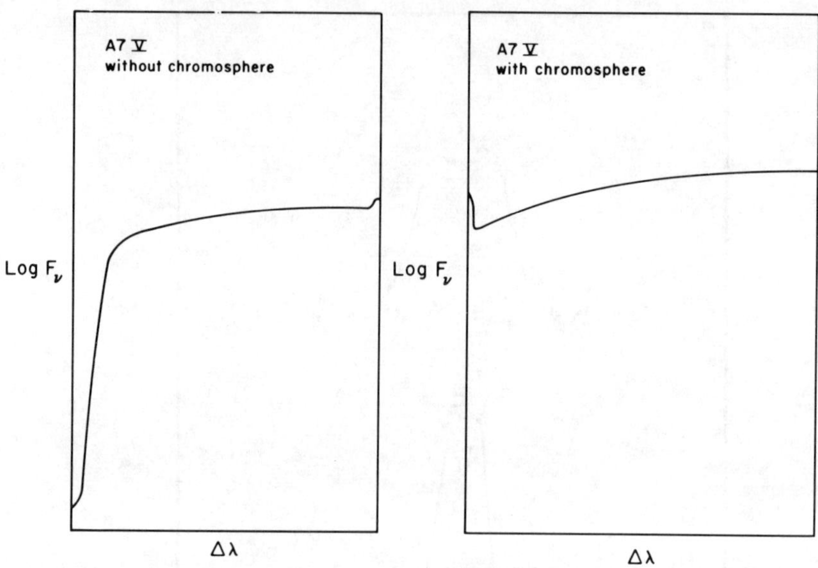

Figure 3a. See caption for Fig. 3b.

chromospheric emission in the Hα absorption cores of active-chromosphere, solar-type stars. The enhanced chromospheric densities in active regions on the stellar surface lead to coupling of the Hα source function to the local chromospheric temperature rise in these regions. As a result, an Hα emission component contributes to the observed absorption profile by slightly filling in the core.

3.2.2. The Hα line in main sequence A stars. The applicability of Hα as a potential chromospheric diagnostic in the main sequence A stars invites both theoretical and observational exploration. Previous attempts to discern the presence of a chromosphere in the A V stars utilized observations of the Mg II or Ca II resonance lines. Of course, these are the most difficult diagnostics to use since they occur near the spectral energy maximum of an A star. The contrast against the background photosphere is therefore highly unfavorable. However, the Hα line is located in the red where the contrast is substantially improved.

I display in Fig. 3(a)-(b) the results of model computations of the Hα line as it appears in two A stars with and without chromospheres. The chromospheric model consists of an *ad hoc* temperature rise similar in form (i.e., in dT/d log m, where m is the column mass density) to a solar plage adjoined to a photospheric model taken from Kurucz (1979). The model computations were performed in collaboration with E. H. Avrett using the PANDORA code. The Hα line transfer consists of a NLTE, CRD treatment involving an 8-level + continuum atomic model. The profile is symmetric and so only the half profiles are displayed.

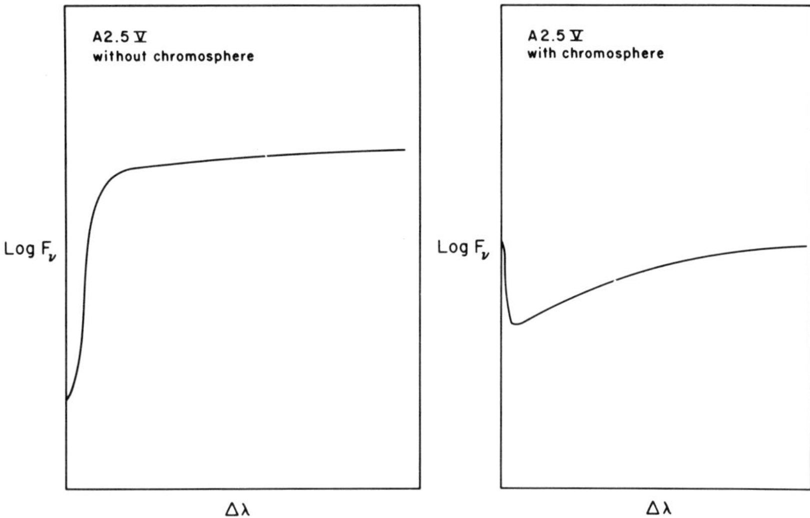

Figure 3b. Computed Hα half-profiles with and without a chromospheric temperature rise adjoined to A V star model photospheres.

The model results in Fig. 3(a)-(b) show the clear presence of Hα core emission and the associated modification of the line wings as compared with the purely photospheric profiles. Of course, rotational broadening would smear out any core emission. Nevertheless, the core would be filled in beyond that which could normally be accounted for by rotational broadening alone.

In brief summary, the aforementioned considerations imply the following:

(a) Any study of the line formation of absorption features, such as Hα or the lines of neutral metal species, must be cognizant of the likely presence of residual core emission components that can arise from stellar surface "active" regions.

(b) Theoretical models should be constructed in which positive temperature gradients are superimposed on model photospheres. Potential line diagnostics of these chromospheric-like thermal structures (e.g., Hα, He I, etc.) should then be calculated in order to predict their appearance in stellar spectra.

3.2.3. Hα in the M dwarf stars. The appearance of Hα in emission in the cool M dwarf stars has been recognized as an indicator of the presence of a relatively dense ($n_e \sim 10^{11}$ cm^{-3}) chromosphere (Fosbury

1974). However, the appearance of a Hα absorption line is also a result of chromospheric heating in the atmospheres of these stars (Cram and Mullan 1979). Recently, Young *et al.* (1984) noted the appearance of Hα absorption in some M dwarfs that was of an intermediate strength. This result can be attributed to the contribution of Hα emitting regions in combination with the absorbing regions on the stellar surface. A naive extrapolation of this result would suggest that the observed absence of any Hα feature indicates the existence of an average chromospheric structure that is intermediate in "strength" between those M dwarfs that exhibit strong Hα absorption and those that show Hα emission.

Is this a correct interpretation? The answer has important ramifications for the understanding of dynamos in the fully convective, very late M dwarf stars (Giampapa and Liebert 1984). Useful insights can be attained by examining the Hα line source function from a two-level + continuum model atom in a region where the Lyman lines are in detailed balance or (Thomas 1957)

$$S_L = \frac{\int \phi_\nu J_\nu d\nu + \varepsilon B_\nu + \eta B^*}{1 + \varepsilon + \eta}.$$

Adopting expressions for the source-sink terms given by Cram and Mullan (1979) yields the results given in Fig. 4. These results demonstrate that the Hα line is photoionization dominated in the very late M dwarfs at electron densities $n_e < 10^8\ cm^{-3}$. The results of single-component, semi-empirical model chromospheres of M dwarf stars, however, reveal that chromospheric electron densities are typically $n_e \geqslant 10^{10}\ cm^{-3}$ (Giampapa, Worden and Linsky 1982). Hence the Hα line is inevitably collision dominated at plausible chromospheric densities and would therefore appear only in emission, or not at all, in the very late, cool M dwarf stars. Thus the absence of a Hα feature in the very cool (i.e., later than dM 5.5) dwarfs means exactly that -- it's not there!

The aforementioned considerations serve as an illustrative example of the utility of single-component models for deciding on the plausibility of the occurrence of specific atmospheric properties (e.g., the range of reasonable electron densities) in stars of various spectral types.

4. MULTI-COMPONENT ATMOSPHERES

4.1 The T Tauri Stars

An interesting example of the implications of a single-component versus a multi-component representation of a stellar atmosphere involves the pre-main sequence T Tauri stars. These pre-cursors of late-type, main sequence stars exhibit a variety of spectral characteristics (e.g., see Giampapa *et al.* 1981). The investigation of the

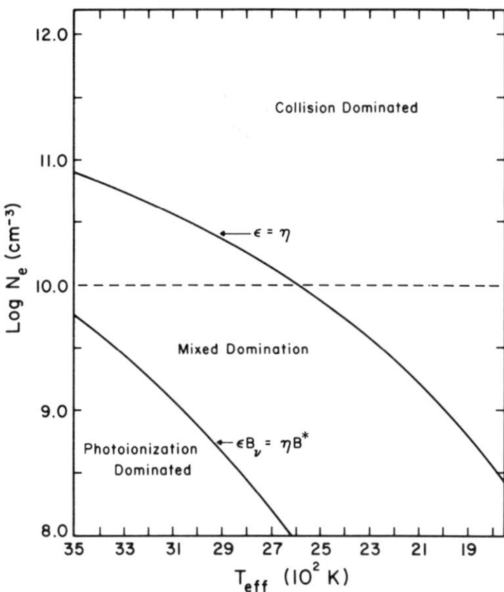

Figure 4. Election density versus photospheric effective temperature for M dwarf stars. The dashed line indicates a typical electron density for the relatively weaker chromospheres of the non-dMe stars.

properties of the T Tauri stars has historically involved a controversy between "compact chromospheres", analogous to an enhanced solar chromosphere, versus an "extended warm region" to account for the observed line spectrum within the context of a single-component model approach. Of course, the adopted approaches have determined the treatment of the line formation problem in these stars.

A single-component schematic representation of a T Tauri atmosphere, as offered by Bertout (1984), is shown in Figure 5. This model can account for the observed spectral characteristics of many known T Tauri stars. But is this single-component model physically realistic? I present an alternative two-component, schematic model in Figure 6. This model consists of a compact chromospheric and coronal region, analogous to magnetic "closed regions " on the Sun, combined with an "open region", analogous to a solar coronal hole, characterized by outflow and extended plasma at temperatures $T \sim 10^5$ K. This latter component is the site of a cool wind where the plasma never attains coronal temperatures (i.e. $T \sim 10^6$K). The evidence that supports this interpretation is extensively discussed by Giampapa (1984b).

The physics of the representations in Figures 5 and 6 is quite different. The single-component approach (Fig. 5) implies a "Parker-type" thermally driven wind while the two-component model (Fig. 6)

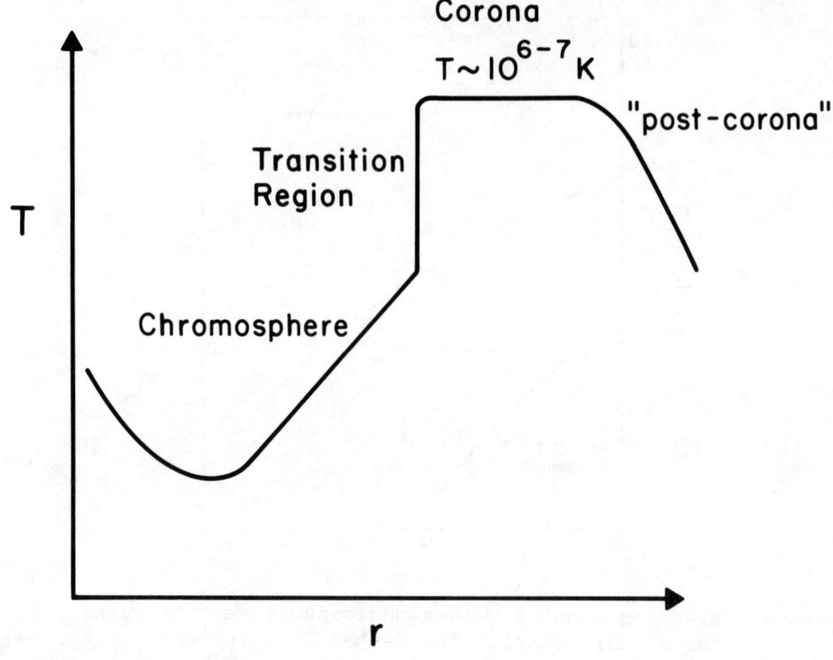

Figure 5. A single-component schematic model of T Tauri atmospheres.

must utilize an Alfvén wave-driven wind mechanism to account for wind temperatures that never attain coronal values. This latter model requires large turbulent velocities that are of the same order as the expansion velocity thus rendering the Sobolev approximation invalid. Consequently, the computed Balmer line profiles are broad, as are the observed profiles, and their widths can be reproduced at significantly smaller mass loss rates than those previously inferred (Hartmann, Edwards and Avrett 1982).

4.2 Magnetic Fields

The investigation of the Sun has clearly revealed that inhomogeneities as defined by magnetic field structures are a fundamental property of the solar atmosphere. Both indirect and some direct evidence suggests that this is true of stellar atmospheres as well. Indeed, the discussion I have given in this paper has mainly been concerned with thermal inhomogeneities on stellar surfaces. These inhomogeneities are presumably associated with, and defined by, emergent and evolving magnetic field configurations.

Given the crucial role of magnetic fields in producing stellar atmospheric structure, it becomes important to measure the range of

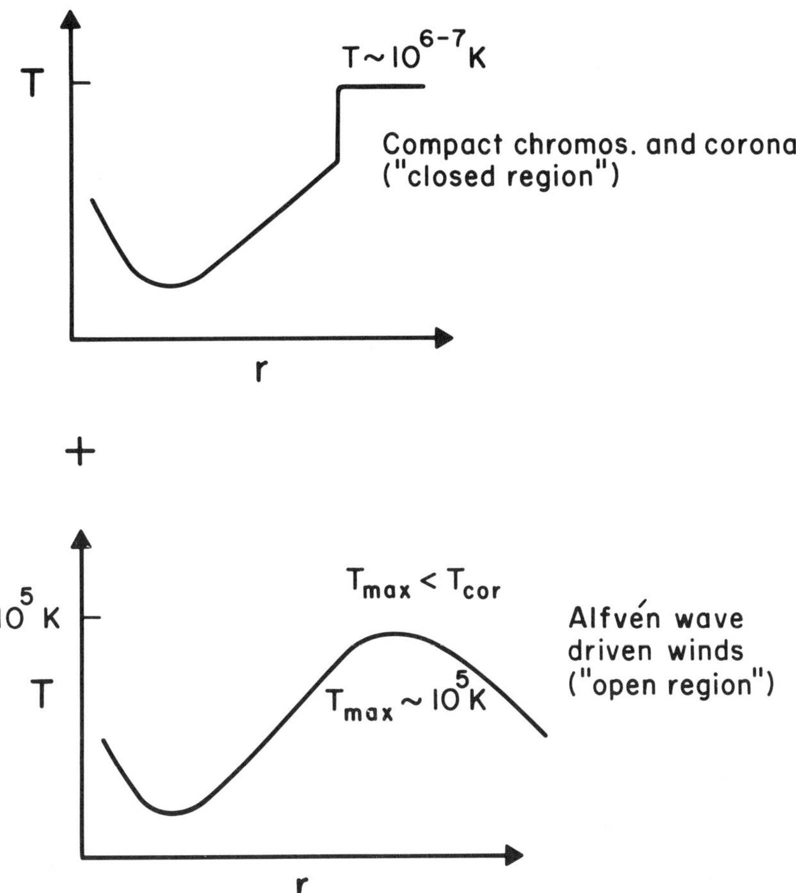

Figure 6. An alternative multicomponent representation of T Tauri atmospheres.

magnetic field strengths and area coverages (or "filling factors") that can occur on stellar surfaces for stars throughout the H-R diagram. The attempts to measure these quantities have recently been reviewed by Giampapa (1984c). Thus far, the explicit treatment of radiative transfer in areas of significant magnetic flux, and possibly very different thermal structure from the quiet photosphere, has been unjustifiably neglected. A single-component approach has thus far been adopted, in the interest of simplicity, to infer magnetic field strengths and filling factors. However, as noted by Giampapa (1984c), the determmination of these quantities is in reality wavelength dependent. This is due to the cooler temperatures of spots, which constitute significant concentrations of magnetic flux, and which will contribute to observed Zeeman line broadening according to the wavelength region observed. For example, the "magnetic signature" of cool spots will be more important in the infrared than in the optical.

In brief summary, the further refinement of these initial attempts to measure magnetic flux on stars must involve an explicit treatment of the radiative transfer in the σ and π components of a simple triplet. The theoretical approach must take into account the differing depths of formation of the components, the modification of the hydrostatic balance by magnetic field gradients, and the potentially different thermal structures of the field and field-free regions on the stellar surface.

4.3 Temporal "Inhomogeneities"

I have concentrated on the <u>spatial</u> structuring of stellar atmospheres that is, in turn, related to magnetic field configurations and manifested in stellar line spectra. However, stars can exhibit "temporal structure" as well. This structure must be considered in the interpretation of stellar line profiles. A vivid example of a time-dependent line profile in the B0.5 V star ε Per is given in Figure 7. This behavior is attributed to the occurrence of non-radial pulsations (Smith 1984). Obviously, this kind of phenomenon could only be observed with high spectral and high temporal resolution.

I note, parenthetically, that ε Per is sufficiently bright visually that previous observations consisted of nightly "snapshots". An intercomparison of the resulting spectra lead investigators to erroneaously conclude that ε Per must be a spectroscopic binary! Moreover, the star was included as a photometric calibration standard for the <u>International Ultraviolet Explorer</u> (IUE) satellite until recent EUV photometry obtained with the Voyager spacecraft disclosed the occurrence of 50% flux variations in this spectral range!

5. SUMMARY

The spatial structuring of stellar atmospheres by emergent and evolving magnetic fields defines thermal inhomogeneities that are, in turn, manifested in stellar spectra. The proper interpretation of stellar spectra must therefore involve a multi-component model approach. Furthermore, the realistic treatment of line transfer in stellar atmospheres will require input from the following kinds of observations:

(a) High spectral resolution
(b) Multi-spectral
(c) High temporal resolution
(d) Monitoring of representative objects within a class

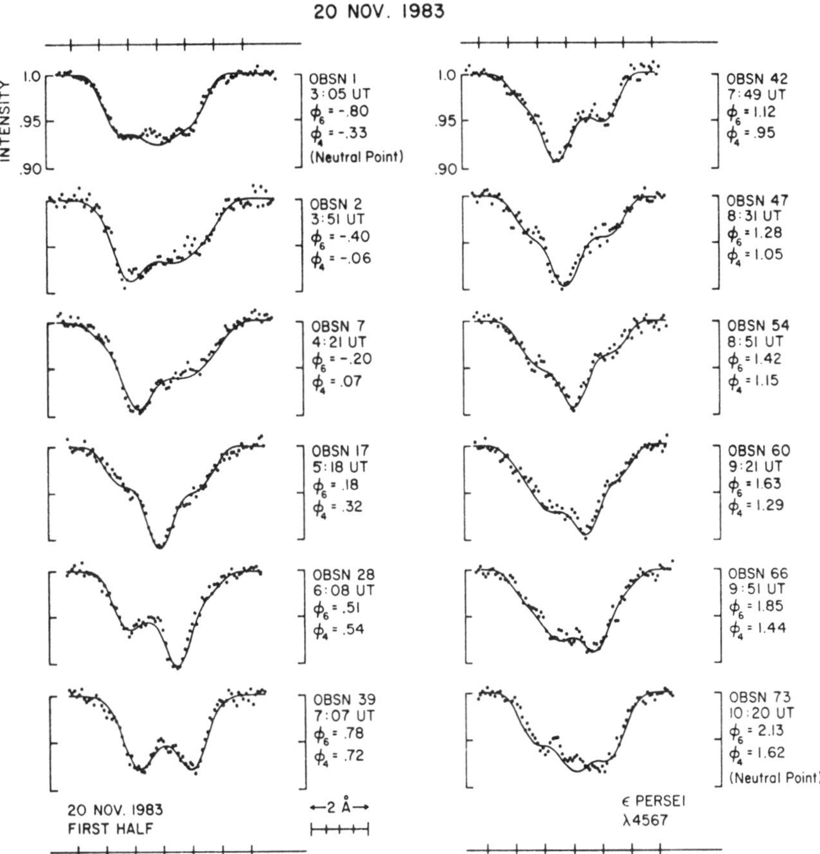

Figure 7. High spectral resolution, time-resolved observations of the pulsating B star ε Persei. Courtesy of M. A. Smith, National Solar Observatory.

The key question to address upon obtaining the aforementioned observations is 'What is the best way to decompose line profiles into the various components that arise from different regions on a star and combine to form the spatially unresolved spectra we actually observe?'

The relevant theoretical questions include the following:

(a) What is the nature of the interaction (if any) between magnetic field and field-free regions in a stellar atmosphere (i.e., are closed field regions constitute "brick walls")?

(b) How do we incorporate magnetic fields and filling factors into model atmosphere calculations?

(c) How realistic are single-component models? What do they physically represent?

The answers to these questions will yield exciting, new perspectives on radiative transfer theory and the structure of stellar atmospheres.

I greatfully acknowledge the support allocated by the North Atlantic Treaty Organization for the Workshop. I am especially grateful to Dr. L. Crivellari and Dr. J. Beckman.

REFERENCES

Bertout, C. 1984, Rept. Prog. Phys., in press.
Cram, L. E. and Mullan, D. J. 1979, Ap. J., **234**, 579.
Fosbury, R. A. E. 1974, M.N.R.A.S., **169**, 147.
Giampapa, M. S., Worden, S. P. and Gilliam, L. B. 1979, Ap. J., **229**, 1143.
Giampapa, M. S., Calvet, N. Imhoff, C. L. and Kuhi, L. V. 1981, Ap. J., **251**, 113.
Giampapa, M. S., Worden, S. P. and Linsky, J. L. 1982, Ap. J., **258**, 740.
Giampapa, M. S. 1984a, Ap. J., **277**, 235.
Giampapa, M. S. 1984b, "Results from Ultraviolet Observations of T Tauri Stars", in The Third Cambridge Workshop on Cool Stars, Stellar Systems and the Sun: Evolution and Structure of Cool Stars, eds. S. L. Baliunas and L. Hartmann, (Berlin-Heidelberg: Springer-Verlag), p.14.
Giampapa, M. S. 1984c, "Direct and Indirect Measurements of Stellar Magnetic Fields", in the Proceedings of the Workshop on Space Research Prospects in Stellar Activity and Stellar Variability, eds. F. Praderie and A. Mangeney, Observatoire de Paris-Meudon, p.309.
Giampapa, M. S. and Liebert, J. W. 1984, Ap. J., submitted.
Hartmann, L. W., Edwards, S. and Avrett, E. H. 1982, Ap. J., **261**, 279.
Kurucz, R. L. 1979, Ap. J. Suppl., **40**, 1.
Linsky, J. L., Worden, S. P., McClintock, W. and Robertson, R. M. 1979, Ap. J. Suppl., **41**, 47.
Linsky, J. L. 1980, Ann. Rev. Astr. Ap., **18**, 439.
Skumanich, A., Lean, J. L., White, O. R. and Livingston, W. C. 1984, Ap. J., **282**, 776.
Smith, M. A. 1984, Ap. J., submitted.
Thomas, R. N. 1957, Ap. J., **125**, 260.
Vernazza, J. E., Avrett, E. H. and Loeser, R. 1981, Ap. J. Suppl., **45**, 635.
Wilson, O. C. 1973, in Stellar Chromospheres, IAU Colloq. No. 19, eds. S. D. Jordan and E. H. Avrett (Washington D. C.: NASA SP-317), p.305.
Young, A., Skumanich, A. and Harlan, E. 1984, Ap. J., **282**, 683.
Zarro, D. M. 1983, Ap J. (Letters), **267**, L61.

THE THEORY OF LINE TRANSFER IN EXPANDING ATMOSPHERES

P. B. Kunasz
Private Consultant, Boulder, Colorado
Boulder, Colorado U.S.A.

ABSTRACT. The observational motivation for the development of line transfer in moving atmospheres is discussed, with an emphasis on problems not solved. This is followed by a review of methods. Special attention is given to techniques developed within the past few years, and to their application and extension to problems of current astrophysical interest, with an emphasis on those involving multi-dimensional geometries, and formidable atomic coupling. Finally, some speculations on future developments are offered.

1. MOTIVATION

1.1. Astronomical Observations

Stellar line profiles broadened by macroscopic atmospheric motions were observed in supergiants at least as early as 1934 by Struve and Elvey.[1] At this time macroscopic motions were not expected, and the broadened lines were interpreted as non-thermal microturbulence with trans-sonic amplitude. As spectroscopic resolution improved, profiles orignally thought to be broadened by microturbulence alone were discovered to be the result of a spectrum of modes of organized motion. A well studied star simultaneously exhibiting many modes of photospheric motion is α Cygni,[2] while the most studied example of photospheric motion is the sun, in which amplitudes on the order of a thermal doppler width have been detected in Fe I and Fe II spectra.

The hot stars exhibit the largest Doppler shifts, which indicate the largest wind speeds. The O- and B-type supergiants are characterized by strong P Cygni profiles, while pure emission lines of unusual strength are produced by many Wolf-Rayet stars. The displacement of the violet edge of a P Cygni absorption feature bounds the zone of line formation for the corresponding transition. Many stars exhibit a progression of violet displacements in the Balmer series, which we take as evidence of an accelerating wind, with lines of different strengths being formed at different heights. The largest displacements are in ultra-violet resonance lines not observable from the Earth. Some of these indicate winds exceeding 1% the speed of light.

J. E. Beckman and L. Crivellari (eds.), Progress in Stellar Spectral Line Formation Theory, 319–333.

The range of morphology of Doppler shifted profiles is quite large. It runs from slightly asymmetric absorption lines in photospheric spectra, to strongly shifted saturated absorption in the early A-type supergiants, and from slightly filled in absorption lines formed near a variety of photospheres to the emission lines in Wolf-Rayet winds, some of which are thought to form as far out as several hundred stellar radii. The P Cygni profiles, observed most frequently in hot stars, assume a wide variety of shapes,[3] and occasionally exhibit more than one emission feature. In Be and Oe stars, we observed broadly self-reversed strong emission lines, and shallow broad emission lines with weak self reversal, respectively, both being variable in time.

The dynamic nature of the processes which cause these large scale motions is revealed by their variations. Time scales for significant changes in line profiles range from minutes to years. Some observations, such as those of Hα in α Cyg, suggest to some of those who study them,[4] blob-like concentrations of matter rising within the surrounding wind. However, it is not only the most dynamic atmospheres which require a theory of line formation which accounts for velocity fields. Organized motion as small as a tenth of a thermal Doppler unit can produce an asymmetric line profile. The presence of small scale features on the sun which produce iron lines shifted half a Doppler width suggest that most stars, if angularly resolved, would exhibit macroscopic velocity fields. Ultimately, highly resolved, time-averaged flux profiles from ordinary main sequence stars may be understood only if motion is included in the transfer calculations used for synthetic profiles. If we could analyze the photospheric Doppler shifts of a star in terms of a spectrum of oscillatory modes, as we do for the sun, we might be able to determine more about the stellar interior than mass, radius, and luminosity; hence, the importance of line formation theory in turbulent and oscillatory media, and of theory accounting for the multidimensional structures which may be found in such media.

In the zoo of stellar profile shapes, most of the simpler ones can be produced synthetically. In fact an atlas of synthetic P Cygni profiles has been compiled[5] for the detailed elucidation of the subject. In general, the Sobolev approximation[6] is valid if the line is primarily formed in the rapidly expanding outer layers of the wind, but it must be relaxed for lines formed in slowly expanding regions. In fact it has been shown that a wind amplitude of ten thermal Doppler units cannot be accurately treated with the Sobolev approximation.[7] For such winds, the entire line profile is significantly distorted by the approximation. If the motions to be modelled are complicated, yet the amplitude of the flow is small, the solution of the transfer equation is most easily found in the laboratory frame. For rapid monotonic expansion the comoving frame technique is the most useful. Both techniques have been used within their limitations to successfully fit most of the frequently observed profile shapes.

However, some line shapes continue to defy analysis. For example, consider the profile known as Beals type III,[3] which has an emission feature blueward of the blue absorption edge of a P Cygni profile. It has been speculated[8] that this profile, in ζ Pup, at λ4686, is formed by a wind in which the inner one or two stellar radii are held in co-rotation with the star by a magnetic field of >70 gauss. Observations of circular polarization in the winds of Hβ, in ζ Pup, do not disallow such a field: however, recent theoretical work[9] indicates that co-rotation is dynamically prohibited, favoring a gradually decreasing tangential velocity component.

If wind rotation must be considered, then line transfer solutions become difficult due to cross-talk between separate regions of the wind. A semi-empiracle approach to such problems may be fruitful, but first we must have the numerical tools to solve the two-level line transfer problem in a differentially rotating wind, perhaps an axially extended one. An accurate solution technique for such flows and geometries has yet to be developed.[10,11,12]

The deep absorption lines of Mg^+ in α Cygni and HR 1040[13] can be fit by a model incorporating supersonic microturbulence,[14] but this explantion may be non-unique and perhaps non-physical, the correct one still waiting to be found. In fact, this theory may have to be replaced by one resembling a recently proposed idea purporting to explain the depth of singlet and double resonance absorption components of P Cygni profiles in O-type stars.[15,16] It is assumed that a train of radiatively driven shocks provide localized deceleration regions in the wind which scatter photons on the violet side of line center, with the net effect that a greater fraction are returned to the photosphere, darkening the absorption line. This model also provides a source of heating to explain the X-ray fluxes observed in early type stars with the Einstein satellite. A rough calculation using the Sobolev approximation confirmed the explanation, but a solid comparison with observational data, and predictions for coupled subordinate lines must await a line transfer solution in a non-monotonic spherical flow, with high spacial resolution around the shock features. In principal, this can be done by use of a laboratory frame calculation with a very fine mesh.[17] In practice, it will be at the very least a painstaking piece of work.

Another example is the broad, shallow, self-reversed emission lines of the Oe stars such as λ Cep. Models with coronae and microturbulence have been somewhat successful[18]; but, again, are they unique? We are not certain. Yet another example of the failure of the line formation theory is the dense winds of the Wolf-Rayet stars, in which a single photon may be Doppler shifted through a succession of line transitions, as it bounces back and forth within the wind. Photons trapped in such a wind may be able to deliver several times their intrinsic momentum by a succession of back-scatterings followed by long flights. The complexity of the problem is such that the only serious attempt to solve it[19] used the Monte Carlo method, and assumed only one scattering event to characterize the interaction between a photon and a resonance region. By this means it has been demonstrated that a Wolf-Rayet wind probably can be driven by line radiation pressure. At least it isn't out of the question, as some have proclaimed.

Another problem which has been treated primarily in static atmospheres but is also present in winds is that of partial frequency redistribution, and of angle-frequency correlations. The effects of partial redistribution are most pronounced in observations with high resolution in both frequency and angle, but it is known that coherent scattering in the frame of the atom can strongly influence such quantities as excited atom density and escape probability, and that atmospheric energy balance can be significantly influenced by the radiative flux carried in the wings of a single line. For monotonic flows, line transfer can be treated in the comoving frame, and accurate representations of partial redistribution can be found.[20] If the flow is non-monotonic, the comoving frame formulation is very awkward,[21,22] and we are forced into the laboratory frame. But in the lab frame an accurate representation of partial redistribution is difficult to obtain.[21,23,24] This seems to be a serious dilemma. Even for monotonic acceleration, partial redistribution has not been studied at the level of angle-frequency correlation. Angle-averaged redistribution functions are generally thought to be adequate for gross properties in planar atmosphere, or regions with more-or-less isotropic radiation. But it has been clearly shown[25,26] that a detailed treatment of angle-frequency coupling is necessary to explain the resonance profiles of Ca^{+} in the solar limb. We also suspect that angle-frequency correlation may be important in extended spherical winds, due to angular peaking.

Finally, even though line formation theory for winds has been rather successful in fitting individual lines, it has been less so for the simultaneous fit to all lines observed in one or more species in a single star. Two notable exceptions to this rule are ζ Pup and τ Sco.[27,28] which were successfully modelled to fit their UV resonance lines. The problem of a comprehensive fit to all lines which can be regarded as indicators of structure is exaggerated by the fact that crucial data in observationally difficult spectral regions are often missing.

Serious efforts should be made to fill in the missing data for a few representative stars in any category of interest. For variable spectra the observations must be made simultaneously, with coordinated ground-based and extraterrestrial observations. Such work will almost certainly require international cooperation and planning.

1.2. Laboratory Spectra

Though astronomical observations have provided the primary motivation for line transfer solutions in moving media, there is also a need for such solutions in laboratory plasmas. In modern applications such as inertially confined fusion, radiative transfer can play a critical role in radiation hydrodynamical phenomena occurring on very short time scales. Also, new techniques for obtaining data from rapidly evolving plasmas have revealed detailed emission spectra in the X-ray band. In turn, laboratory work may be able to test the theories of line broadening, and those of redistribution, so that we may proceed into the interpretation of stellar spectra on a more solid footing.

The general problem of radiation hydrodynamics, in both laboratory and astrophysical applications, has not been solved. Here, the transfer of radiation occurs in media which not only is moving, but assumes a time sequence of shocks and arbitrary geometrical shapes. In addition, extremely complex networks of atomic levels and ion charge states are present in this regime, and time-coupling may be important. From this point of view, line transfer technique is in a relatively primitive state, as are the computers used to do it.

2. LINE TRANSFER IN EXPANDING ATMOSPHERES

Let us turn now to a discussion of the hierarchy of the problems solved and see why those of the future are so difficult. The early analytic work in this area was limited to assumptions such as the square absorption profile, homogeneous media, winds with constant velocity gradients, and the Eddington approximation. Fundamental effects were found, but interpretation of observations and the treatment of realistic cases required more sophistication.

A great deal of work has been done on the differential equations of physics, but these equations usually contain only local coupling. The biggest obstacle to the development of line transfer theory is the global nature of radiative coupling, or in mathematical terms, the presence of a scattering source term in the line transfer equation. This term, which represents emission into a given frequency and direction of photons scattered from all other coupled frequencies, and directions is mathematically represented, for complete redistribution, as the convolution of the frequency-dependent mean intensity and the line absorption profile. When the problem includes time coupling or multi-dimensional geometry, or when the transfer equation is solved in the comoving frame of an expanding medium, the global line transfer equation becomes partial as well. If the multiplicity of partial derivations is entirely spatial a solution along characteristics can usually be constructed. But if a time or frequency derivative is included in the operator, then specialized techniques for P.D.E.'s must be woven into the global solution technique.

Leaving for the moment the question of the physical coupling among the variables, we can appreciate the complexity of the general transfer problem by examining the logical domain of the itensity function, assuming the Feautrier[29] approach, with adequate sampling on all coordinates. Table 1 illustrates the dimensionality of the dependence of the specific intensity on the variables of space, frequency, angle, and time, for a sequence of progressively more coupled and complicated problems. Here, we are omitting the transfer of Stokes parmeters, important for solar magnetic field determinations, and that of phase coherence, which is important in laser cavities. The number of mesh points shown in the right-hand-column quickly goes beyond our ability to cope as the problems get more complicated. Similarly, the operation count soars with the complexity factor. An example of this can be found by comparing the solution for a two-level atom in spherical geometry, and in two-dimensional Cartesian geometry, assuming both are solved by second order finite difference methods. The spherical case runs in a few second on a Cray 1a if 60 mesh points are used, while the 2D case requires a few minutes, assuming 60 points on each axis, and optimal vectorization. Here, we have assumed a small velocity amplitude. Since the 2D problem is solved in the laboratory frame, a larger velocity amplitude would be correspondingly more costly. The complexity of the solution is partly due to complexity of global coupling in two dimensions, but not entirely. A lambda step, in which the internal mean intensity is calculated from a known source function, requires about 25% of the time for a full solution[30]; and yet it is a non-global solution. For 2D and 3D models a larger portion of the overall numerical complexity comes from the difficulty of representing the full angular dependence of the intensity at every spatial mesh point than in 1D models, in which rays may be constructed from one boundary to another without missing any relevant mesh points. For this reason, a lambda operation is relatively more formidable in multi-dimensional geometries. Therefore, rapid methods which depend on fast formal (lambda) solutions, such as the integral operator perturbation techniques, may have to accept approximate formal solutions, which results in approximate final solutions. No one has seriously attempted the 3D case by the Feautrier method for obvious reasons, even though six years have elapsed since the 2D case was worked out in x-y coordinates.[31]

Nevertheless, looking over the various areas within the field of line transfer, we observe that steady progress has been made, and this progress has followed the development of the digital computer. Of course, we cannot measure progress along only one axis. For example, while the quest for accurate numerical solutions procedes along one line, that for better approximate solutions procedes along another. In fact, the problems we face in the general hydrodynamic medium are so severe, that a new interest in approximate methods has recently been kindled.

Let us categorize five major areas in which advances are being made as geometry, complexity of flow, redistribution, atomic couplings, and approximate methods. In some respects, the challenges of each of these areas can be viewed as independent of the others. In developmental work, at least, this is the view we usually have to take.

2.1. Geometry

The simplest geometry is obviously that of the plane parallel slab; and the most complicated is the arbitrary 3D medium. There are exactly three 1D geometries: the plane slab, the sphere, and the infinitely long cylinder. The first two have only one angle coordinate, but differ in difficulty due to the degree of angular peaking in the radiation field. The radiation field in the 1D cylinder requires both angle coordinates for its specification, and also may be quite peaked about the axial direction. Line transfer in planar geometry has been extensively studied in both the laboratory and comoving frames,[20] while spherical geometry has been treated almost exclusively in the comoving frame. The 1D cylinder has been studied only in the comoving frame,[32] and only for a limited number of cases. This is not surprising in light of the greater expense and complexity of the cylindrical solution, and the greater applicability of planar and spherical models to moving stellar atmospheres. Large scale flows are almost always, if not always, extended from the photosphere. Such protrusions as solar flares are best modelled in planar geometry, while global extension, through probably asymmetric in active chromospheres and variable atmospheres, has most easily been modelled in spherical geometry. It is also worthy of note that we have no evidence of global extension without differential motion. That motion be steady expansion, explosion, collapse, or steady infall. We cannot state that an atmosphere with moving components must be extended as a whole, though it is likely that the moving structures themselves are multi-dimensional or periodic; any free-standing structure is apt to be in a dynamic state. Certainly, any technique for line transfer in two or three dimensions should include the possibility of differential motion.

The 1D problems have been well studied for models of practical value. In 2D, several papers have been written for the x-y coordinate case, including a Rybicki-style solution in the laboratory frame of a slow flow,[31] which has also been incorporated into an ETLA solution for species with many levels and charge states.[30] The basic method holds promise for generalization to 2D axisymmetric geometry as well. For the analysis of line profiles from axisymmetric atmospheres with differential velocity fields in both radial and tangential directions, a formal solution for emergent fluxes has also been developed,[33] but the parameter space has not yet been explored. It is worth pointing out that almost all line transfer theory for stellar spectroscopic applications has assumed that the atmosphere is semi-infinite, that with integration to a deep enough layer, all processes represented, if represented with physical realism, will thermalize to the local Planck function. In fact, this has not only provided a convenient analytic check on solutions, but it also provides stability for the numerical procedure, for the thermalizing photosphere translates mathematically into a region in which the transfer equation is dominated by the thermal driving term. Free-standing atmospheres, such as solar flares, or certain laboratory gasses, must be treated without this convenience. Similarly, there is an important difference between the effects of periodic and non-periodic boundary conditions on free-standing or multi-dimensional structures. For 2D slabs in x-y geometry, it was found that a 1D representation is completely adequate for the determination of the source function in a periodic 2D slab. However, the 2D representation was necessary for accurate emergent fluxes. Such a model is called the 1 1/2-D slab.[31]

Finally, though no full 3D solution has been attempted, global properties of line transfer in a 3D medium have received some attention,[34] and 3D deviations from a 1D slab have been treated.[35]

2.2. Complexity of Flow

Complexity of flow usually goes hand in hand with complexity of geometry because the kinematics of a flow usually determine the shape of the object. However, within a given model geometry, the flow may, in principle, be simple or complicated, monotonic or non-monotonic, small or large in amplitude, and may be confined to optically thin layers, or penetrate the thermalizing depths of the atmosphere. The major divisions among models assumed for line profile calculations have been between slow and fast flows, and monotonic and non-monotonic ones. By "slow flow" we mean one which does not Doppler shift the medium beyond one or two line widths of the absorption profile. Slow flows can be treated in the laboratory frame by a straight-forward extension of techniques for static atmospheres. Non-monotonic motion fits well into this formulation, but partial redistribution is difficult to represent accurately, for all but the slowest flows.[21,23,24] The laboratory frame method may be applied to larger flows, but only if the spatial mesh resolves the spatial variation of the profile function, and the Doppler shifted frequency band is entirely covered by numerical frequency points on a mesh of adequate density. Lines formed entirely in regions with large velocity gradients compared with the other physical gradients, may be treated by the escape theory of Sobolev,[6] if boundaries are not involved, and certain other restrictions are satisfied. It is the intermediate flow cases, especially those which are not monotonic, which are the most difficult. A solution of the transfer equation in the comoving frame of the flow is economical, and can support the treatment of partial redistribution.[20] Non-monotonic flows have been treated in the comoving frame, for planar[21] and spherical[22] geometries, but such solutions are awkward and lose much of the original economy of the comoving frame method. Only a few test cases have been calculated, and the question of accuracy has plagued some of this work. Fast non-monotonic flows in planar or spherical geometry can be treated in the laboratory frame, for only a few times the cost of the comoving solution, if a computer calculation is written with vectorization over parameters that do not have to be treated recursively.[17] Again, almost no exploration has been done in this area, even though astrophysical applications, such as the interpretations of inverse P Cygni profiles in T Tauri stars,[38] apparently exist.

2.3. Partial Redistribution

Partial redistribution in moving atmospheres has not been extensively investigated, though it may be important for analysing highly resolved profiles with broadening not dominated by the wind. The comoving frame seems to be the natural frame in which to represent the details of redistribution without loss of accuracy,[20] and any new formulation of frequency redistribution may be directly implemented in the comoving frame. To study redistribution, we are prohibited from using the Rybicki variant[39] of the Feautrier technique,[29] because the Rybicki variant achieves its superior performance by taking advantage of the fact that in complete redistribution the scattering integral doesn't depend on frequency. The cost of the Feautrier method is proportional to the cube of the number of angle points, and the cube of the number of frequency points. The problem this presents in geometries which produce angular peaking and require many angle points may be avoided by use of the variable Eddington factor technique.[40] This method has been used in spherical geometry to study atomic and electron scattering frequency redistribution in an expanding spherical atmosphere.[20] It has also been used in expanding one-dimensional cylindrical atmospheres,[32] assuming complete frequency redistribution.

Certain problems in PRD are ripe for investigation. For example, a method has been suggested[36] for treating angle-frequency coupling in spherical flows in the comoving frame, using an extension of the variable Eddington factor technique. The intense angular peaking in an extended atmosphere may accentuate the importance of angle-frequency coupling, in the atomic scattering process. In addition, as shown by heuristic arguments, angle-frequency coupling in electron scattering in an expanding wind can generate a prominant redward tail in a line profile.[37] This phenomenon should be explored quantitatively so that it can be used, in conjunction with atomic redistribution, to infer electron densities in stellar atmospheres or the expanding regions of Seyfert galaxies, and to calculate diagnostic flux profiles.

2.4. Atomic Coupling

Most line transitions do not form in isolation from the surrounding atomic network. A few resonance lines, such as those of Mg^+ and C^{+++}, may be safely treated as isolated two-level transitions, but the majority of lines must be treated in their atomic context. Unfortunately, this means that to obtain results for one line you must effectively calculate them all, for a given species, along with the continuum radiation fields. However, this is not all bad because a good fit to all the lines observed in a given species is much more convincing than a good fit to one, especially considering the number of free parameters available in semi-empiracle wind models. One of the two primary methods, the equivalent two level atom (ETLA) technique, is intrinsically adaptable to whatever advance can be made for isolated two level atom calculations because the transitions in the network are sequentially isolated and solved as independent two-level atoms, in each interation. The ETLA technique has now been extensively used to model the hydrogen, helium and metal spectra in a broad range of atmospheres, including slowly and rapidly[41] expanding ones. It has not been used in decelerating winds even though the two-level atom problem has been solved for non-monotonic winds in spherical geometry.[17] Problems of atomic interlocking which can plague ETLA convergence in static atmospheres are reduced in rapidly moving atmospheres due to the desaturating effects of differential motion. In addition, new methods to overcome the interlocking problem are under study. These include the implementation of the non-local extrapolation method of Ng, which has been used with success by Buchler and Auer[42] in a two-level atom context.

The second method, complete linearization,[43] is easily adaptable to laboratory frame solutions in slow flows, and has been used to interpret such lines as the Ca^+ infrared triplet in solar spectra. It has been formulated in the comoving frame for spherical geometry,[41] and has been used in this context for the analysis[44] of HD 50896, a WN7 star.

2.5. Approximate Methods

Approximate methods are useful for line transfer problems being solved within a parameter space too poorly known to justify a numerically rigorous procedure, or with a computer too small to treat the problem accurately by known methods. The latter category includes super computers along with micros. In fact, it is partly due to the existence of important problems too complex for the super computers of the present and near future that there is a renewal of interest in faster methods for accurate solutions, and very fast methods for approximate ones.

One of the most used approximate methods is that of escape probability theory.[45] Here, the line source function is assumed to be locally coupled. This justifies its separation from the kernel function in the integral representation of the transfer equation. In recent years, escape theory has been generalized to a second order formulation with greatly improved accuracy near open boundaries.[46] It has also been extended to include arbitrary velocity gradients,[47] to bridge the gap between static escape theory and Sobolev theory, and to media with overlapping sources and sinks of radiation.[48] However, it seems limited to schematic problems because of its restrictions to radiation fields dominated by local coupling. Significant depth variation of the absorption profile, or the assumption of coherent scattering in line wings, is not consistent with the approximation.

For large velocity gradients, the radiation transport process becomes local, for monotonic flows. The invariance of the line source function over photon interaction zones allows us to make the approximation of Sobolev.[6] Sobolev theory has also been generalized considerably. It can now be applied to multi-level atoms,[49] certain special cases of non-monotonic flows,[50] and to media with overlapping background opacity.[51] It is usually applied to obtain a stellar line source function and again, in different form, to calculate emergent line profiles. It has been shown that the approximation breaks down for typical stellar wind models with wind amplitudes less than ten thermal Doppler units. Most of the numerical degradation occurs in the calculation of the emergent radiation.[7]

One of the oldest approximations, which has allowed an analytic solution for many otherwise intractable problems is the Eddington approximation. Here, the Eddington closure factor, the ratio of the second to the zeroth angle moments of the intensity, is set to 1/3, its value in an isotropic radiation field. This is also an immediate consequence of the Gaussian two-stream approximation. This approximation allows the angle moment form of the transfer equation to be greatly simplified. An improvement on accuracy is made if a shrewd estimate of the Eddington factor, as a function of depth and displacement from line center, can be made. In this case, the moment equation cannot be solved analytically due to the variation of its coefficients, but the result is more accurate. This method can also be used in spherical and cylindrical geometry.[32] In the cylindrical case, where two angle coordiantes are required, the Eddington closure factor becomes a sparse tensor of closure factors,[52,32] all of which must be evaluated or estimated for the specification of the second order moment equation. Some of these factors vary through a wide range of values, as in the spherical case, due to angular peaking in curved geometries. Unfortunately, in the comoving frame of a stellar wind, the closure factors are very difficult to estimate because the radiation sources within the local profile are difficult to locate as a function of direction. The estimation of ratios of flux-like moments, which can pass through zero, is particularly difficult.

The method of Scharmer,[53] which was partly inspired by the angle and frequency quadrature perturbation techniques of Cannon,[54,55] can also be used as an approximation, if limited to a single iteration, or if implemented with an approximate lambda-type operator. The transfer equation is split into an approximate integral equation for a source function correction, and a solution for the driving term of this equation. This solution is structurally the same as the lambda formal soution, and may be evaluated accurately or approximately. In semi-infinite slabs the method has been quite successfull. One-dimensional models with a variety of complications have been separately treated. These include the 1D slab with: 1) a velocity field, 2) a strong temperature gradient, 3) partial frequency redistribution, and 4) a ten level hydrogen atom with all lines included. The method runs roughly a factor of ten faster than the standard second order method. If an approximate lambda operator allows a large savings in the operation count, or the running time of ones computer, then a relatively inexpensive approximate method exists which cannot itself be iterated to an accurate solution without a substantially greater numerical effort. This is an important consideration because multi-dimensional problems may be characterized, as mentioned above, by lambda solutions which are costly if implemented with high (second) order accuracy, and cheap if implemented as approximate low (first) order procedures.

3. FUTURE DIRECTIONS

Line transfer becomes most challenging in time dependent flows. If a hydrodynamic structure depends on the radiation field, or the emergent radiation spectra are sufficiently time resolved, then the transfer

DESCRIPTION	SPECIFIC INTENSITY	NUMBER OF VARIABLES	SPACE	ANGLE	FREQUENCY	TIME ≐ PRODUCT	
1D, gray, planar	$I(\mu,r)$	2	40	2	1	1	100
1D, line, planar	$I(\mu,\nu,r)$	3	40	2	6	1	500
1D, line, spherical, expanding	$I(\mu,\nu,r)$	3	50	30	15	1	20 K
1D, line, cylindrical, expanding	$I(\mu,\psi,\nu,r)$	4	50	150	15	1	100 K
2D, line, planar, expanding	$I(\mu,\psi,\nu,x,z)$	5	50^2	15	15	1	500 K
3D, line, expanding	$I(\mu,\psi,\nu,x,y,z)$	6	50^3	25	15	1	50 M
3D, line expanding, Voigt profile, time dependence	$I(\mu,\psi,\nu,x,y,z,t)$	7	50^3	25	30	100	10 G

TABLE 1. Dimensionality and complexity of the dependence of specific intensity on its free variables for a hierarchy of increasingly complex models.

equation must be included among the equations of hydrodynamics. The time dependence of hydrodynamical structures can enter the transfer equation at different levels of coupling. For many cases, the radiation field and the microkinetic processes adjust so fast to structural changes that the radiation field is almost instantaneously consistent with the fluid. For these cases, the relevant transfer problem is solved, as in a static atmospheres, at each time step of the hydro calculation. For shoter hydro time scales, the radiation field adjusts almost instantaneously to the detailed population densities, but these do not adjust immediately to the bulk state of the matter. Here, the rate equations and line source functions are time coupled over a time step. For even shorter time scales, or for very long distances, the non-zero time for propagation of radiation across the medium becomes relevant, and the intensity time derivative must be included in the transfer equation.[56] A possible example of this is an extended atmosphere in the presence of a rapidly varying X-ray intensity from the accretion disk of a compact object. Both these processes must be considered for laboratory experiments in which a resonant vapor is suddenly illuminated, or de-illuminated, by an external light source such as a laser, and time-resolved measurements of the diffuse emission must be analyzed. However, these experiments are usually done in static media. Moving media are encountered in the study of X-ray spectra from expanding metal plasmas, and of the spectra produced by the target in inertially confined fusion experiments.

At the present time we are forced to make physical approximations which are too restrictive for some of the stellar wind phenomena we study and the high quality data we can now gather. Looking down the road to a time when large numbers of solutions will be needed for complicated ions in rather arbitrary hydrodynamical geometries, we predict the need for theory and methods which greatly exceed what we presently have. We must find ways of representing the global coupling which controls optically thick line transfer without solving for all the descretized intensities of the Feautrier method. These new methods must account for non-monotonic velocity fields as well as complicated atomic coupling. A start in this direction has been made in 1D geometries by the integral operator perturbation method mentioned above. If these methods, or similar ones, can be adapted to free-standing structures, especially those in two and three dimensions, without loosing their intrinsic computational advantages, we will take an important step toward line transfer in hydrodynamic media.

REFERENCES

1. Struve, O., and Elvey, C. 1934, Ap. J., 79, 409.
2. Lucy, L. B. 1976, Ap. J., 206, 499.
3. Beals, C. S. 1951, Pub. Dominion Ap. Obs., 9, 1.
4. Inoue, M. O., and Uesugi, A. 1977, Pub. A. S. P., 29, 149.
5. Castor, J. I., and Lamers, H. J. G. L. M. 1979, Ap. J. Suppl., 39, 481.

6. Sobolev, V. V. 1957, Sov. Astron., 1, 678.
7. Hamman, W.-R. 1981, A. A., 93, 353.
8. Mihalas, D., and Conti, P. S. 1980, Ap. J., 235, 515.
9. Barker, P. K., and Landstreet, J. D., Marlborough, J. M., and Thomson, I. 1981, Ap. J., 250, 300.
10. Bertout, C. 1978, Mitt Astron. Ges. Nr., 43, 176.
11. Bertout, C. 1979, AA., 80, 138.
12. Friend, D., and McGregor, K. 1984, Ap. J.
13. Praderie, F., Talavera, A., and Lamers, H. J. G. L. M. 1980, A.A, 86, 271.
14. Kunasz, P. B., and Praderie, F. 1981, Ap. J., 247, 949.
15. Lucy, L. B. 1982, Ap. J., 255, 278.
16. Lucy, L. B. 1983, Ap. J., 274, 372.
17. Mihalas, D. 1980, Ap. J., 238, 1042.
18. Grady, C., Kunasz, P. B., and Snow, T. 1985, Ap. J., in press.
19. Abbott, D., and Hummer, D. G. 1985, Ap. J., in press.
20. Mihalas, D., Kunasz, P. B., and Hummer, D. G. 1976, Ap. J., 210, 419.
21. Mihalas, D., Shine, R. A., Kunasz, P. B., and Hummer, D. G. 1976, Ap. J., 210, 419.
22. Marti, F., and Noerdlinger, P. D. 1977, Ap. J., 215, 247.
23. Vardavas, I. M. 1974, J.Q.S.R.T, 14, 909.
24. Vardavas, I. M. 1976, J.Q.S.R.T., 16, 781.
25. Milkey, R. W., and Mihalas, D. 1973, Ap. J., 185, 709.
26. Milkey, R. W., Shine, R. A., and Mihalas, D. 1975, Ap. J., 199, 718.
27. Hamann, W.-R. 1980, A. A., 84, 342.
28. Hamman, W.-R. 1981, A. A., 100, 169.
29. Feautrier, P. 1964, C. R. Acad. Sci. Paris, 258, 3189.
30. Kunasz, P. B., and Mihalas, D. 1985, in preparation.
31. Mihalas, D., Auer, L. H., and Mihalas, B. W. 1978, Ap. J., 220, 1001.
32. Kunasz, P. B. 1984, Ap. J., 276, 677.
33. Kunasz, P. B. 1985, in preparation.
34. Rybicki, G. B., and Hummer, D. G. 1983, Ap. J., 274, 380.
35. Nordlund, A. 1985, "NLTE Spectral Line Formation in a 3D Atmosphere with Velocity Fields" in Progress in Stellar Spectral Line Formation Theory.
36. Mihalas, D. 1980, Ap. J., 238, 1034.
37. Auer, L. H., and Van Blerkom, D. 1972, Ap. J., 178, 175.
38. McGregor, K., Solar Wind Five, NASA Conference Publication #2280.
39. Rybicki, G. B. 1971, J.Q.S.R.T., 11, 589.
40. Auer, L. H., and Mihalas, D. 1970, M.N.R.A.S., 149, 65.
41. Mihals, D., and Kunasz, P. B. 1978, Ap. J., 219, 635.
42. Buchler, J. R., and Auer, L. H. 1985, "Iterative Solution to the Non-LTE Transport Problem," in The Second International Conference and Workshop on Radiative Properties of Hot Dense Matter.
43. Auer, L. H., and Heasley, J. 1976, Ap. J., 205, 165.
44. Hillier, D. J., Ph.D. Thesis, Australian National University.
45. Bibermann, L. M. 1948, Dokl. Akad. Nuak. S. S. S. R., 49, 659.
46. Hummer, D. G., and Rybicki, G. B. 1983, JILA preprint.

47. Hummer, D. G., and Rybicki, G. B. 1982, Ap. J., 254, 767.
48. Noerdlinger, P. D. 1984, private communication.
49. Castor, J. I. 1970, M.N.R.A.S., 149, 111.
50. Rybicki, G. B., and Hummer, D. G. 1976, Ap. J., 219, 654.
51. Hummer, D. G., and Rybicki, G. B. 1984, JILA preprint.
52. Weaver, R., Mihalas, D., and Olson, G. 1982, Los Alamos Report LA-UR-82-961.
53. Scharmer, G. B. 1982, preprint, to apper in Methods of Radiative Transfer, ed. W. Kalkofen (Cambridge University Press).
54. Cannon, C. J. 1973, Ap. J., 185, 621.
55. Cannon, C. J. 1973, J.Q.S.R.T., 13, 627.
56. Kunasz, P. B. 1983, Ap. J., 271, 321.

Computed He II spectra for Wolf-Rayet stars

W.-R. Hamann
Institut für Theoretische Physik und Sternwarte
Olshausenstr. 40 - 60
D-2300 Kiel
Federal Republic of Germany

ABSTRACT. Synthetic spectra of spherically symmetric, expanding stellar atmospheres of pure helium are calculated in non-LTE for a given atmospheric structure. The radiation transfer is solved in the "comoving frame", and consistency with the multilevel atom rate equations is achieved iteratively by an "equivalent two level atom approach". A small grid of models is calculated. He II emission lines as typically observed in WR stars are obtained, if the effective temperature of the star lies above 35 kK and the mass loss rate exceeds 10^{-5} $M_{\odot}$/yr.

I. Introduction

Due to the lack of adequate model calculations, the spectral analysis of WR stars is a more or less unsolved problem of astrophysics. Classical models fail, because the emission lines as well as the continuum are formed in the expanding stellar envelope. Consequently, the basic parameters of WR stars (luminosity, effective temperature, chemical composition) are poorly known (see, e.g., the review by Conti, 1982). Structure and physics of the expanding atmospheres are also not yet understood.

Very simplified models have been used so far for analyzing WR stars. An adequate method for treating the line formation in expanding atmospheres is the "comoving frame" technique (Mihalas et al., 1975). With its restriction to two-level atoms, this method was applied to the UV resonance lines from hot stars of various types (Hamann, 1980, 1981; Hamann et al., 1981, 1982, 1984). An "equivalent two level atom approach" to the multilevel problem was presented by Mihalas and Kunasz (1978). However, this technique has not been used so far to calculate WR spectra. The present paper is based on a new code which follows the method of Mihalas and Kunasz (1978).

II. Radiative transfer calculations

Synthetic spectra of spherically symmetric, expanding envelopes are calculated in non-LTE for a given atmospheric structure (density- and

J. E. Beckman and L. Crivellari (eds.), Progress in Stellar Spectral Line Formation Theory, 335–342.

tempertature stratification). The radiation transfer problem is solved by means of the "comoving frame" formalism. The consistency with the multilevel-atom rate equations is achieved iteratively by an "equivalent two level atom (ETLA) approach". Hence our method follows in principle the pioneering work of Mihalas and Kunasz (1978). For details, the reader is referred to that paper or to the textbook of Mihalas (1978). In the following, we will only summarize how we have applied or modified the quoted techniques.

1. Continuum

From a set of approximate population numbers, the continuum opacity and emissivity are calculated and the continuous radiation transfer (spherically, but static) is solved. From the resulting radiation field, the Eddington factors are evaluated. It is not necessary to recalculate the Eddington factors in each iteration cycle. Hence a formal solution can be obtained with less computational effort from the moment equations, keeping the Eddington factors fixed over several iterations.

Optionally, we provide the possibility that distinct continua are calculated with the ETLA source function, which explicitely displays the scattering term. Hence the solution satisfies not only the equation of radiative transfer, but simultaneously also the rate equations, with all other rates fixed except the considered transition. The scattering term couples all frequencies where the considered continuum dominates. The solution is obtained by means of a Feautrier elimination scheme.

2. Lines

Each line transition is considered separately. From a set of approximate population numbers, the line opacity and emissivity is evaluated. Pure Doppler braodening and complete redistribution are assumed. The radiation field is obtained from a formal solution of the comoving-frame equation of transfer. For each impact-parameter separately, the calculation proceeds from the blue wing of the line to the red. The blue wing boundary condition is specified by the (angle-dependent) radiation field of the ambient continuum, as stored from previous continuum calculations. The resulting radiation field is integrated over the impact parameters to obtain mean intensities and Eddington factors.

Optionally, now the same line may be recalculated with the ETLA line source function, which explicitely displays the scattering term. Using the Eddington factors from the formal solution, the moment equations are solved. The scattering term couples all line frequencies, and a Feautrier elimination scheme is applied. The resulting radiation field hence satisfies simultaneously the eq. of radiative transfer and the rate equations, for all other transition rates being fixed.

3. Formal solution in the observer's frame

With the converged set of population numbers, the emergent flux profiles of the lines are calculated finally by a separate program. The intensities along each ray (specified by impact parameter and observer's frame frequency) are integrated straightforward by a simple quadrature sum, where special care is taken for a suitable spacing of mesh points in the zone of interaction with the line.

III. The models

1. The atmospheric model

The stratifications of temperature and density must be specified in advance. The following assumptions allow to construct the model from a few basic parameters:

With the mass loss rate $\dot{M}$ being given, the density stratification $\varrho(r)$ and the velocity law $v(r)$ are related by the equation of continuity:

$$\dot{M} = 4\pi r^2 \varrho(r) v(r) \quad (1)$$

In the high velocity region, we specify the velocity field by the usual law

$$v(r) = v_\infty (1 - r_o/r)^\beta \quad (2)$$

In the low velocity part of the atmosphere we assume a hydrostatic density stratification. The scale height is calculated from a prefixed temperature parameter T_*, the effective gravity g_{eff} and the mean particle mass. The connection point of both domains is defined by requiring that the transition should be smooth, and lies in practice

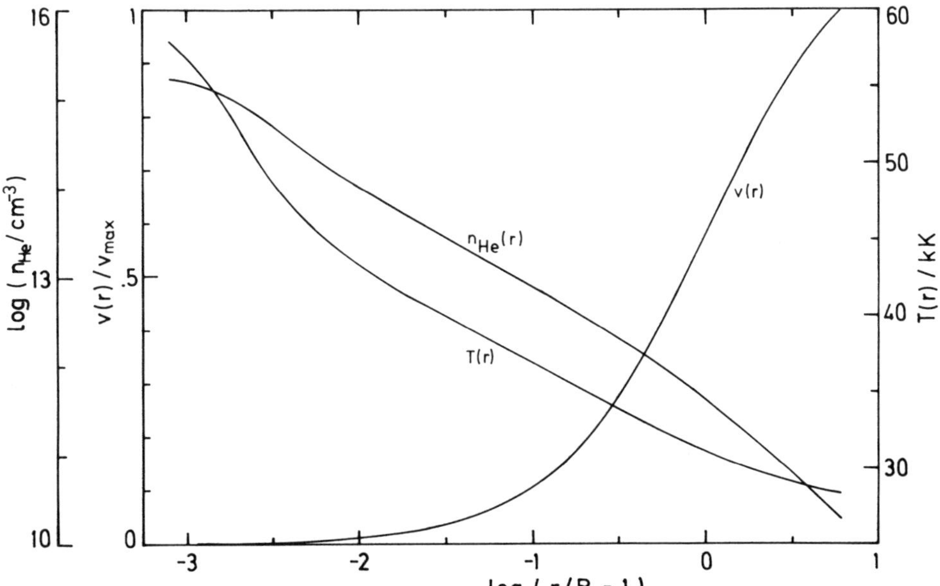

Fig. 1: Velocity field, density stratification and temperature structure for the model with T_* = 35 kK and $\dot{M}$ = 4 10^{-5} $M_\odot$/yr

roughly at the sonic point. At the inner boundary, the velocity v_{min} is chosen small enough to secure a sufficient Rosseland optical depth (around 10). By specifying the velocity v_{max} at the outer boundary (given by r_{max}), now the velocity field and density structure are completely defined.

The parameter T_* defines the temperature stratification as follows: With the Rosseland mean opacity calculated in LTE, we use the formula of a planeparallel grey atmosphere with $T_{eff} = T_*$. For the continuum forming region, the planeparallel case hardly deviates from the spherical result, because the spherical extension is still small. In the wind, our model temperature approaches the planeparallel boundary value, which seems to be more realistic than the radial temperature decrease which results in a spherical grey atmosphere. In reality, nonradiative heating may even raise the temperature in the envelope. Note that T_* equals roughly the effective temperature of the star, related to that radius where the Rosseland optical depth reaches unity.

In the present model series, we have only varied T_* and $\dot{M}$. All other parameters were set to values considered to be "typical"; their variation should be subject to future work. The maximum velocity (at $r_{max} = 7\ R_*$) is set to 2500 km/s, which corresponds to the typical line widths observed. The velocity law in WR stars is speculative, unless some

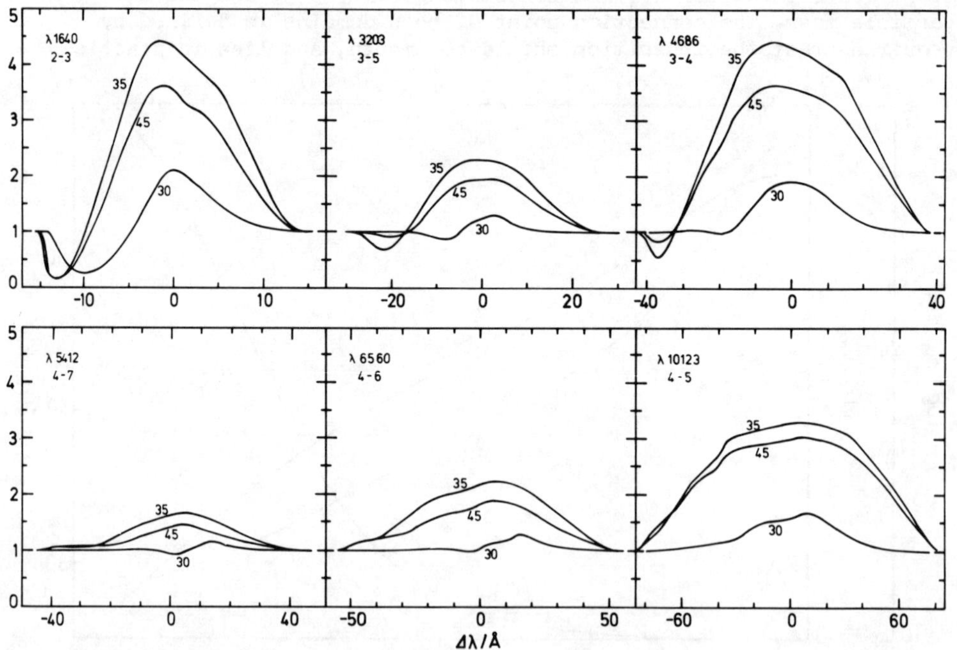

Fig. 2: Line profiles of the first series of models with common mass loss rate of 4 10^{-5} $M_\odot$/yr. The labels denote the temperature T_* in kK (for definition see Sect. II.1)

empirical estimate is made. We use the velocity field described above with $\beta = 1$, which was found to be adequate for many stellar winds studied so far. The photospheric density stratification is defined by $\log g_{eff} = 3.5$ and a mean particle mass corresponding to fully ionized helium. The inner boundary radius is set to $R_* = 5\ R_\odot$. Another free parameter not mentioned so far is the Doppler-broadening velocity v_D. We assume $v_D = 100$ km/s, because similar high random velocites ("microturbulence") are commonly observed in stellar winds of early-type stars (Hamann, 1980, 1981; Hamann et al., 1981, 1982, 1984).

2. The atomic model

The model atmospheres consist of pure helium. The lowest seven bound levels of He II are taken into account. Removing the seventh level hardly changes the remaining population numbers, and hence we feel this number of levels being sufficient for the present purpose. He I is included with its ground state, because its continuum opacity may not always be negligeable.

IV. Results and discussion

A first series of models is computed for a mass loss rate of $4\ 10^{-5}\ M_\odot$/yr, which may be a typical value for WR stars. The stellar temperature T_* (for definition see Sect. III.1) is set 30, 35 or 45 kK, respectively. The resulting line profiles are plotted in Fig. 2, and the equivalent widths are compiled in Table 1.

The 30 kK model does not resemble a WR star, because only weak emission lines are present. At 35 kK, however, the situation has changed drastically, as very bright emission lines do occur now. At higher temperature (45 kK), this general feature remains the same, but the

Table 1

Model	1640	4686	3202	10128	6563	5414
$T_*/\dot{M}$	2–3	3–4	3–5	4–5	4–6	4–7
30/4	-3.97	-21.8	-2.48	-39.1	-6.27	-1.63
35/4	-43.8	-147.	-30.7	-248.	-74.2	-26.3
45/4	-29.7	-114.	-21.7	-224.	-48.6	-16.4
35/4	-43.8	-147.	-30.7	-248.	-74.2	-26.3
35/1	-12.2	-65.9	-9.4	-151.	-22.3	-6.0
35/.4	0.16	-7.07	0.25	-15.5	-1.69	-0.04

Equivalent widths of He II lines in Å. The models are labelled by T_* in kK (first number) and the mass loss rate in units of $10^{-5}\ M_\odot$/yr (second number). The lines are identified by their wavelength in Å and by the principle quantum numbers of the corresponding transition emissions decrease gradually again. Obviously, there exists an optimum temperature of about 35 kK for the He II emission lines.

The line emissions are caused mainly by the geometrical extension of the line forming region. Hence they depend essentially on the corresponding line opacities. (Note the difference to the formation of photospheric absorptions, which are determined by the line/continuum opacity ratio.) In the models of the first series, the mass loss rates and the density stratifications are the same. Therefore the strengths of the emission lines reflect the corresponding level populations. Going from the 35 kK model to 30 kK, the main stage of ionization switches from He III to He II. The excited He II levels, which are responsible for all observable He II lines, lie closer to the ionization energy than to the He II ground state and thus are depopulated at 30 kK. Going from 35 kK towards higher temperatures (45 kK), the gradual decrease of the emissions reflects a corresponding depopulation of all He II bound states in favour of He III.

In a second series of models, we keep T_* at 45 kK, but vary the mass loss rate (Fig. 3 and Table 1). A reduction by a factor of 4 (to 1 10^{-5} $M_\odot$/yr) diminishes the emissions by about the same factor. Hence the star still resembles a WR star. A further decrease of mass loss to 0.4 10^{-5}

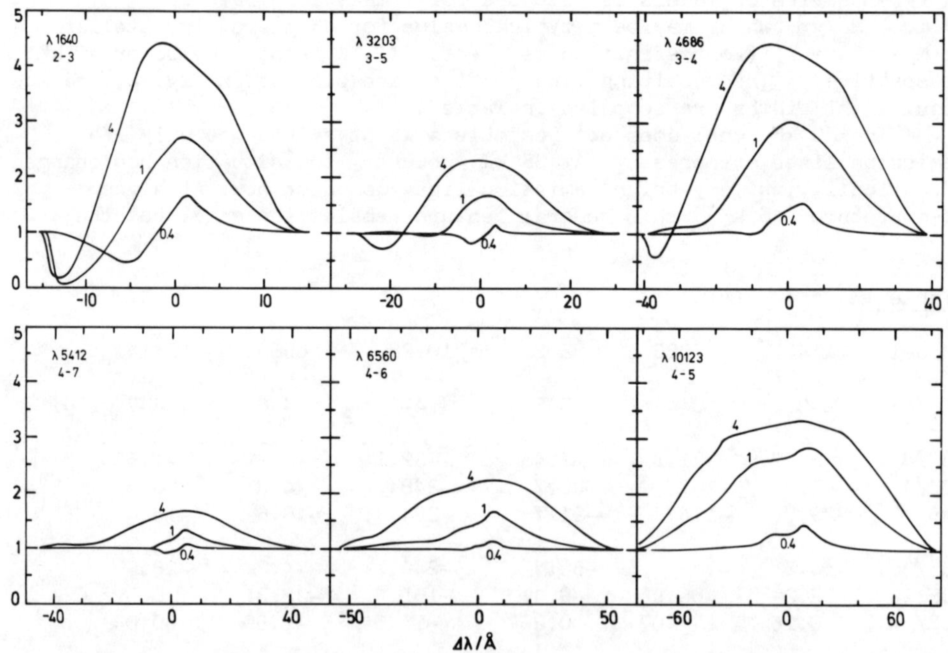

Fig. 3: Line profiles of the second series of models with the common temperature T_* = 35 kK. The labels denote the mass loss rate in units of 10^{-5} $M_\odot$/yr

$M_\odot$/yr (i.e. by another factor of 2.5) leads to a sudden breakdown of the emissions.

It is obvious that the line emitting region becomes smaller if one proceeds to models with smaller mass loss rates (second series of models). However, the non-linearity of this effect, i.e. the breakdown of the emissions between 1 10^{-5} and 0.4 10^{-5} $M_\odot$/yr, is striking.It is caused by an additional decrease in population numbers of the excited He II levels. The reason can be traced back to reduced pumping in the resonance lines, which in turn is caused by the He I continuous absorption that becomes prominent only at small mass loss rates (see below).

The emergent flux continua of all computed models are displayed in Fig. 4. The WR-like models 45/4 and 35/4 (Fig. 4a) show strong UV excess between 228 Å and 911 Å, and also strong IR excess which already begins in the visible. At $T_* = 30$ KK, the He I absorption becomes important and prevents the UV excess. The IR excess also nearly vanishes because of reduced bound-free and free-free emission.

The second series (Fig. 4b) demonstrates that the continuum strongly depends on the mass loss rate. At the lowest rate (model 35/0.4), the

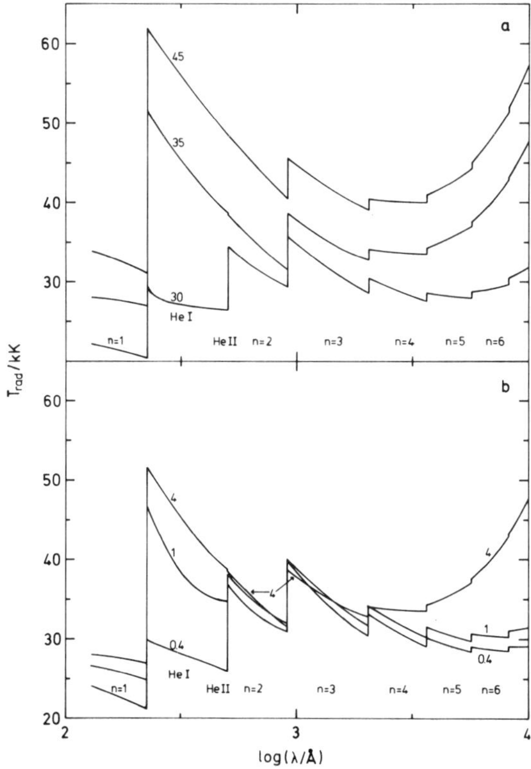

Fig. 4: Continuum flux of both model series, expressed by radiation temperatures T_{rad}. The labels denote the stellar temperature T_* in kK, or the mass loss rate in 10^{-5} $M_\odot$/yr, respectively

spectrum already resembles that of a planeparallel, static atmosphere. With higher mass loss, the He I absorption edge at 504 Å vanishes, and the IR excess also increases.

As a preliminary conclusion from this small grid of models we may state that the occurrence of He II emission lines which are typical for WR stars is restricted to (effective) temperatures higher than or about 35 kK, and (at 35 kK) to mass loss rates greater than or about 10^{-5} $M_\odot$/yr. However, this result may still depend on further model properties (e.g. the stellar radius). A systematic study of all free parameters which specify a model will be subject to future work, as well as some refinements of the model calculations. Detailed line fits should allow then to analyze observed WR spectra with respect to the stellar parameters, envelope structure and chemical composition.

References

Conti,P.S.: 1982, in C.W.H. de Loore and A.J. Willis (eds.), "Wolf-Rayet Stars: Observation, Physics, Evolution", IAU Symp. No. 99, p.3

Hamann,W.-R.: 1980, Astron. Astrophys. **84**, 342

Hamann,W.-R.: 1981, Astron. Astrophys. **100**, 169

Hamann,W.-R., Gruschinske,J., Kudritzki,R.P., Simon,K.P.: 1981, Astron. Astrophys. **104**, 249

Hamann,W.-R., Schönberner,D., Heber,U.: 1982, Astron. Astrophys. **116**, 273

Hamann,W.-R., Kudritzki,R.P., Méndez,R.H., Pottasch,S.R.: 1984, Astron. Astrophys. (in press)

Mihalas,D.: 1978, "Stellar Atmospheres", 2nd edition, San Francisco, Freeman

Mihalas,D., Kunasz,P.B., Hummer,D.G.: 1975, Astrophys. J. **202**, 465

Mihalas,D., Kunasz,P.B.: 1978, Astrophys.J. **219**, 635

PARTIAL REDISTRIBUTION IN THE WINDS OF RED GIANTS

K. Hempe
Hamburger Sternwarte
University of Hamburg
Gojenbergsweg 112
D-2050 Hamburg 80
Federal Republik of Germany

ABSTRACT. Most late-type giants and supergiants are losing mass in cool stellar winds. The only evidence for mass loss in single stars is the presence of a characteristic asymmetry deep in the core of photospheric lines. These asymmetries can be observed with modern high resolution coude echelle spectrographs and can be used for a study of mass loss rates and the wind structure in red giants. In this paper, we study the effects of partial redistribution functions R_{I-A} and R_{II-A} and complete redistribution on line profiles. The equation of transfer has been solved for a two level atom in the comoving frame.

1. INTRODUCTION

Partial redistribution effects in expanding atmospheres have been studied by Cannon and Vardavas (1974) who found large differences between complete redistribution (CR) and partial redistribution (PR). Magnan (1974), Hamann and Kudritzki (1977) and Mihalas (1976) showed that these effects are due to an incorrect angle averaging of the redistribution function in the observers frame. They all found that the differences of CR and PR are negligible at high expansion velocities. Peraiah (1979) and Wehrse and Peraiah (1979) studied the effects of the angle averaged R_{II-A} function in the observers frame for a spherical, moving atmosphere. Drake and Linsky (1983) used the comoving frame method to compare PR with CR for chromospheric lines in spherical geometry. They found strong differences in profiles calculated with CR and PR. These differences are mainly caused by the high cromospheric densities, which favour photon diffusion into the line wings.

In this paper we concentrate our interest upon lines which are formed in the extended, expanding shell of late type supergiants. From the work of Sanner (1976), Weymann (1962), Deutsch (1956), Reimers (1981), and Che et al. (1983) we know that the wind velocities in late type giants and supergiants are low, while the observed non-thermal turbulence velocities are high. The velocity ratio is in the range of 1 - 3. Therefore we can expect that the differences of CR and PR are of

J. E. Beckman and L. Crivellari (eds.), Progress in Stellar Spectral Line Formation Theory, 343–349.

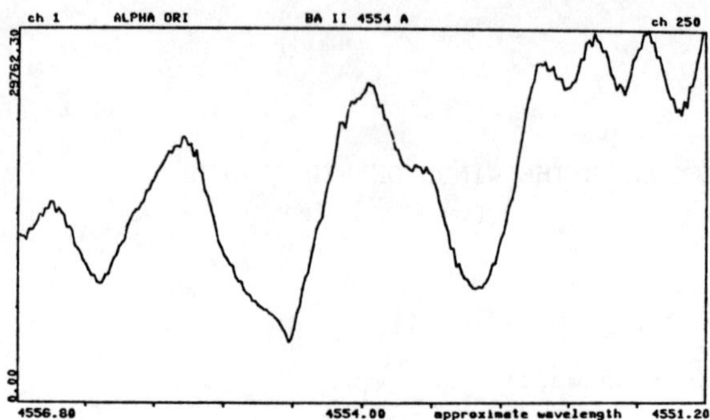

Figure 1. Photospheric Ba II line of α Ori with P-Cygni profile in the line core. The spectral resolution is 44.8 milli-Angstrom. The scan was obtained with the ESO CES scanner.

the same order as in static atmospheres at the same densities.

The wind lines will be formed in a spherical shell, which is detached from the photosphere. The inner radius is of the order of several photosphere radii. At the inner boundary of this shell a diluted photospheric line will be seen as incident radiation. This radiation will be redistributed in the expanding, spherical shell. The resultant profiles can be observed with modern high-resolution coude-echelle spectrographs at a resolving power of 100 000 to 200 000. Line profiles of α Ori, α Sco and a few other bright giants and supergiants at such a high resolution have been obtained by Sanner (1976) using the double pass echelle scanner at the coude focus of the 107 inch telescope at McDonald Observatory and by Hempe using the ESO coude echelle spectrograph at the 1.4 m coude auxiliary telescope at La Silla. In the cores of strong photospheric lines like Ba II λ 4934, λ 4554 Å and Sr II λ 4216 Å, and 4078 Å one finds only a small asymmetry which results from the redistribution of the photospheric radiation (fig.1). At the high resolving power which has been used in these observations even small differences of CR and PR profiles might be important in a quantative analysis.

2. SOLUTION OF THE TRANSFER EQUATION

The equation of radiative transfer has been solved numerically for spherical symmetry in the comoving frame. We use a computer code which is based on the method developed by Mihalas (1975) and modified by Hamann (1977). We can use the static redistribution functions as given by Hummer (1962) because in the comoving frame there is no macroscopic velocity. The emergent flux profiles have been calculated from the formal solution of the transfer equation after a final transformation of the source function to the observers frame.

The source function of a two level atom for complete redistribution and partial redistribution is given by eq. (1) and (2) respectively (cf. Mihalas, 1978).

$$S = (1-\varepsilon)\int \phi(x)\, J(x)\, dx + \varepsilon B \quad (1)$$

$$S_x = (1-\varepsilon)\phi^{-1}\int R(x,x')\, J(x')\, dx' + \varepsilon B \quad (2)$$

x is the frequency in Doppler units ($\nu_0 v_n/c$). v_n is determined by the thermal and stochastic (micro turbulence) velocity of the flow.

$$v_n^2 = v_{th}^2 + v_{sto}^2 \quad (3)$$

Here we assume pure scattering, i.e ε =0.

In the case of PR we start with the CR solution of the radiation field and iterate the source function. Convergence is achieved with an accuracy of .01 percent after a few iterations depending on optical depth. At the inner boundary of the envelope we assume that the radiation field is prespecified by a photospheric line which can be written as

$$F_x = 1 - a\, \mathrm{EXP}\,(-x^2/b) \quad (4)$$

The temperature and the ionization degree in the envelope is assumed to be constant. The density is given by the equation of continuity while the velocity is described by

$$v = v_\infty \sqrt{1 - .9/r} \quad (5)$$

The value of the velocity at the inner boundary is already a large fraction of the terminal velocity, because we consider a shell with an inner radius of several photospheric radii.

For the study of partial redistribution versus complete redistribution we use the R_{I-A} and R_{II-A} functions as given by Hummer (1962),

$$R_{I-A}(x,x') = .5\, \mathrm{erfc}\, (|\bar{x}|) \quad (6)$$

$$R_{II-A}(x,x') = \pi^{-3/2}\int_{|\underline{x}-\bar{x}|/2}^{\infty} \mathrm{EXP}(-u^2)\left[\tan^{-1}\frac{\underline{x}+u}{\sigma} - \tan^{-1}\frac{\bar{x}-u}{\sigma}\right]du \quad (7)$$

where $\underline{x}$ and $\bar{x}$ are the larger and smaller of $|x|$ and $|x'|$. The R_{II-A} has been calculated numerically by the simpson rule on the frequency grid used for the solution of the equation of transfer.

3. RESULTS

The models have been calculated for a velocity ratio of 1.5 as can be observed for α Sco. The damping constant σ has been set to 10^{-3}.

The comparison of CR and R_{I-A} in fig.2a shows for models with total

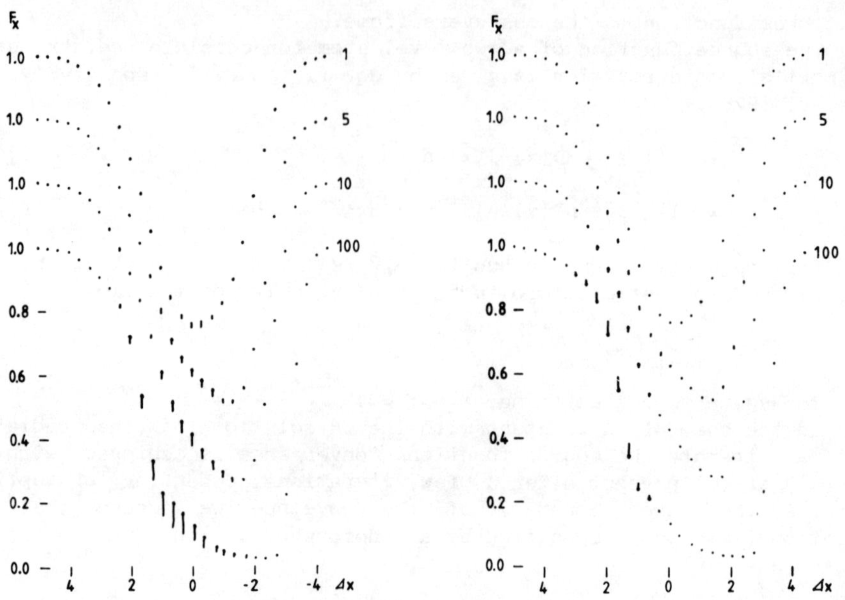

Figure 2a (left). Line profiles calculated with the assumption of CR and PR (R_{I-A}). At the right side the profiles are labeled with the total optical depth in the line. The profiles calculated with CR are marked by the dots at the end of the strokes. The length of the strokes are a measure of the differences between CR and PR. The velocity law is $v=1.5(1-.9/r)^{.5}$ and the parameters of the photospheric component are a=.9, b=2.
Figure 2b (right). Same as fig.2a, but for R_{II-A}. In the model with total optical depth τ =100, the CR solution is below the R_{II-A} solution.

optical depth of 1,5,10, and 100 small differences within 1 Doppler frequency in the emission part of the profile, i.e. the red wing of the photospheric line. In general the difference will be enhenced with increasing optical depth. Using the R_{II-A} function (fig.2b) we find the same behaviour but the differences are nearly negligible. The largest differences occur at 1 to 2 Doppler units outside of the core at the red wing of the photospheric line. To study the effect of the frequency dependent background continuum we calculated three models (fig.3) with different widths of the photospheric components. The parameter b (eq.4) has been chosen to 1, 2, and 4. Again redistribution effects will be largest for the R_{I-A} function compared to CR and the differences depend strongly on the background continuum. The larger the frequency derivative of the background continuum the larger the differences. The sensitivity of CR and PR to the velocity has been checked with velocity ratios of 0.5, 1.5, and 3. We find the well known result that the

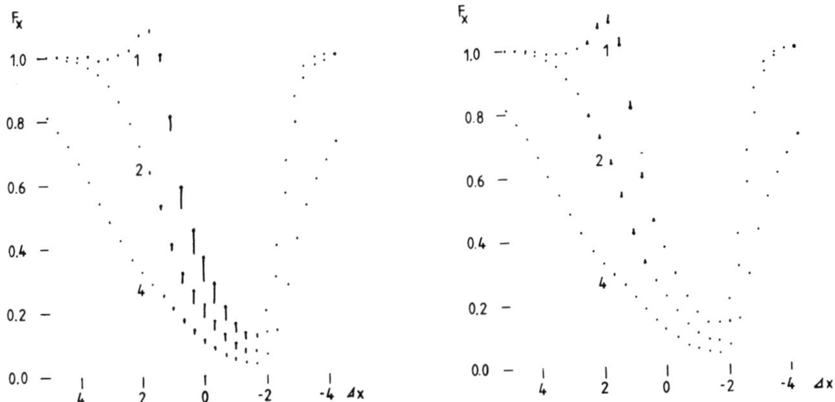

Figure 3a(left). The profiles are labeled with the width parameter b of the photospheric line. The CR solution is marked by the dots at the end of the strokes. The lengths of the strokes are a measure of the differences between the CR and the R_{I-A} solution.
Figure 3b (right). The same as fig. 3a, but for R_{II-A}. Here the CR solution is below the PR solution for b=1,2.

differences become smaller and smaller with increasing velocity.

The shapes of the line profiles can be understood by the run of the source function. In the line core the PR source functions (fig.4) are always below the CR source function. This effect is caused by the frequency gradient of the background continuum and the coherence of the scattering in the line wing. If we assume as a first approximation that no point in the flow is radiatively coupled to any other point in the flow than the local absorption profile sees only the blue part of the photospheric line. Therefore the radiation field in the blue wing of the local profile is stronger than in the line core and the red wing. In the case of the R_{II-A} function we have strong coherence in the wing and non-coherence in the core region. Thus after redistribution the source function will be less in the core, but higher in the wing compared to CR. This explains why we will find less emission in lines of low and moderate optical depth compared to CR. At low optical depth we see the source function at the line core, and at high optical depth we see the source function at the blue wing of the local profile. The radiation field in the line core obtained by CR and R_{II-A} are very similar, because the scattering in the line core is non-coherent. Therefore the profiles are similar too. The scattering described by the R_{I-A} function is only non-coherent for photons scattered exactly at the line center. All other photons at frequencies x will be redistributed with equal probability over an region $\Delta x=\pm x$. Therefore the radiation field in the line core is smaller than in the case of CR. Consequently the line source function is below the CR solution in the line core.

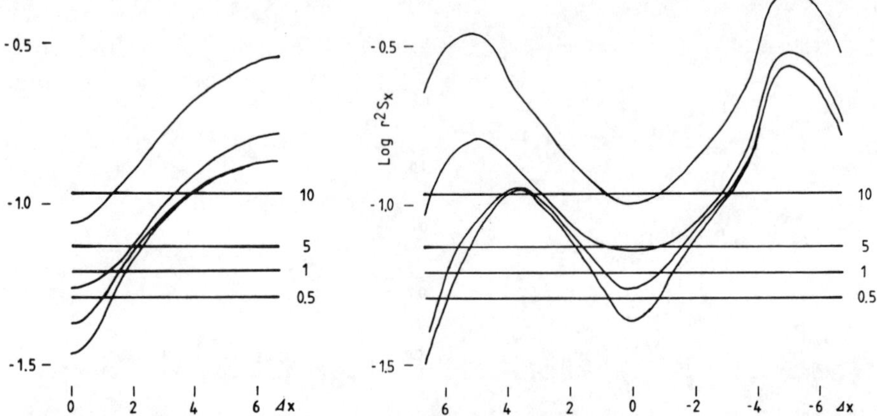

Figure 4a (left). Source functions in an isothermal, spherical expanding atmosphere with ε =0, assuming pure Doppler redistribution. Straight lines:CR solution. Abscissa:displacement from line center in Doppler units. The curves are labeled by the optical depth.
Figure 4b (right). Same as fig. 4a, but for a line with $\sigma = 10^{-3}$, assuming coherent scattering in a broadened profile in the atoms frame with Doppler redistribution in the laboratory frame. Straight lines:CR solution. Abscissa:displacement from line center in Doppler units.

4. CONCLUSIONS

Partial redistribution with broadening by radiation damping (R_{II-A}) compared to complete redistribution shows only small differences for lines formed in a typical late type giant envelope. The differences depend mainly on the frequency gradient of the background continuum, the optical depth and the wind velocity. In the case of coherence in the rest frame of the atom (R_{I-A}) the differences are larger.

ACKNOWLEDGEMENT

The scanner observations were made at the European Southern Observatory, La Silla, Chile. We thank ESO for the appropriate allocation of observing time.

REFERENCES

Cannon,C.J.,Vardavas,I.M.:1974, Astron.Astrophys. 32, 85
Che,A.,Hempe,K.,Reimers,D.:1983, Astron.Astrophys. 126, 225

Deutsch,A.J.:1956, Astrophys.J. **123**, 210
Drake,S.A.,Linsky,J.L.:1983, Astrophys.J. **273**, 299
Hamann,W.-R.,Kudritzki,R.-P.:1977, Astron.Astrophys. **54**, 525
Hamann,W.-R.:1980, Astron.Astrophys. **84**, 342
Hummer,D.G.:1962, Monthly Notices Roy.Astr.Soc. **125**, 21
Magnan,C.:1974, Astron.Astrophys. **35**,233
Mihalas,D.,Kunasz,P.B.,Hummer,D.G.:1975, Astrophys.J. **202**, 465
Mihalas,D.,Kunasz,P.B.,Hummer,D.G.:1976, Astrophys.J. **205**, 492
Mihalas,D.:1978,Stellar Atmospheres (San Francisco:Freeman), p. 438
Peraiah,A.:1979, Kodaikanal Obs.Bull.Series A, **2**, 203
Reimers,D.:1981, Physical Processes in Red Giants (ed. I.Iben, A. Renzini), Reidel, p. 269
Sanner,F.:1976, Astrophys.J.Suppl.Series **32**, 115
Wehrse,R.,Peraiah,A.:1979, Astron.Astrophys. **71**, 289
Weymann,R.:1962, Astrophys.J. **136**, 844

MODELING LINES FORMED IN THE EXPANDING CHROMOSPHERES OF RED GIANTS

Stephen A. Drake
Joint Institute for Laboratory Astrophysics, University of Colorado and National Bureau of Standards, Boulder, Colorado 80309

ABSTRACT. In this paper, I discuss the application of radiative transfer techniques to the study of the physical conditions in the extended, expanding chromospheres of cool giants and supergiants in the spectral ranges K2-M5 (for luminosity type III stars) and G0-M5 (for supergiants). The important diagnostic feature indicating the outflow of chromospheric material in such stars is the presence of blue-shifted absorption components in the h and k resonance lines of Mg II. To model these lines, I use a spherically symmetric co-moving frame solution of the equation of radiative transfer that takes proper account of partial redistribution effects. I give, as a specific example, my preliminary results of the study of the Mg II lines in the K2 III star α Boo.

1. INTRODUCTION

There is a growing body of evidence that late-type giants and supergiants in a wide region of the H-R diagram have geometrically extended, outflowing chromospheres (Stencel 1982; Carpenter et al. 1985). For MK class III stars, the spectral range is approximately from K2 to M5, with giants earlier than K2 having geometrically thin chromospheres and coronae, and giants later than M5 having cool, dusty winds. For MK class I and II stars, the spectral range of stars showing this phenomenon is broader, extending perhaps from G0 to M5. The two basic observational techniques available to study these "warm" (5×10^3 K $\lesssim T_e \lesssim 10 \times 10^3$ K) winds are (a) observations in the radio continuum at centimeter wavelengths and (b) observations of resonance lines of dominant stages of ionization of abundant atoms, e.g., Mg II h and k lines at λ2796 Å and 2803 Å, Ca II H and K lines at λ3933 Å and λ3968 Å, Mg I λ2852 Å.

The winds of K2-4.5 III stars are particularly difficult to study because only the Mg II lines show the blue-shifted absorption components indicating outflow, presumably because the mass loss rates are so small ($\dot{M} \lesssim 10^{-10}$ $M_\odot$ yr^{-1}) that the opacity in the wind for the other lines such as Ca II H and K is insufficient to produce similar

J. E. Beckman and L. Crivellari (eds.), Progress in Stellar Spectral Line Formation Theory, 351–357.

features. Also, only one such star (α Boo, K2 IIIp) has been detected as a weak radio continuum source (Drake and Linsky 1984), indicating the limited applicability of the first approach for determining the physical parameters in the wind.

Analysis of the profiles of the Mg II h and k lines in a particular cool star should in principle provide specific information on both the mass flux of the wind <u>and</u> the variation with radius, r, of quantities such as the electron temperature, T_e, and the density, ρ, from the photosphere up to that point in the wind where the lines become optically thin. In practice, it is difficult to "deconvolve" this information uniquely from <u>one</u> observed line profile, and any physical quantities inferred from such an analysis, while "sufficient" to reproduce the observed profile may not "necessarily" be the actual ones. Thus, such an analysis should be cross-checked as much as possible with all other pertinent information known about the chromosphere of the star being studied.

If reliable parameters can be established for cool giant winds, however, the resultant information would be valuable in helping to discriminate between the various mechanisms that have been proposed for generating such winds (for example, by comparison of theoretical and observationally inferred $T_e(r)$ distributions). Another important application would be in the field of stellar evolution: the observationally inferred mass loss rates of stars in the red giant phase can be compared with the integrated mass loss rates deduced from comparison of theoretical and observational cluster diagrams, and could provide a new constraint or test on the validity of the present evolutionary calculations.

In Sec. 2, I discuss how the previously deduced physical conditions in the Mg II line formation region of cool giants, and the physics of radiative transfer in these resonance lines, determine the particular analytical tools needed to study them. In Sec. 3, I apply these techniques to the Mg II emission lines of the K2 IIIp star α Boo, and, in the final section, I present the conclusions of this study and some suggestions for future work in this area.

2. RADIATIVE TRANSFER METHODOLOGY ADOPTED FOR THIS STUDY

The proper method for solving a line radiative transfer problem in a moving atmosphere is clearly dependent on both the macroscopic properties of the wind and on the atomic properties of the particular line being analyzed. The winds of K and M giants of luminosity class III are known to have the following properties:

(i) The winds are not isothermal -- they originate in the lower chromosphere where $T_e \gtrsim T_{eff}$, further out the temperature reaches a maximum value of $\lesssim 1.5 \times 10^4$ K, and then at greater distances from the star (in the M giants) the temperature must decrease to $\lesssim 5 \times 10^3$ K.

(ii) The winds have maximum outflow velocities, V_{wind}, that are typically in the range 10-50 km s^{-1}.

(iii) The microturbulence, V_{Dopp}, deduced for the chromospheric lines is of order 5-20 km s^{-1}.

(iv) The chromospheres extend out to at least a few stellar radii.

The low ratio of systematic to random velocities, which is commonly true for the winds of cool giants, means that approximate radiative transfer solutions of the Sobolev type (valid in the high velocity limit) will not be reliable, and that an exact method must be used. As discussed by Drake and Linsky (1983a), for these cases it is also not appropriate to assume complete frequency redistribution in the radiative transfer for a resonance line, and the correct redistribution function should be used, viz.:

$$R(\nu',\nu) = \gamma\ R_{II}(\nu',\nu) + (1-\gamma)\ \phi_\nu\phi_\nu' \quad ,$$

where I have used the standard notation [cf. Mihalas 1978, eq. (13-72)].

I have therefore adopted a spherically symmetric co-moving frame (CMF) technique for the radiative transfer, based on an original program kindly provided by Paul Kunasz, and modified it to include partial redistribution of this particular type (see Drake and Linsky 1983a for further details). I have used a two-level atom to represent the upper and lower level of the resonance line.

I have used angle-averaged redistribution functions in this analysis, but as Milkey *et al.* (1975) point out, this approximation may be inaccurate for the case of extended chromospheres with velocity fields. To the best of my knowledge, no detailed wind model calculations have yet been done using the full angle-dependent partial redistribution functions to verify this supposition. Another limitation of the particular CMF formulation used is that it can correctly handle only monotonic velocity laws such as accelerating outflows, but this restriction is probably not too important for this particular study.

3. MODELING THE Mg II k LINE OF ARCTURUS

I have adopted a semi-empirical approach in applying the techniques discussed in Sec. 2 to fitting an observed Mg II profile. I have not attempted to adopt $T_e(r)$ and $V_{wind}(r)$ laws predicted by specific theoretical wind models as Hartmann and Avrett (1984) did in their study of the wind of α Orionis (M2 Iab). For the photosphere and temperature minimum region of the atmosphere of α Boo where $V_{wind} \ll V_{Dopp}$, I have adopted the Ayres and Linsky (1975) plane parallel model atmosphere. This assumption has little effect on the modeling of the Mg II k line emission core. For the outer region of the chromosphere where the outflow velocity is significant, I choose on a trial and error basis the following quantities: $T_e(r)$, $V_{wind}(r)$, $V_{Dopp}(r)$, and $\dot{M}$. (The density $\rho(r)$ is implicitly fixed, given $\dot{M}$ and $V_{wind}(r)$, by the equation of continuity.) Since all of these parameters help determine the resultant line profile, it is difficult to determine the actual atmospheric structure uniquely and more than one valid solution that fits the observed profile well might exist.

Given this input model atmosphere, the radiative transfer in the specified line is carried out in the co-moving frame (CMF) or fluid frame, and the emergent line profile calculated after a transformation into the observer's frame (Mihalas et al. 1976). The theoretical line profile of say, the k line, is then compared with the observed line profile (see Fig. 1), and the nature of the discrepancies between the two usually helps one to produce a new "improved" set of input parameters. This procedure is repeated for many iterations until "optimal" agreement is reached between the two profiles. Because of the number of free parameters and the wide range of parameter space that has to be explored, this iterative process requires a large amount of time and the decision as to when the "optimal" or best fit has been reached is clearly subjective. Since this process may not lead to unique atmospheric parameters, I believe that it is crucial to compare the predicted properties of the "optimal" fit model(s) with as many other observational constraints as possible.

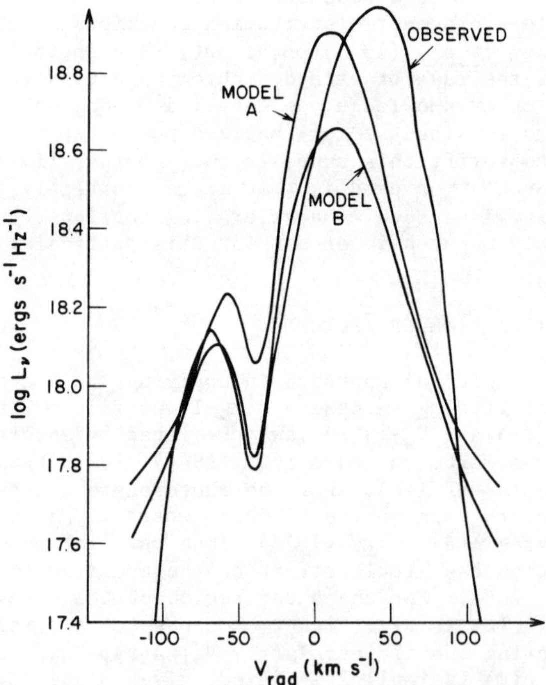

Figure 1. The observed profile of Mg II K in α Boo is compared with two calculated profiles. The ordinate is monochromatic luminosity in ergs s^{-1} Hz^{-1} and the abscissa is radial velocity relative to the stellar photosphere in km s^{-1}. The models are discussed in the text.

In Figure 1, I compare the observed Mg II k profile in α Boo with that predicted by two model atmospheres. The agreement is not perfect: Model A has about the right peak emission but the blue-shifted absorption is weaker than is observed, while model B fits the absorption well but underestimates the peak emission by a factor of 2. In Figure 2, the temperature and velocity structure of model B is shown. The other properties of this particular model atmosphere can be summarized as follows:

(i) The wind velocity and electron temperature climb steeply to their maximum values of 40 km s^{-1} and 8400 K, respectively, in a fairly short distance above the photosphere.

(ii) There is a broad high temperature plateau with $T_e \gtrsim 7 \times 10^3$ K extending from 1.2 to ~13 r_*.

(iii) There is a region further out, extending from 13 to 50-100 r_*, where the temperature is rather cooler ($T_e \sim 5 \times 10^3$ K).

(iv) The mass loss rate is 2×10^{-10} $M_\odot$ yr^{-1}, and the ionization fraction of the outer atmosphere is about 50%.

(v) The maximum microturbulence reached is 5 km s^{-1}.

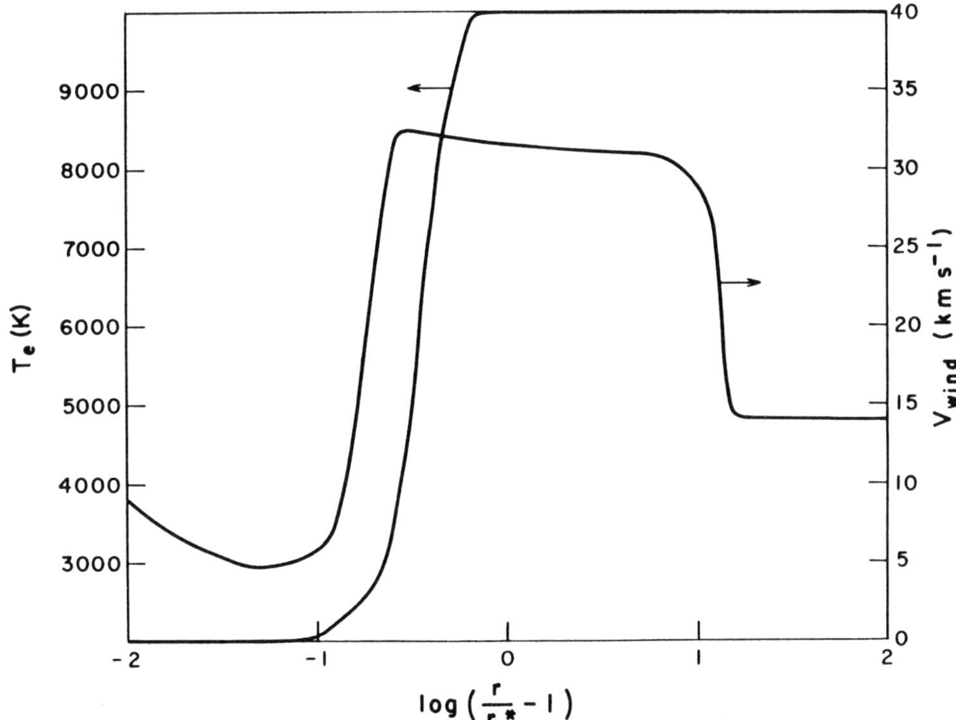

Figure 2. The wind velocity and electron temperature for Model B are shown as a function of radial distance above the photosphere (see text).

The major difference between Model B and Model A is that the latter has an even more extensive high temperature plateau region extending as far as 35 r_* (required to produce the stronger Mg II peak emission of this model). Specific predictions that can be made based on Model B are:

(i) The angular size of Arcturus in the Mg II k line is large: assuming a photospheric angular diameter of 0".021, the predicted size of α Boo in the integrated Mg II emission is ~0".27.

(ii) The free-free emission from the ionized wind of α Boo should make it a radio continuum source of ~0.4 mJy at 6 cm.

(iii) The angular size of Arcturus imaged in the blue absorption component of Mg II k is ~1".

The only one of the above predictions that has been tested to date is that α Boo should be radio source: Drake and Linsky (1983b) have detected α Boo as a 0.25 mJy source at 6 cm, in good agreement with the predicted value. The remaining predictions should be easily verifiable or disproven by the Faint Object Camera on Space Telescope.

An independent estimate of the chromospheric temperature can be obtained from the ratio of the C II resonance and intersystem lines. Brown and Carpenter (1984) estimated $\langle T_e \rangle \simeq 8 \times 10^3$ K (±500 K) for α Boo, which is also consistent with the models discussed above. Using the same PRD program, I also plan to calculate the Lyman-α profile for α Boo and to compare it with the observed profile (e.g., McClintock *et al*. 1978). Although Lyα is strongly affected by both interstellar absorption and geocoronal emission, any realistic Mg II model atmosphere should be able to reproduce approximately the observed Lyα strength and shape.

One final property of these models is that the line profiles calculated assuming CRD are nearly the same as those shown in Fig. 1 (where PRD is assumed). This is presumably due to the fairly high V_{wind}/V_{Dopp} ratio in these particular models. It may also be due in part to the Gaussian smearing by 25 km s^{-1} FWHM that has been performed on the theoretical profiles in order to reproduce the instrumental resolution of the observed *IUE* profile.

4. SUMMARY

I have described a technique for modeling the chromospheric lines of late-type stars and applied it to the K2 IIIp giant α Boo. The models which come closest to reproducing the observed Mg II k profile are not inconsistent with other evidence available concerning the mass flux and temperature of the expanding chromosphere of this star. However, additional observational evidence is still needed for complete verification.

As for future directions of research, I think that it would be fruitful to apply this technique to later type giants like α Tau (K5 III) or β And (M0 III), where there are additional lines showing evidence for expansion at chromospheric temperatures, and there will thus be much tighter constraints on any model atmosphere. I also aim to compare my best fit α Boo models with the atmospheric structure predicted by _theoretical_ models of the winds from the stars, e.g., the Hartmann and MacGregor (1980) Alfvén-wave driven winds. Finally, it would be most valuable as a consistency check, if two independent groups were to model the same star using these techniques. Any differences between the "best-fit" model atmospheres would provide invaluable insight as to the reproducibility of these techniques.

REFERENCES

Ayres, T. R. and Linsky, J. L. 1975, _Ap. J._, _200_, 660.

Brown, A. and Carpenter, K. G. 1984, _Ap. J. (Letters)_, _287_, in press.

Carpenter, K. G., Brown, A. and Stencel, R. E. 1985, _Ap. J._, in press.

Drake, S. A. and Linsky, J. L. 1983a, _Ap. J._, _273_, 299.

_____. 1983b, _Ap. J. (Letters)_, _274_, L77.

_____. 1984, in preparation.

Hartmann, L. and Avrett, E. H. 1984, _Ap. J._, in press.

Hartmann, L. and MacGregor, K. B. 1980, _Ap. J._, _242_, 260.

McClintock, W., Moos, H. W., Henry, R. C., Linsky, J. L. and Barker, E. S. 1978, _Ap. J. Suppl._, _37_, 223.

Mihalas, D. 1978, _Stellar Atmospheres_ (Freeman: San Francisco).

Mihalas, D., Kunasz, P. B. and Hummer, D. G. 1976, _Ap. J._, _210_, 419.

Milkey, R. W., Shine, R. A., and Mihalas, D. 1975, _Ap. J._, _202_, 250.

Stencel, R. E. 1982, in _Second Cambridge Workshop on Cool Stars, Stellar Systems, and the Sun_, SAO Special Report No. 392, Vol. 1, p. 137.

TRANSFER OF LYMAN-α RADIATION IN SOLAR CORONAL LOOPS

P. GOUTTEBROZE, J.-C. VIAL and G. TSIROPOULA
L.P.S.P., PO Box 10
F-91370 Verrieres-Le Buisson, France

ABSTRACT The emission and scattering of Lyman-α radiation within the loop-like structures of the solar corona are investigated, for a large range of physical conditions within these objects. Different methods are used, which apply to different physical cases and allow different insights into the problem:

i Statistical equilibrium of a 3-level plus continuum atom and single scattering, in the optically thin case.

ii 2-level Monte-Carlo technique in cylindrical geometry, for moderate optical depths ($\tau \lesssim \sim 1000$).

iii 3-level plus continuum, plane-parallel, difference-equation methods for large optical depths.

Results from partial and complete redistribution computations are compared. A series of predictions, concerning line profiles, integrated intensities, and directional diagrams are given for observation diagnosis. Prospects for more general (multi-level, multidimensional) methods are presented.

1. INTRODUCTION

Lyman-α filtergrams of the Sun with high angular resolution, such as those obtained with the "Transition Region Camera" (TRC; cf. Bonnet et al., 1980), show that the regions of the solar atmosphere where the Lyman-α line is formed have a very complex structure. Thus, a substantial part of the Lyman-α emitting (or absorbing) material appears

J. E. Beckman and L. Crivellari (eds.), Progress in Stellar Spectral Line Formation Theory, 359–372.

to be concentrated in loop-like objects, probably due to magnetic fields. In order to use these pictures for diagnosing the structure of the high chromosphere, as well as preparing future observations with improved resolution (ATRC2/SPARTAN), we examine in details the emission of Lyman-α radiation by structures of finite size lying at low altitude above the solar surface (i.e. embedded in the solar corona).

Since there is no numerical method to treat the general problem of objects with arbitrary shapes and opacities, we limit ourselves to three particular, but representative, cases. The first one is that of objects, with arbitrary shapes, which are optically thin in every transition of the hydrogen atom. The second case is that of cylindrical objects, simulating coronal loops, that are optically thick in Lyman-α , but optically thin for bound-free transitions (especially the Lyman continuum). The third case is that of plane-parallel slabs, with arbitrary optical thicknesses. In every case, we assume that temperature and pressure are uniform throughout the object.

In the following, to allow easier comparisons with observation, we often express the intensity integrated over frequency as "LRI" ("Lyman-α Relative Intensity"). It is defined as the ratio of the intensity emitted by the object, integrated over a 100 nm wavelength bandpass around line center, to the corresponding quantity emitted by the average solar disk.

2. THE OPTICALLY THIN CASE

We consider here some object with an arbitrary shape. Its optical thickness at Lyman-α , along any solar radius, is lower than 1, so that it is also optically thin for every other transition of the hydrogen atom. The state variables and the radiation fields being constant (and known) throughout the object, it is sufficient to solve the statistical equilibrium to obtain the hydrogen ionization ratio, absorption coefficient and source function. The emitted intensities are then obtained by integration along the line-of-sight of observation.

Our hydrogen model atom includes three levels and one continuum. For the three lines ($L\alpha$, $L\beta$ and $H\alpha$), the incoming intensities from the solar disk are taken from observations. The radiation temperatures for the Lyman, Balmer and Paschen continua are set to 6500 K, 5000 K and 4800 K, respectively. More details about these computations are given in the paper of Tsiropoula et al. (1984).

The analysis of statistical equilibrium shows that, among the processes that can populate the level 2 of hydrogen atoms and thus produce Lα emission, two are dominant: excitation from the ground level by electron collision and absorption of Lα photons from the solar disk (i.e. scattering). The relative importance of these two processes depends principally on the pressure and, but weakly, on the temperature. Roughly speaking, the Lα emission is dominated by scattering at low pressures ($p \leq 0.5$ dyn cm^{-2}) and by collisional creation at high pressures ($p \leq 1$ dyn cm^{-2}).

Figure 1 summarizes the results of these calculations for an object whose geometrical thickness is 1 Mm, as a function of pressure and temperature:

a- the optical thickness of the object is approximately proportional to pressure (indicated on the curves in dyn cm^{-2}), and decreases rapidly with temperature, as a result of ionization (and also of density).

b- this figure shows Lyman-α line profiles (the intensity is in erg s^{-1} cm^{-2} sr^{-1} Hz^{-1} and $\Delta\nu$ in THz). The solid curve represents the profile emitted by the solar disk. The dotted and the dashed lines represent the profiles emitted (when seen outside the solar disk) by a low pressure, low temperature object (0.1 dyn cm^{-2}, 70 000 K), with the assumptions of complete and partial frequency redistribution, respectively. Complete redistribution produces line wings somewhat higher than partial redistribution, but the consequences on integrated intensities are negligible. The dot-dashed line represents the profile emitted by a high pressure, high temperature object (10 dyn cm^{-2}, 500 000 K). In this case, there is no significant difference between complete and partial redistribution profiles, since scattering is dominated by collisional excitation.

c- the relative intensity (LRI) for an object seen outside the solar disk decreases with temperature, as a result of the decrease of opacity, and increases with pressure. This increase of LRI is approximately linear for low pressures (scattering being proportional to density), and quadratic for high pressures, when collisions dominate.

d- this figure represents the LRI for objects observed on the solar disk, and shows the different aspects of low and high pressure objects. The first ones appear darker than the surrounding solar disk, since the scattering, which absorbs radiation within $2\times\Pi$ sr and reemits it within 4Π sr, produces a decrease of intensity. On the contrary, high pressure features, dominated by collisional excitation, appear brighter than the disk.

3. INTERMEDIATE OPTICAL THICKNESSES: EMISSION BY CYLINDRICAL OBJECTS

The optically thin case treated above applies to a number of structures seen on Lα filtergrams, such as faint objects visible near the solar limb. Nevertheless, the existence of dark structures on the solar disk of bright objects at the limb indicates that the optically thick case should also be considered. The study of optically thick objects implies the treatment of radiative transfer effects and needs the definition of the geometry of these objects. Since this study is principally aimed at coronal loops, we treat now the case of cylindrical objects lying at low altitude above the solar surface. This case can be easily treated by the Monte-Carlo method, provided that a two-level atom may be reasonably assumed, and that the optical thickness is not too high. Since the two dominant processes for Lα emission (scattering and collisional creation) involve only levels 1 and 2, a 2-level atom is a possible assumption for this line. The restriction on optical thickness comes both from considerations on computational efficiency, and departures from "optically thin ionization equilibrium" which appears when the opacity in the Lyman continuum cannot be neglected. Firstly, the computer time needed to process each photon, which is roughly proportional to the number of scatterings, grows rapidly with optical thickness. Secondly, the ionization equilibrium is principally controlled by radiative processes in the Lyman and Balmer continua. While the Balmer continuum remains optically thin even at large opacieties in Lyman-α , the ratio between the absorption coefficient at the head of the Lyman continuum, and the Lα mean absorption coefficient lies between 10^{-4} and 10^{-3} for temperatures considered here. We thus restricted these computations to Lα mean optical depths lower than 10^{3}.

The principle of the calculation is summarized on Fig. 2a: there are two sources of primary photons: emission by the solar disk (in a 2Π steradian solid angle) and internal creation by electron collisions and other processes, except Lα absorption, that can populate level 2. The ponderation between these two sources is defined by temperature and pressure, as seen in the preceding section. On the other hand, processes other than Lα emission which depopulate level 2 result in a "destruction probability" which is applied after each absorption. Photons escaping from the cylinder, after a number of scatterings, are recorded by different counters: firstly, directional counters record their polar coordinates (θ , ϕ). We also define two exit "windows": an upper window ($\theta \leq 30°$) simulates observation near the center of the solar disk, while a side window ($60° \leq \theta \leq 120°$, $-30° \leq \theta \leq 30°$) represents observation near the limb, perpendicularly to the axis of the cylinder. For photons escaping by each of these windows, we record their frequency ($\Delta\nu$) and their position relative to axis of the cylinder (ξ), in order to have a transversal picture of the object. Some similar counters record primary photons emitted by the disk, to obtain the reference flux for the calculation of the LRI.

On Figures 2b to 2d, we show a sample of typical results, for a cylinder with a diameter of 1 Mm, a temperature of 30 000 K, and a pressure of 0.1 dyn cm^{-2} (the mean optical thickness along a diameter is about 180). This object is observed through the side window, i.e. out of the solar disk. Figure 2b compares the line profiles emitted by the sun and by the cylinder: the second one is narrower than the first (an effect of optical thickness), brighter in the line core and darker in the wings. Figure 2c represents the variation of intensity (integrated over frequency) along a line crossing perpendiculary the cylinder. The upper curve represents the reference flux of the Sun, which should be constant: the variations are due to the sampling noise pertaining to the Monte-Carlo method, and give an information about the precision of the results. It appears that the brightness of the object is somewhat lower than that of the average Sun, and that the sides are darker than the middle. Figure 2d is a polar diagram of emission in an horizontal plane. For this relatively high optical thickness, the emission is almost isotropic. When one approaches the optically thin case, a reinforcement of emission appears along the axis of the cylinder, as shown on Figure 2e, which represents the same diagram for a similar, but hotter, loop ($T = 50\ 000$ K, $\tau \sim 5$).

Using several numerical experiments of this kind, we can complement, towards low temperatures, the diagram of Fig. 1c giving the variations of LRI with pressure and temperature. The result is shown on Fig. 3, for a cylinder with a diameter of 1 Mm (intensities are taken near the middle of the object).

4. LARGE OPTICAL DEPTHS

For optical depths larger than 10^3, the preceding method is no longer valid: the Lyman continuum opacity cannot be neglected and important changes of ionization occur within the object. So that we return to a multilevel atom (three levels plus continuum) and solve simultaneously the equations of equilibrium for atomic level populations, ionization state and pressure, and radiative transfer for lines and continua. For the sake of tractability, we limit ourselves to a plane-parallel, horizontal slab and solve the transfer equations using a finite-difference scheme. The lower face of this slab receives the same solar radiation field as in the preceding sections. This geometry does not permit the computation of intensities emitted in an horizontal plane, so that this method will be restricted to the determination of intensities of objects observed on the solar disk.

A set of computations using this last method is used to complement the "contrast diagram", which gives the LRI for objects seen on the solar disk (near its center), as a function of temperature and pressure. The results obtained by the three methods for a slab geometrical thickness 1 Mm (or a cylinder with the same diameter) are summarized on Figure 4.

Towards high temperatures, since all objects become optically thin, the LRI tends to 1. When the temperature decreases, high pressure objects become brighter than the solar disk (the threshold is about 10^5 K for a pressure of 1 dyn cm^{-2}, and 4×10^5 K for 10 dyn cm^{-2}). When their optical depth exceeds 1, low pressure objects become darker than the solar disk. For very low pressures (0.01 dyn cm^{-2}), their LRI remains lower than 1. For intermediate pressures (0.1 dyn cm^{-2}), the variation of LRI temperature is more complex. Below 70 000 K, it begins to decrease, since scattering causes a dilution of radiation (as seen in the preceding section). About 40 000 K, the intensity becomes to increase, probably as a result of radiation trapping inside the object, which causes an increase of the local radiation field. A maximum is reached near 20 000 K. At low temperatures ($T \lesssim$ 15 000 K), the object becomes darker than the solar disk again.

For the same pressure, plane-parallel and cylindrical objects have similar variations. The differences between the corresponding curves arise both from the differences between the geometries and between the model atom (2 and 3 levels).

5. PROSPECTS

For the treatment of loop-like objects of large optical depth, and especially for the determination of their emission in an horizontal plane, the development of a radiative transfer method in cylindrical coordinates, using either an integral or a difference equation scheme, would be very useful. The radiative transfer problem considered here ia s full two-dimension problem, since the incident radiation varies with the angular coordinate. Thus, the ionization ratio, and consequently the line and continuum opacities, will vary as a function of the two (radial and angular) coordinates. This transfer method should be designed to be included within a loop with the other equations of the problem (statistical, pressure and ionization equilibria), in order to solve them iteratively. Such a method, using a difference equation scheme, is presently under study.

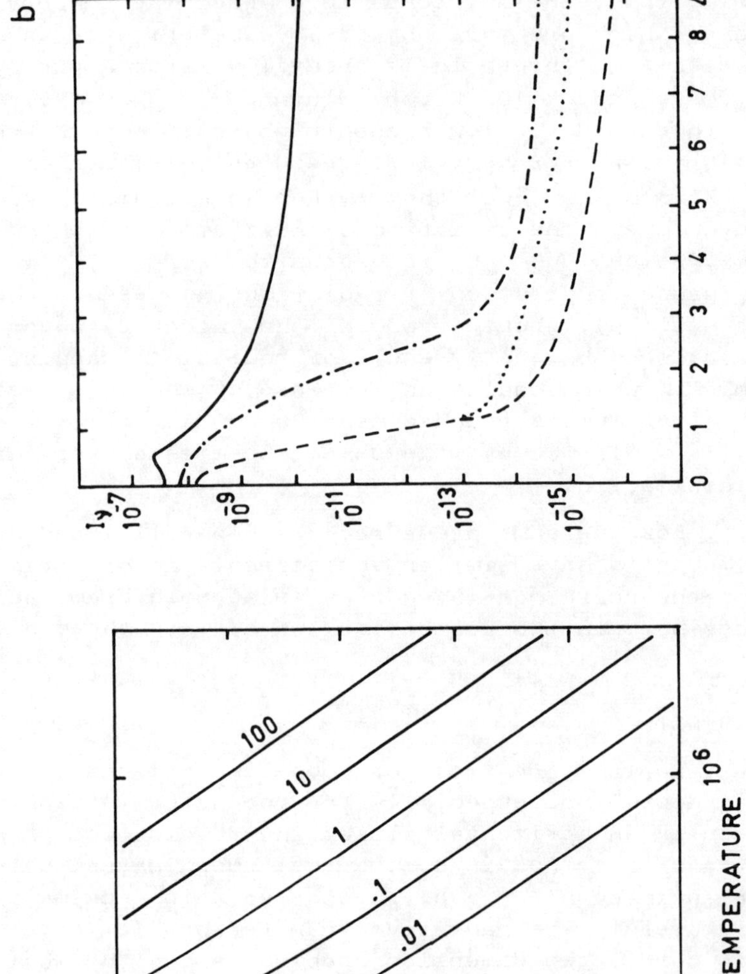

Fig. 1
a
b
OPTICAL THICKNESS
TEMPERATURE
100
10
1
.1
.01
10^0
10^{-1}
10^{-2}
10^{-3}
10^{-4}
10^5
10^6
I_ν
10^{-7}
10^{-9}
10^{-11}
10^{-13}
10^{-15}
0 1 2 3 4 5 6 7 8
$\Delta\nu$

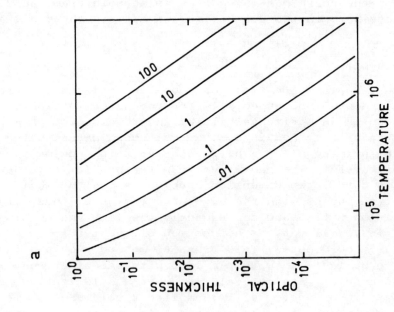

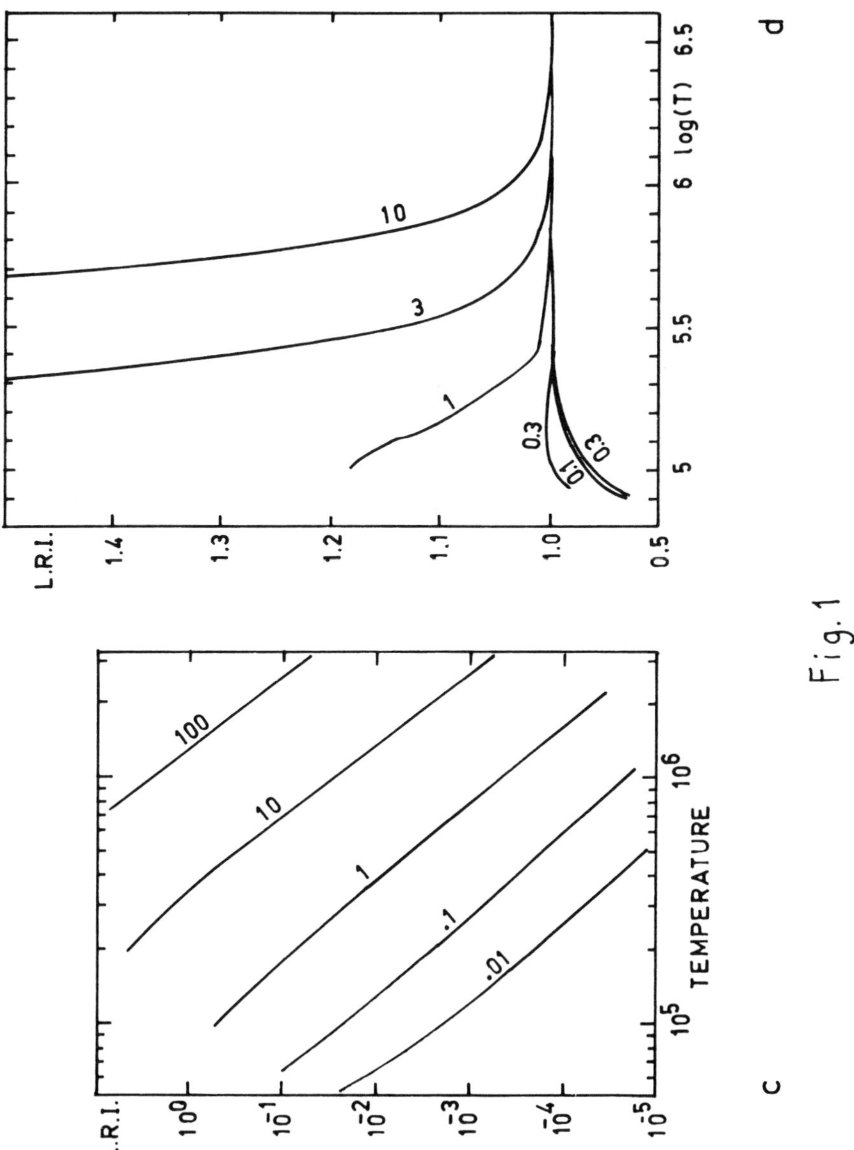

c
L.R.I.
10^0
10^{-1}
10^{-2}
10^{-3}
10^{-4}
10^{-5}
10^5
TEMPERATURE
10^6
100
10
1
.1
.01
d
L.R.I.
1.4
1.3
1.2
1.1
1.0
0.5
5
5.5
6
log(T)
6.5
10
3
1
0.3
0.1
0.3

Fig. 1

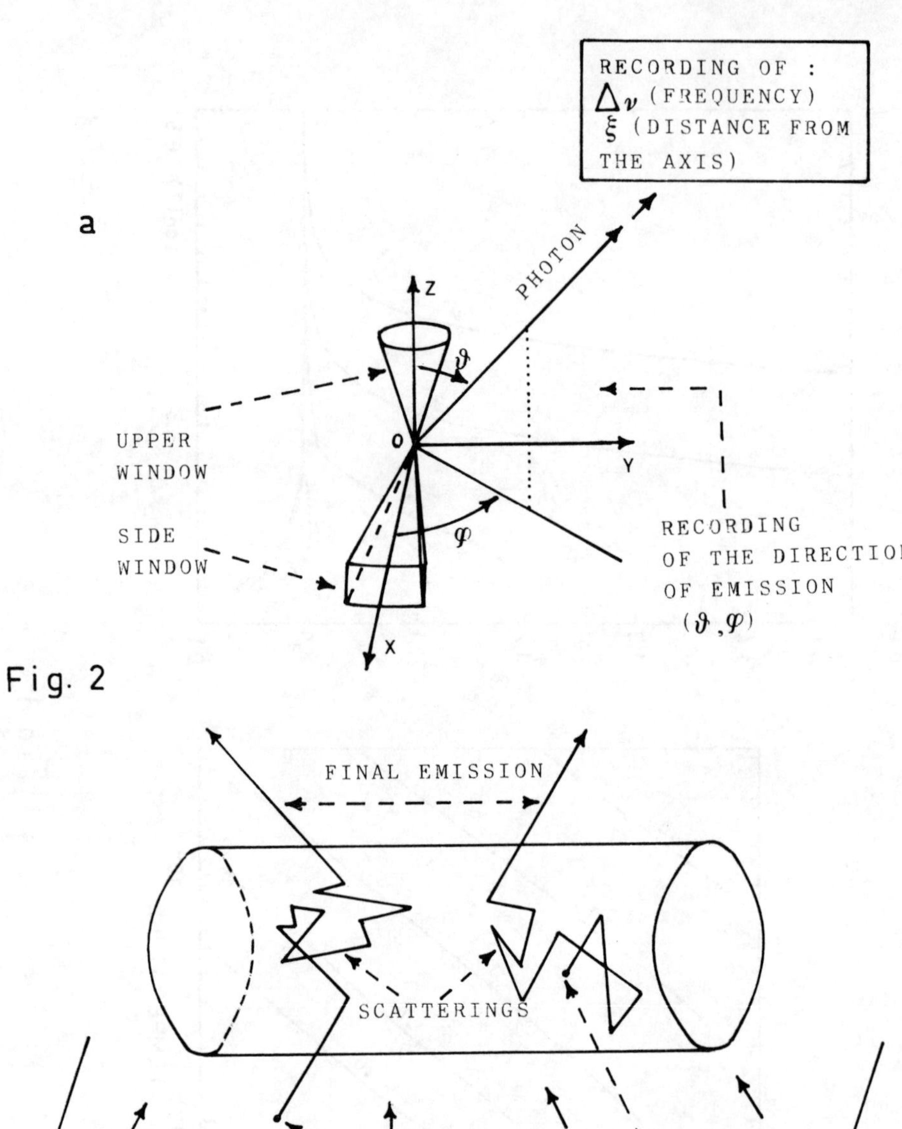
RECORDING OF :
$\Delta\nu$ (FREQUENCY)
ξ (DISTANCE FROM THE AXIS)
a
PHOTON
Z
ϑ
UPPER WINDOW
O
Y
SIDE WINDOW
φ
RECORDING OF THE DIRECTION OF EMISSION (ϑ, φ)
X
FINAL EMISSION
SCATTERINGS
SOURCE 1 EMISSION BY THE SUN
SOURCE 2 INTERNAL CREATION BY COLLISIONS
SOLAR SURFACE

Fig. 2

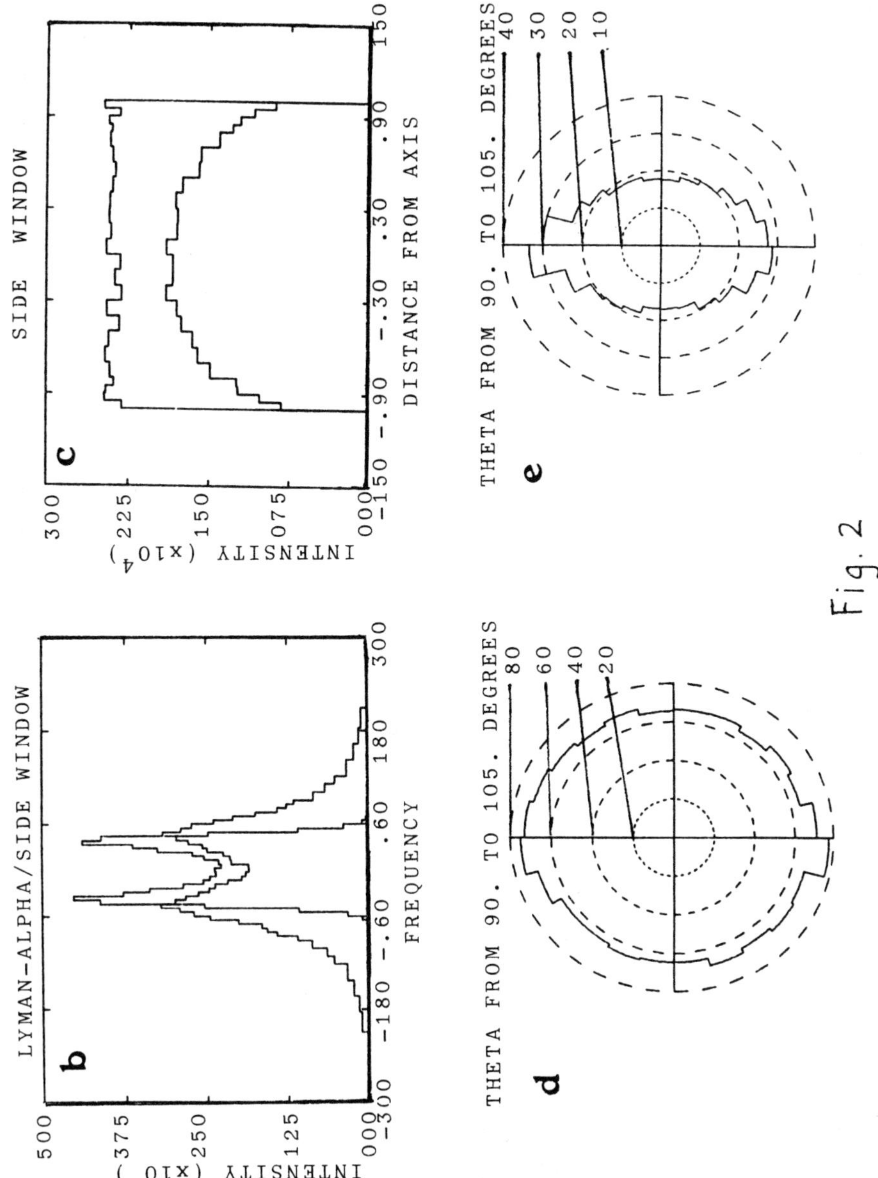
b
LYMAN-ALPHA/SIDE WINDOW
INTENSITY (x10^4)
500
375
250
125
000
-300
-180
-.60
.60
180
300
FREQUENCY
c
SIDE WINDOW
INTENSITY (x10^4)
300
225
150
075
000
-150
-.90
-.30
.30
.90
150
DISTANCE FROM AXIS
d
THETA FROM 90. TO 105. DEGREES
80
60
40
20
e
THETA FROM 90. TO 105. DEGREES
40
30
20
10

Fig. 2

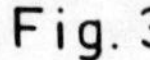

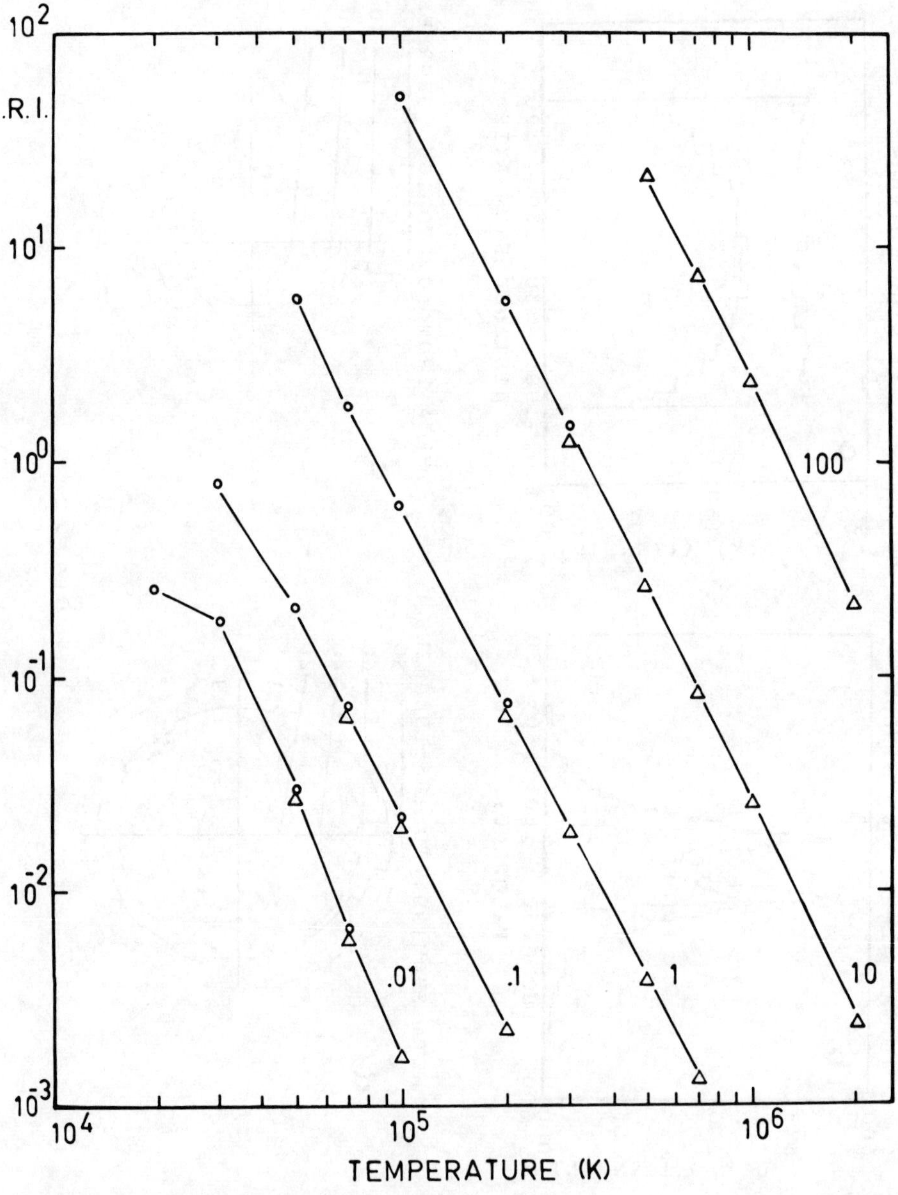

∘ ∘ ∘ cylinder (Monte-Carlo method)

△ △ △ optically thin computations

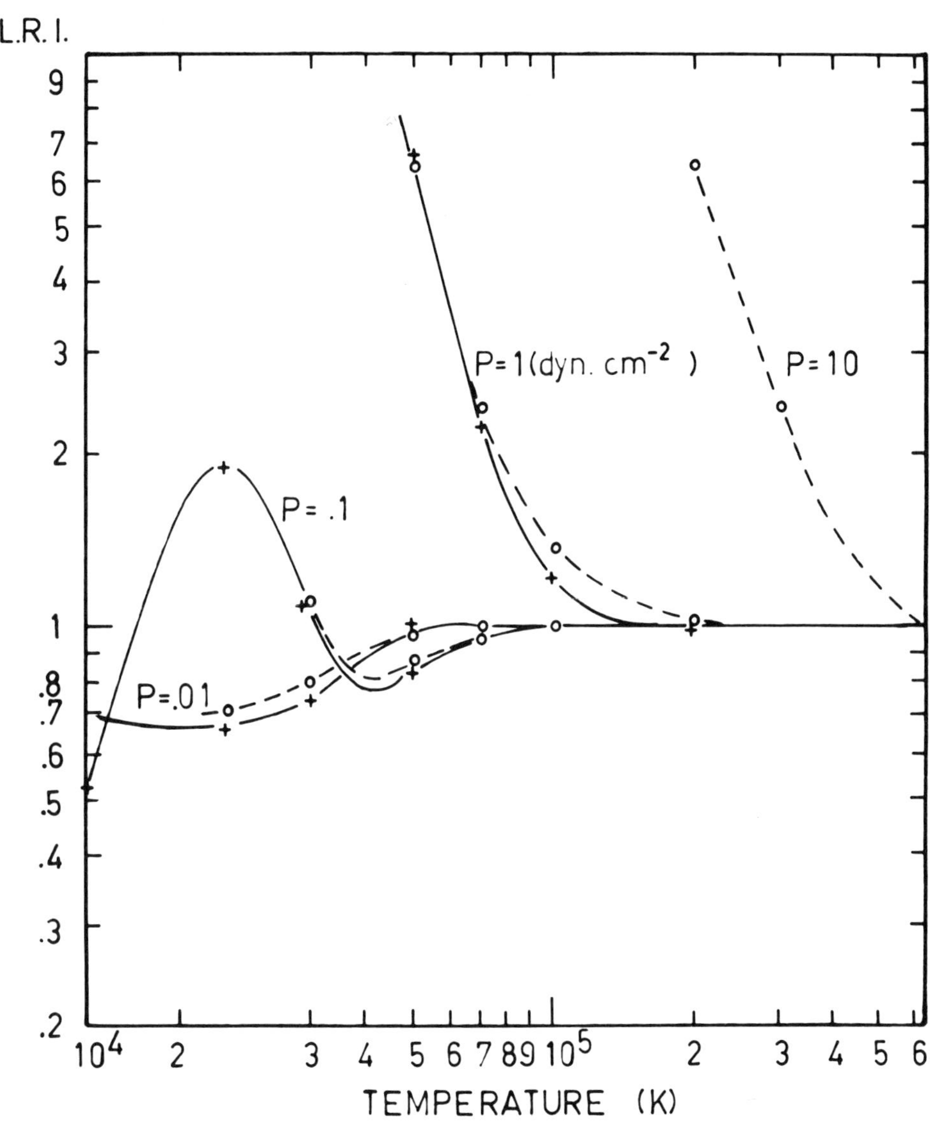

+—+—+ slab o---o---o cylinder

REFERENCES

- Bonnet, R.M., Bruner, E.C. Jr., Acton, L.W., Brown, W.A. and Decaudin, M. (1980), Ap. J. 237, L47

- Tsiropoula, G., Gouttebroze, P., Alissandrakis, C.E. and Bonnet, R.M., (1984), Submitted to Astronomy and Astrophysics.

PRESSURE BROADENING AND SOLAR LIMB EFFECT

Ištvan Vince, Milan S.Dimitrijević and
Vladimir Kršljanin
Astronomical Observatory
Volgina 7
11050 Beograd
Yugoslavia

ABSTRACT. The pressure broadening contribution to the Solar limb effect has been discussed for the case of a spectral series using Na I 3p-ns and 4p-ns series as an exemple. We have found that the influence of the pressure broadening on the Solar limb effect is reduced due to different signs of hydrogen and electron impact shifts in the case considered. We have discussed also the behaviour of pressure broadening contribution to the Solar limb effect within a spectral series and we have compared our results with simple radial-current theory predictions.

1. INTRODUCTION

Careful measurements of the Fraunhofer lines show a small but systematic red shift across the Solar disk (as compared with their wavelength at the centre) reaching a maximum on the limb. One of attempts to explain completely or partialy this effect is so called pressure broadening hypothesis (see e.g. Hart, 1974 or Vince et al., 1984). According to this hypothesis the observed line shifts are the consequence of collisions between the absorbing atoms and surrounding particles. In this paper, we will study the behaviour of pressure broadening contribution to the Solar limb effect within a spectral series.

2. RESULTS AND DISCUSSION

Recently, it was shown (Dimitrijević and Sahal-Bréchot, 1984a,b) that for lines belonging to a spectral series, electron and proton impact width increases gradually with increasing principal quantum number of the upper state. The electron and proton impact shift changes gradually

J. E. Beckman and L. Crivellari (eds.), Progress in Stellar Spectral Line Formation Theory, 373–380.

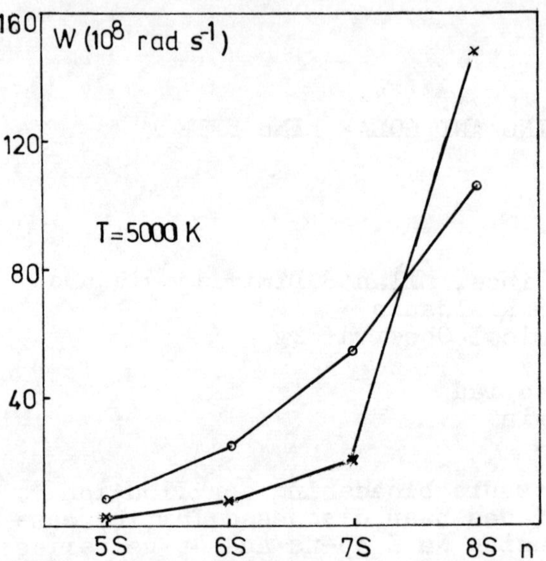

Figure 1. The full halfwidth due to collisions with electrons (x) and hydrogen atoms (o) for Na I 3p-ns lines as a function of n for T = 5000 K, $N_e = 10^{13}$ cm^{-3} and $N_H = 10^{16}$ cm^{-3}.

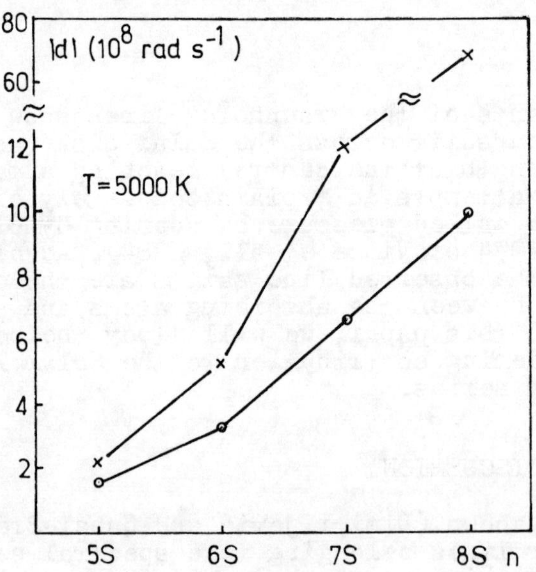

Figure 2. The same title as in Fig. 1 applies but for shift

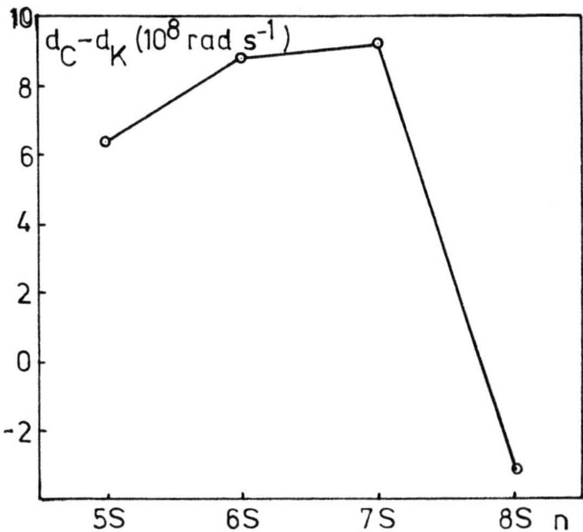

Figure 3. Differences between line shifts in the centre of the Solar disk and at the limb for Na I 3p-ns lines as a function of n.

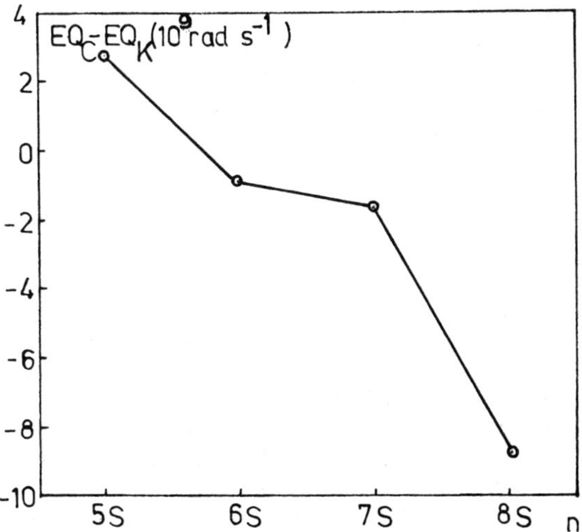

Figure 4. Differences between equivalent widths at the centre of the Solar disk and at the limb for Na I 3p-ns lines as a function of n.

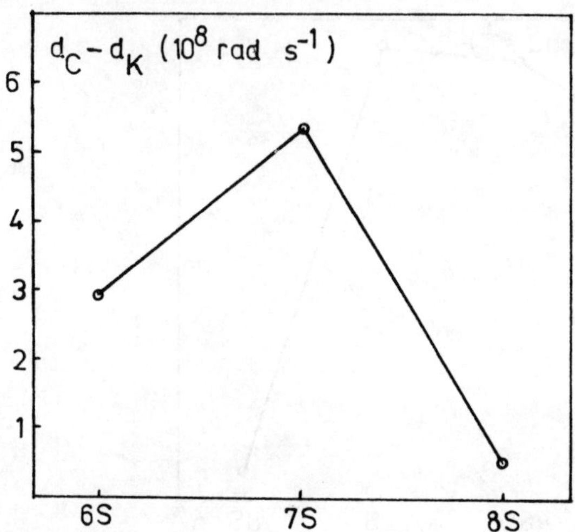

Figure 5. Differences between line shifts at the centre of the Solar disk and at the limb for Na I 4p-ns lines as a function of n.

within a spectral series if the Debye cut-off is larger than the strong collisions cut-off for the shift. If this is not the case, the shift is zero. If the shift is negative (blue) for lower members of a series due to larger polarization of the lower level of the transition, for higher members of the series it becomes positive (red) owing to the gradual increase of the upper level contribution.

For the calculation of the H-impact contribution to the line profiles, we have used Smirnov-Roueff (Roueff, 1972) exchange potential, which takes into account the overlap at intermediate distances of the electronic orbitals. This method was adapted by Roueff (1975, 1976) to the broadening of Solar Na I lines with upper s-states, with neutral hydrogen as the perturber. Within a spectral series, the difference between various line profiles is produced only because the upper energy level is different. Since this value changes gradually within a spectral series, we can expect a gradual change of the line broadening parameters.

The hydrogen and electron impact full halfwidths and shifts (in angular frequency units) for the Na I 3p-ns series as a function of the principal quantum number of the upper level (n) are shown in Figs. 1 and 2. In all cases we can see a gradual change of the line broadening parameters. We can notice also that the hydrogen impact width increases gradually within the series considered (as well as Stark width and shift) while the shift is negative (blue) and decreases. This is the consequence of the fact that the Smirnov-Roueff Na-H potential leads to a repulsive interaction resulting in the blue line shift.

Differences between line broadening parameters at the centre of the Solar disk and at the limb for Na I 3p-ns and 4p-ns lines as a function of n, are presented in Figs. 3-5. We can see that the pressure broadening contribution to the limb effect increases for lower members of the series due to the dominant rôle of the H-impact contribution. For n=8 this difference decreases due to the large contribution of the electron impact red shift. For n>8 the difference considered changes the sign due to the larger Stark shift. Finaly, for higher n, when the Stark shift becomes zero due to the Debye screening (see Dimitrijević and Sahal-Bréchot, 1984b), the sign of this difference is again determined by the hydrogen impact contribution. We can see also from Fig. 4 that the difference between equivalent widths at the centre of the Solar disk and at the limb decreases with n within Na I 3p-ns series.

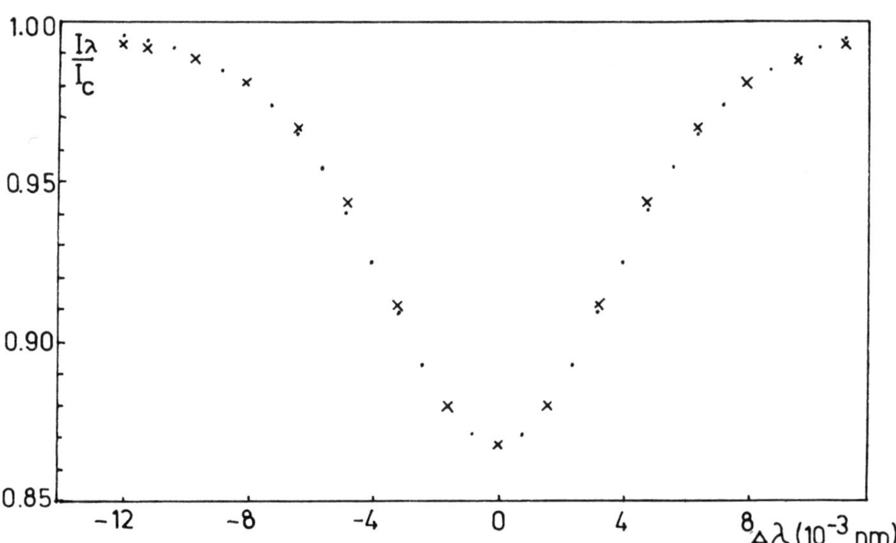

Figure 6. Calculated profile (·) for Na I 3p-6s (λ =514.88 nm) line (μ =1) compared with observational points (x) of Pierce and Slaughter (1982).

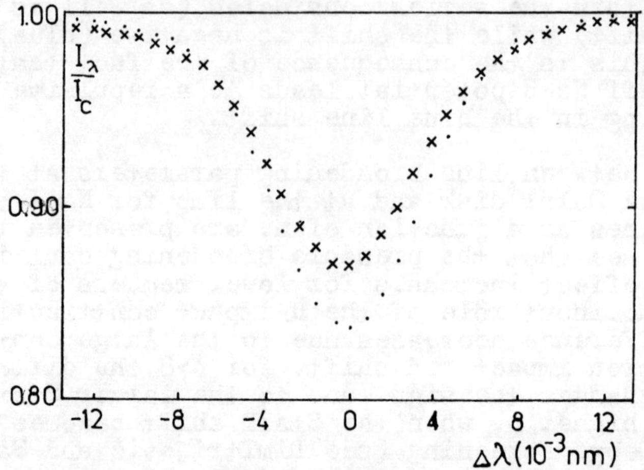

Figure 7. Absorption profile for Na I 3p-6s (λ = 514.88nm) line at the centre of the Solar disk (x) (μ =1) and for μ = 0.1 (·).

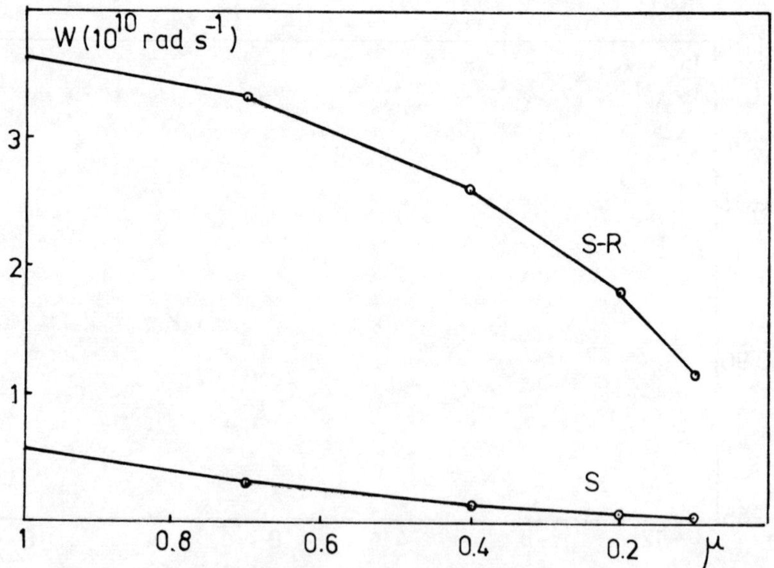

Figure 8. The full halfwidth for Na I 3p-6s line due to collisions with neutral (S-R) and charged particles (S) as a function of the heliocentric angle (μ = cosθ).

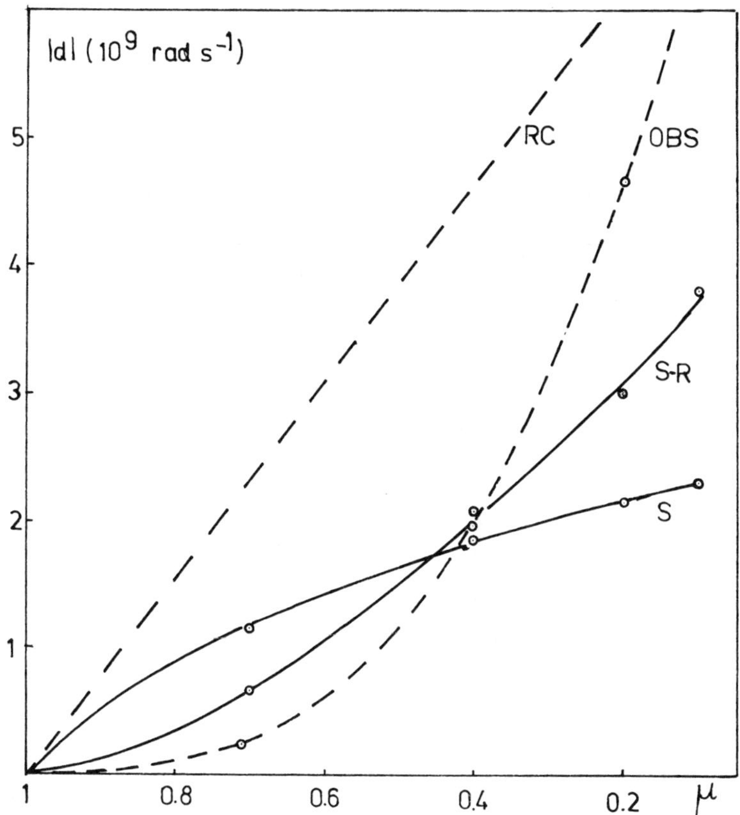

Figure 9. Relative Na I 3p-6s line shifts (shift at the Solar disk centre is taken as zero) as a function of the heliocentric angle ($\mu = \cos\theta$) of observed points; RC - the shift according to the radial current hypothesis (see e.g. Hart, 1974); OBS - averaged observation for Fe I 525.02 nm line ((Labonte and Howard, 1982); S-R - Relative shift due to Na-H collisions using Smirnov-Roueff potential; S - Relative shift due to collisions with charged particles.

We calculated line profiles within Na I 3p-ns and 4p-ns series, across the Solar disk using HSRA model of the Solar atmosphera. The neutral atom broadening contribution is calculated using the approach of Roueff (1975, 1976). The basis for calculation of Stark broadening contribution is the computer code which evaluates electron and ion impact broadening parameters of isolated spectral lines, using the semiclassical perturbation approach (Sahal-Bréchot, 1969ab).

Our results for 3p-6s line are presented in Fig. 6 and are in good agreement with observational points of Pierce and Slaughter (1982). In Fig. 7 are presented pressure broadened profiles for Na I 3p-6s line at the centre of the Solar disk and for μ = 0.1, where $\mu = \cos\theta$ and θ is the heliocentric angle. We can see that pressure broadening effects change the whole profile and in the case considered this contribution to the limb effect is measurable.

Our calculations of line widths and relative line shifts across the Solar disk for Na I 3p-6s line are presented in Figs. 8 and 9. Relative line shifts are compared also with simple radial current theory predictions (Hart, 1974). As an illustration, in Fig. 9 are given also averaged observations for Fe I 525.02nm line (Labonte and Howard, 1982). Since H-impact shift is blue (as well as the radial current and observational ones) and the Stark shift is red, the resulting pressure shift across the Solar disk, agree better with averaged observations for Fe I line than with the shift according to the radial current hypothesis. We can notice also that within the pressure broadening hypothesis each particular line has its own limb effect and comparison with averaged data for Fe I line give only an illustration of disagreement with simple radial current theory predictions.

REFERENCES

Dimitrijević, M.S., and Sahal-Bréchot, S.: 1984a, Astron. Astrophys. 136, 289

Dimitrijević, M.S., and Sahal-Bréchot, S.: 1984b, J.Quant. Spectrosc.Radiative Transfer 31, 301

Gingerich, O., Noyes, R.W., and Kalkofen, W.: 1971, Solar Phys. 18, 347

Hart, M.H.: 1974, Astrophys.J.187, 393

Labonte, B.J., and Howard, D.R.: 1982, Solar Phys. 80, 361

Pierce, A.K., and Slaughter, Ch.S.: 1982, Astrophys.J.Suppl. 48, 73

Roueff, E.: 1972, J.Phys.B 5, 279

Roueff, E.: 1975, Astron.Astrophys. 38, 41

Roueff, E.: 1976, Astron.Astrophys. 46, 149

Sahal-Bréchot, S.: 1969a, Astron.Astrophys. 1, 91

Sahal-Bréchot, S.: 1969b, Astron.Astrophys. 2, 322

Vince, I., Dimitrijević, M.S., and Kršljanin, V.: 1984, in *Spectral Line Shapes* Vol. III (ed. F.Rostas) W. de Gruyter, Berlin, in press

HYDROGEN LINE FORMATION IN DENSE PLASMAS IN THE PRESENCE OF A MAGNETIC FIELD

G. Mathys
Institute of Astronomy, Swiss Federal Institute of Technology
ETH-Zentrum
CH-8092 Zurich
Switzerland

ABSTRACT. The transfer equation of arbitrarily polarized radiation in the hydrog lines is formulated. The effect of a large scale, uniform magnetic field, the linear Sta effect due to the charged particles of a surrounding dense plasma and the Dopp broadening due to the thermal motions of the radiating atoms are included. The Sta effect is taken into account through a modified version of the Unified Theory; the theo is suitable to conditions typical of atmospheres of magnetic CP stars or of active so regions. It is shown from consideration of spontaneous emission intensity profiles of l that it is unsafe to use the "classical" nonmagnetic hydrogen line profiles to derive t physical parameters (electronic density, $\log g$) in magnetic CP stars.

1. INTRODUCTION

To get a good insight into the physical properties of stellar atmospheres where a m: netic field is present, one needs to take properly into account the influence of this fi upon radiation. Two effects of particuliar importance are the polarization induced the spectral lines of all elements and, for the hydrogen lines, the coupling between t Zeeman effect due to the magnetic field and the Stark broadening by the surroundi charged particles. This latter point is of considerable interest to people studying the u per main sequence chemically peculiar (CP) stars, or Ap stars — the only nondegener stars known to have a large organized field, up to now —, where the hydrogen lines widely used as a diagnostic tool for deriving fundamental stellar parameters such $\log g$.

The transfer equation in the spectral lines for the Stokes vector, describing state of an arbitrarily polarized light beam, has been obtained in various ways by seve authors since Unno's (1956) pioneering work and there now exist formulations of t equation that are valid in non-LTE and fully account not only for the emission a absorption processes but also for the anomalous dispersion effects (see e.g. Mathys, 1! = Paper 1, and Landi Degl'Innocenti, 1983). As to the combined effects on the hydro line profiles of a large scale magnetic field and of the electric microfields due to charged particles surrounding the emitting and absorbing atoms in a plasma, the o detailed studies are those of Nguyen-Hoe *et al.* (1967) and of Mathys (1983a, 198 1984b = Papers 2 to 4); all these works are only dealing with spontaneous emissior optically thin media, paying no attention to transfer and polarization questions.

Our purpose here is to show how the formalisms developed in Papers 1 to 4

J. E. Beckman and L. Crivellari (eds.), Progress in Stellar Spectral Line Formation Theory, 381–388.

be used to study the transfer of polarized radiation in the hydrogen lines, in a stellar atmosphere, when a large scale uniform magnetic field is present. In order to save some space, the reader is referred to Papers 1 to 4 for the discussion of the conditions of validity of the various expressions presented here; let us just mention that the present treatment seems appropriate in the conditions encountered in the atmospheres of magnetic CP stars and in active regions on the Sun.

2. TRANSFER OF POLARIZED LIGHT IN HYDROGEN LINES IN THE PRESENCE OF A MAGNETIC FIELD

The state of an arbitrarily polarized light beam can conveniently be described by a 4-vector S, known as the Stokes vector, the elements of which are the Stokes parameters. (The definitions adopted for the Stokes parameters are given in Paper 1.) Let us consider an hydrogen line corresponding to a transition between an upper level characterized by the principal quantum number n, and a lower level n' (throughout this paper, a prime is used to distinguish quantities pertaining to the lower level). The transfer equation of S in such a line is, in a plane parallel, static atmosphere:

$$\mu \frac{dS}{dm} = \chi S - \eta,$$

where $\mu = \cos\Theta$, Θ being the angle between the propagation and vertical directions, and where we have used the mass per unit area m as the depth variable. χ and η are respectively the opacity matrix and the emissivity vector:

$$\chi = \chi_c \mathcal{I} + \chi_l,$$

$$\eta = \eta_c \mathcal{J} + \eta_l,$$

where $\mathcal{I}$ is the unit matrix of rank 4, $\mathcal{J}$ is a vector with 4 components, the first of which is 1 and the others 0, and χ_c and η_c are the usual emissivity and opacity of the continuum. χ_l and η_l are the line opacity matrix and emissivity vector; they are explicited as follows:

$$\chi_l = N' \mathcal{B}_{abs} - N \mathcal{B}_{em},$$

$$\eta_l = N \mathcal{A},$$

where N and N' (cm^{-3}) are the populations of levels n and n'. Assuming provisionnally that the line is infinitely narrow, the 4×4 stimulated emission matrix $\mathcal{B}_{em}$, at frequency ω, can be written under the form:

$$\mathcal{B}_{em}(\omega) = \frac{2\pi^2 \omega}{\hbar \rho c} b_{em}(\omega).$$

ρ is the mass density. The elements of b_{em} are given in Table 1. a and a' denote substates of levels n and n', respectively, whose energies are ϵ_a and $\epsilon_{a'}$; $\omega_{aa'} = (\epsilon_a - \epsilon_{a'})/\hbar$. $\mathbf{D}_{aa'}$ is the dipole moment taken between the states a and a', and ρ_a is the probability of finding the atom in the initial state a. $\delta(x)$ is the Dirac function and $\mathcal{P}/x$ denotes the principal value of $1/x$ (see e.g. Heitler, 1954). $\mathbf{e}_1$ and $\mathbf{e}_2$ are two unit vectors; they are orthogonal

Table 1

General form of the matrix b_{em}

$\sum_{aa'} \rho_a \delta(\omega_{aa'} - \omega)$ $(\lvert\mathbf{e}_1 \bullet \mathbf{D}_{aa'}\rvert^2 + \lvert\mathbf{e}_2 \bullet \mathbf{D}_{aa'}\rvert^2)$	$\sum_{aa'} \rho_a \delta(\omega_{aa'} - \omega)$ $(\lvert\mathbf{e}_1 \bullet \mathbf{D}_{aa'}\rvert^2 - \lvert\mathbf{e}_2 \bullet \mathbf{D}_{aa'}\rvert^2)$	$2\sum_{aa'} \rho_a \delta(\omega_{aa'} - \omega)$ $\mathrm{Re}[(\mathbf{e}_1 \bullet \mathbf{D}_{aa'})(\mathbf{e}_2 \bullet \mathbf{D}_{aa'})^*]$	$2\sum_{aa'} \rho_a \delta(\omega_{aa'} - \omega)$ $\mathrm{Im}[(\mathbf{e}_1 \bullet \mathbf{D}_{aa'})(\mathbf{e}_2 \bullet \mathbf{D}_{aa'})^*]$
$\sum_{aa'} \rho_a \delta(\omega_{aa'} - \omega)$ $(\lvert\mathbf{e}_1 \bullet \mathbf{D}_{aa'}\rvert^2 - \lvert\mathbf{e}_2 \bullet \mathbf{D}_{aa'}\rvert^2)$	$\sum_{aa'} \rho_a \delta(\omega_{aa'} - \omega)$ $(\lvert\mathbf{e}_1 \bullet \mathbf{D}_{aa'}\rvert^2 + \lvert\mathbf{e}_2 \bullet \mathbf{D}_{aa'}\rvert^2)$	$2\sum_{aa'} \rho_a \mathcal{P}/(\omega_{aa'} - \omega)$ $\mathrm{Im}[(\mathbf{e}_1 \bullet \mathbf{D}_{aa'})(\mathbf{e}_2 \bullet \mathbf{D}_{aa'})^*]$	$-2\sum_{aa'} \rho_a \mathcal{P}/(\omega_{aa'} - \omega)$ $\mathrm{Re}[(\mathbf{e}_1 \bullet \mathbf{D}_{aa'})(\mathbf{e}_2 \bullet \mathbf{D}_{aa'})^*]$
$2\sum_{aa'} \rho_a \delta(\omega_{aa'} - \omega)$ $\mathrm{Re}[(\mathbf{e}_1 \bullet \mathbf{D}_{aa'})(\mathbf{e}_2 \bullet \mathbf{D}_{aa'})^*]$	$-2\sum_{aa'} \rho_a \mathcal{P}/(\omega_{aa'} - \omega)$ $\mathrm{Im}[(\mathbf{e}_1 \bullet \mathbf{D}_{aa'})(\mathbf{e}_2 \bullet \mathbf{D}_{aa'})^*]$	$\sum_{aa'} \rho_a \delta(\omega_{aa'} - \omega)$ $(\lvert\mathbf{e}_1 \bullet \mathbf{D}_{aa'}\rvert^2 + \lvert\mathbf{e}_2 \bullet \mathbf{D}_{aa'}\rvert^2)$	$\sum_{aa'} \rho_a \mathcal{P}/(\omega_{aa'} - \omega)$ $(\lvert\mathbf{e}_1 \bullet \mathbf{D}_{aa'}\rvert^2 - \lvert\mathbf{e}_2 \bullet \mathbf{D}_{aa'}\rvert^2)$
$2\sum_{aa'} \rho_a \delta(\omega_{aa'} - \omega)$ $\mathrm{Im}[(\mathbf{e}_1 \bullet \mathbf{D}_{aa'})(\mathbf{e}_2 \bullet \mathbf{D}_{aa'})^*]$	$2\sum_{aa'} \rho_a \mathcal{P}/(\omega_{aa'} - \omega)$ $\mathrm{Re}[(\mathbf{e}_1 \bullet \mathbf{D}_{aa'})(\mathbf{e}_2 \bullet \mathbf{D}_{aa'})^*]$	$-\sum_{aa'} \rho_a \mathcal{P}/(\omega_{aa'} - \omega)$ $(\lvert\mathbf{e}_1 \bullet \mathbf{D}_{aa'}\rvert^2 - \lvert\mathbf{e}_2 \bullet \mathbf{D}_{aa'}\rvert^2)$	$\sum_{aa'} \rho_a \delta(\omega_{aa'} - \omega)$ $(\lvert\mathbf{e}_1 \bullet \mathbf{D}_{aa'}\rvert^2 + \lvert\mathbf{e}_2 \bullet \mathbf{D}_{aa'}\rvert^2)$

Table 2

The matrix $(b_{em})_0$, accounting for Stark broadening

$(1+\cos^2\theta)I_{xx} + \sin^2\theta I_{zz}$	$\sin^2\theta(I_{zz} - I_{xx})$	0	$2\cos\theta R_{xx}$
$\sin^2\theta(I_{zz} - I_{xx})$	$(1+\cos^2\theta)I_{xx} + \sin^2\theta I_{zz}$	$2\cos\theta I_{xx}$	0
0	$-2\cos\theta I_{xx}$	$(1+\cos^2\theta)I_{xx} + \sin^2\theta I_{zz}$	$\sin^2\theta(R_{zz} - R_{xx})$
$2\cos\theta R_{xx}$	0	$\sin^2\theta(R_{xx} - R_{zz})$	$(1+\cos^2\theta)I_{xx} + \sin^2\theta I_{zz}$

to each other and to the direction of propagation of the radiation, and represent two (arbitrary) directions of polarization.

The absorption matrix $\mathcal{B}_{abs}$ can be put under the same form as $\mathcal{B}_{em}$; $b_{abs}(\omega)$ is derived from $b_{em}(\omega)$ by replacing ρ_a by $\rho_{a'}$ and $(\omega_{aa'} - \omega)$ by $(\omega - \omega_{aa'})$. Finally, the spontaneous emission vector $\mathcal{A}$ is obtained from $\mathcal{B}_{em}$:

$$\mathcal{A} = \frac{2\hbar\omega^3}{(2\pi c)^2} \mathcal{B}_{em} \mathcal{J}.$$

3. STARK BROADENING

We want to show how our previous studies (Papers 2 to 4) of the Stark broadening of hydrogen lines in the presence of a magnetic field can be modified to derive more realistic expressions for χ_l and η_l. As usual, we only consider those parts of χ_l and η_l that vary rapidly with frequency, i.e. essentially the elements of the b matrices. These are of two types and can formally be written (here, for emission):

$$I(\omega) = \sum_{aa'} \rho_a \delta(\omega_{aa'} - \omega) F(\mathbf{e}_1 \bullet \mathbf{D}_{aa'}, \mathbf{e}_2 \bullet \mathbf{D}_{aa'}),$$

$$R(\omega) = \sum_{aa'} \frac{\rho_a}{\pi} \frac{\mathcal{P}}{\omega_{aa'} - \omega} F(\mathbf{e}_1 \bullet \mathbf{D}_{aa'}, \mathbf{e}_2 \bullet \mathbf{D}_{aa'}),$$

where $F(\mathbf{e}_1 \bullet \mathbf{D}_{aa'}, \mathbf{e}_2 \bullet \mathbf{D}_{aa'})$ is a real function. Radiation is now emitted (or absorbed) by hydrogen atoms (which we call radiators) immersed in a hot, dense plasma; we regard this plasma as divided in a large number of independent cells, each of which contains one radiator, interacting with a vast number of charged particles (ions and electrons), or perturbers. Within that framework, the states a and a' appearing in the expressions of $I(\omega)$ and $R(\omega)$ are the upper and lower states of the cell undergoing the radiative transition.

The Unified Theory is used to account for the Stark broadening. The perturbing ions are treated dynamically as in Paper 4, because this method is very simple and seems quite relevant (mainly when a magnetic field is present) in the density range encountered in stellar atmospheres. The assumptions and approximations that we make are described and discusseed in Papers 2 to 4.

As in most studies of hydrogen line Stark broadening, $I(\omega)$ is advantageously regarded as the Fourier transform of a function $C(t)$ (which in the nonmagnetic, unpolarized case is the autocorrelation function):

$$I(\omega) = \frac{\mathrm{Re}}{\pi} \int_0^{\infty} \exp(i\omega t) C(t)\, dt,$$

with

$$C(t) = \int_{-\infty}^{+\infty} \exp(-i\omega t) \sum_{aa'} \rho_a \delta(\omega_{aa'} - \omega) F(\mathbf{e}_1 \bullet \mathbf{D}_{aa'}, \mathbf{e}_2 \bullet \mathbf{D}_{aa'})\, d\omega.$$

It is easily shown (Mathys, 1983b) that a similar relation holds for $R(\omega)$:

$$R(\omega) = -\frac{\mathrm{Im}}{\pi} \int_0^\infty \exp(i\omega t) C(t)\, dt,$$

so that, in all instances, one only needs to compute $C(t)$. As a matter of fact, it may be useful to introduce a "complex profile":

$$J(\omega) = \int_0^\infty \exp(i\omega t) C(t)\, dt.$$

The only difference between the function $C(t)$ used here and the analogous quantity in our previous papers is the explicit form of $F(\mathbf{e}_1 \bullet \mathbf{D}_{aa'}, \mathbf{e}_2 \bullet \mathbf{D}_{aa'})$, which plays no part in the derivation of a general expression of the profiles. Thus assuming a quite general (and possibly complex) form for F:

$$F = (\mathbf{e}_j \bullet \mathbf{D}_{aa'})(\mathbf{e}_k \bullet \mathbf{D}_{aa'})^*,$$

with j and k equal to 1 or 2, we draw, by analogy with Eq. (3.2) of Paper 4:

$$\begin{aligned} J_{jk}(\omega) = i \sum_{\{ll'mm'\}} & \langle nl_a m_a | \rho_A | nl_a m_a \rangle \times \\ & \times \langle nl_a m_a | \mathbf{e}_j \bullet \mathbf{D} | n'l'_a m'_a \rangle \langle n'l'_b m'_b | \mathbf{e}_k \bullet \mathbf{D} | nl_b m_b \rangle \times \\ & \times \langle n'l'_b m'_b ; nl_b m_b | [\Delta\omega - \mathcal{L}(\Delta\omega_0)]^{-1} | nl_a m_a ; n'l'_a m'_a \rangle. \end{aligned}$$

ρ_A and $\mathbf{D}$ are now the density and dipole momentum operators of the radiator. [Refer to Paper 4 for the other notations and the detailed expression of $\mathcal{L}(\Delta\omega_0)$.]

To get a more explicit formulation of the matrices b, it is convenient to choose $\mathbf{e}_1$ and $\mathbf{e}_2$ respectively parallel and perpendicular to the plane defined by the direction of propagation and the magnetic vector. Taking into account the fact that $[\Delta\omega - \mathcal{L}(\Delta\omega_0)]$ differs from zero only when $m'_a - m_a = m'_b - m_b$, we easily deduce the expression $(b_{em})_0$ of b_{em} for the particular geometry under consideration, in a reference system where the z-axis is parallel to the magnetic field. The elements of $(b_{em})_0$ are given in table 2; θ is the angle between the magnetic vector and the direction of propagation of the radiation and

$$I_{jj}(\omega) = \frac{\mathrm{Re}}{\pi} J_{jj}(\omega),$$

$$R_{jj}(\omega) = -\frac{\mathrm{Im}}{\pi} J_{jj}(\omega).$$

(Notice that the only forms of F that are used are real.) The general form of b_{em} (for an arbitrary choice of $\mathbf{e}_1$) is readily obtained from the expression of $(b_{em})_0$ by the relation (Auer *et al.*, 1977):

$$b_{em} = \mathcal{R}(b_{em})_0 \mathcal{R}^T,$$

where the rotation matrix $\mathcal{R}$ is:

$$\mathcal{R} = \begin{pmatrix} 1 & 0 & 0 & 0 \\ 0 & \cos 2\alpha & \sin 2\alpha & 0 \\ 0 & -\sin 2\alpha & \cos 2\alpha & 0 \\ 0 & 0 & 0 & 1 \end{pmatrix},$$

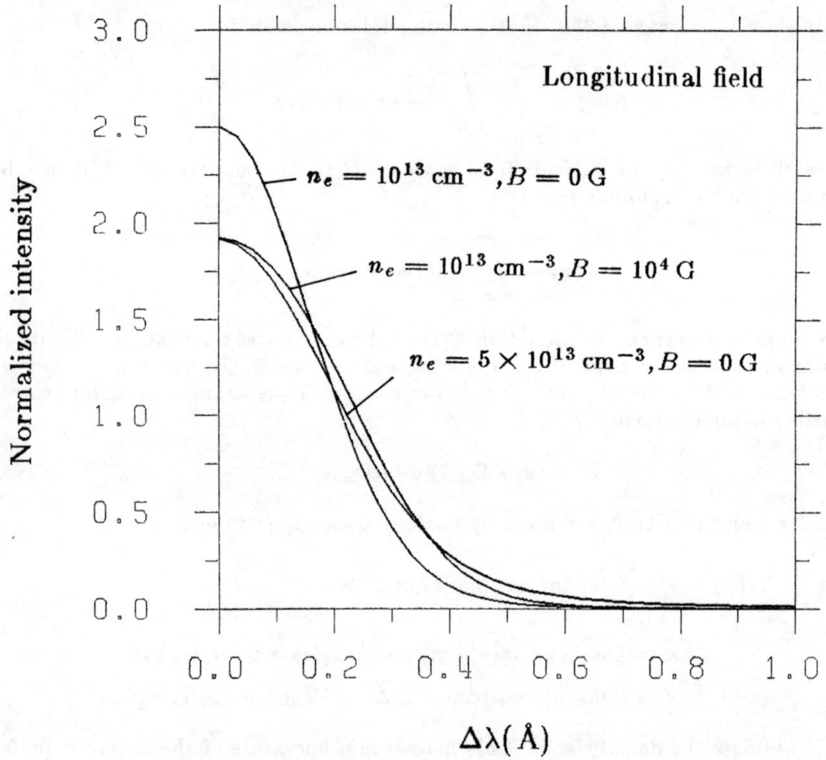

Fig. 1. — Stark and Doppler broadened Hβ profiles at a temperature of 10^4 K, for electron densities $10^{13}\ \mathrm{cm}^{-3}$ (with a longitudinal magnetic field of 10^4 G and without magnetic field) and $5 \times 10^{13}\ \mathrm{cm}^{-3}$ (without magnetic field). Notice how the magnetic profile at $10^{13}\ \mathrm{cm}^{-3}$ closely resembles the nonmagnetic one at $5 \times 10^{13}\ \mathrm{cm}^{-3}$.

α being the angle between the plane defined by the polarization vector $\mathbf{e}_1$ and the direction of propagation, and the plane defined by the magnetic and propagation vectors. $\mathcal{R}^T$ is the transpose of $\mathcal{R}$.

Doppler broadening due to the thermal motions of the radiators can be accounted for by convolving the elements of b_{em} with a Doppler broadening function; this, of course, does not affect the form of the emission matrix, and we shall in the rest of this paper continue to use the notations $I_{jj}(\omega)$ and $R_{jj}(\omega)$ for the Doppler broadened profiles.

The discussion of Sect. 4 of Paper 4 about the symmetries in the expression of the profiles is still relevant here; as a matter of fact, we have already made use of some of these properties to derive the explicit form of the elements of b_{em}. Let us point out that, while the profiles $I_{jj}(\omega)$ are symmetrical about the line center as shown in Paper 4, one can easily prove in an analogous way (for details, see Nguyen-Hoe *et al.*, 1967; Paper 3 and Mathys, 1983b) that the profiles $R_{jj}(\omega)$ are antisymmetrical about the line center. Consequently, in all instances, one only needs to compute half of the profiles.

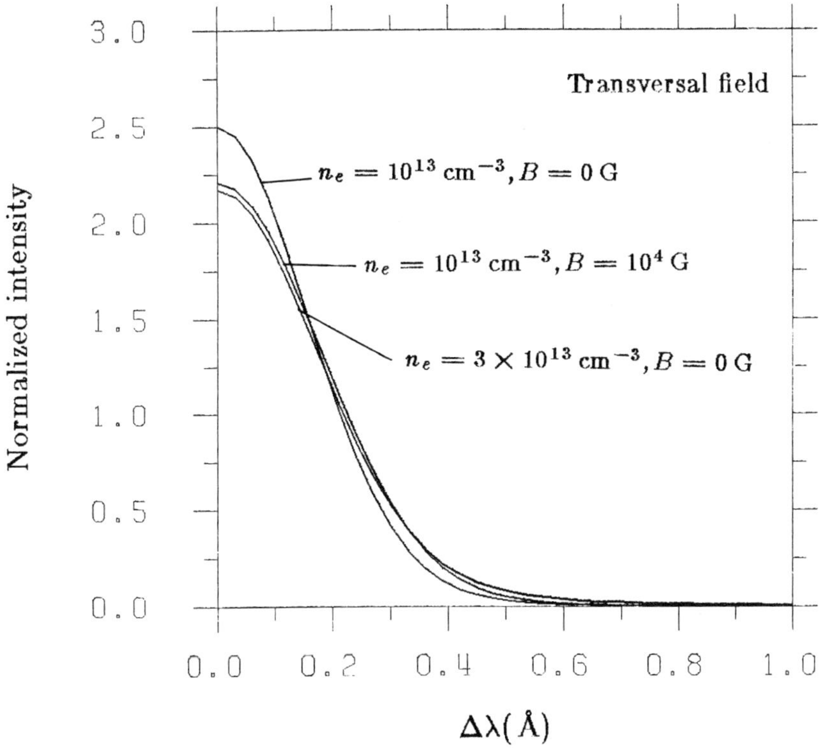

Fig. 2. — Stark and Doppler broadened Hβ profiles at a temperature of 10^4 K, for electron densities $10^{13}\,\mathrm{cm}^{-3}$ (with a transversal magnetic field of 10^4 G and without magnetic field) and $3 \times 10^{13}\,\mathrm{cm}^{-3}$ (without magnetic field). Notice how the magnetic profile at $10^{13}\,\mathrm{cm}^{-3}$ closely resembles the nonmagnetic one at $3 \times 10^{13}\,\mathrm{cm}^{-3}$.

4. PRELIMINARY RESULTS

It was hoped that a profile obtained by solving the complete transfer equation for the Stokes vector in a realistic model atmosphere could be shown at this workshop; unfortunately, this goal could not be achieved in time, and we only present a couple of spontaneous emission intensity profiles of Hβ, in order to give an idea of how a magnetic field can influence the shapes of hydrogen lines. Those profiles have been obtained under the following assumptions: ion and electron densities and temperatures are the same, the perturbing ions are hydrogen ions, and the density matrix of the radiator is constant over the initial states of the transition. This latter hypothesis, called by Auer *et al.* (1977) "complete disalignment", is probably the most questionable one in actual situations, but the question can only be satisfactorily settled by a full non-LTE treatment of the transfer problem, including statistical equations such as those derived by Landi Degl'Innocenti (1983).

As can be seen from the expression of $(b_{em})_0$, all the possible configurations are comprised between the extreme cases of longitudinal ($\theta = 0$) and transversal ($\theta = \pi/2$) magnetic field, which have been represented in Figs. 1 and 2. In each figure, we have plotted the normalized intensity of Hβ vs. the distance $\Delta\lambda$ from the line center at a temperature $T = 10^4$ K, for an electron density $n_e = 10^{13}$ cm^{-3}, in the presence of a magnetic field of 10^4 G and in the absence of a magnetic field, and for $n_e = 5 \times 10^{13}$ cm^{-3} (Fig. 1) or 3×10^{13} cm^{-3} (Fig. 2), without magnetic field. It is obvious that the magnetic field increases the broadening of the lines and makes them look as if they were formed at higher electron densities.

5. CONCLUSION

We have derived the expression of the emission matrix for the transfer of polarized radiation in hydrogen lines, in the presence of a large scale, uniform magnetic field, accounting for Stark and Doppler broadening; the absorption matrix and the spontaneous vector can be easily deduced from this expression. We have shown that, as far as the spontaneous emission intensity profile is concerned, a magnetic field typical of those encountered in magnetic CP stars can significantly contribute to the broadening of a line such as Hβ, and must thus be taken into account when using the Balmer lines to determine physical parameters of atmospheres of magnetic stars. The next step is the resolution of the transfer equation in a realistic model atmosphere. This is under way and we hope to be able to publish the results soon.

REFERENCES

Auer, L.H., Heasley, J.N., House, L.L.: 1977, *Astrophys. J.* **216**, 531

Heitler, W.: 1954, *The Quantum Theory of Radiation*, Clarendon Press, Oxford

Landi Degl'Innocenti, E.: 1983, *Solar Phys.* **85**, 3

Mathys, G.: 1982, *Astron. Astrophys.* **108**, 213 (Paper 1)

Mathys, G.: 1983a, *Astron. Astrophys.* **125**, 13 (Paper 2)

Mathys, G.: 1983b, unpublished Ph. D. thesis, Université de Liège, Belgium

Mathys, G.: 1984a, *Astron. Astrophys.*, in press (Paper 3)

Mathys, G.: 1984b, *Astron. Astrophys.*, in press (Paper 4)

Nguyen-Hoe, Drawin, H.-W., Herman, L.: 1967, *J. Quant. Spectrosc. Radiat. Transfer* **7**, 429

Unno, W.: 1956, *Publ. Astron. Soc. Japan* **8**, 108

A REVIEW OF LINE FORMATION IN MOLECULAR CLOUDS.

John E. Beckman
Instituto de Astrofisica de Canarias
La Laguna, Tenerife, Spain

ABSTRACT A summary of the physical considerations governing line formation in molecular clouds with kinetic temperatures in the range 5K-100K, and hydrogen number densities in the range 10^2-10^6 cm^{-3} is presented. An exposition of the classical Sobolev formulation due to Goldreich and Kwan (1973) is followed by results of a simple microturbulent model with partial redistribution due to Deguchi and Kwan. It is explained how line formation theory in molecular clouds is in a somewhat unsatisfactory state. Large velocity gradient models can account for the observed line shapes, but imply collapse times that are unreasonably short ($<10^6$yr). Microturbulent models give unrealistic self-absorbed CO profiles, which are scarcely improved using transfer theory wich includes partial redistribution. It is hoped that this review will serve to draw the attention of practicioners of line transfer theory in stellar atmospheres to a key problem in interstellar physics requiring further attention.

1. Large Velocity Gradient Models.

The classical way to compute line intensities due to rotational emission from molecules in molecular clouds stems from the seminal paper of Goldreich and Kwan (1974). By that year it had already become apparent that the most striking features of the lines formed in these low temperature, low pressure systems presented some difficulties for immediate physical explanation. Although kinetic temperatures within the bulk of the giant molecular clouds (GMC's) were known to be very low (typically $5K < T_K < 50K$), typical line widths were in the range 5km s^{-1} and would, if attributed

J. E. Beckman and L. Crivellari (eds.), Progress in Stellar Spectral Line Formation Theory, 389–405.

to thermal effects, indicate temperatures at least an order of magnitude higher. These widths cannot be ascribed to pressure or radiative broadening, which fall short by even more than one order of magnitude. The other key observable parameter which gave rise to theoretical difficulty was the dependency of lines of intrinsically very different strengths formed within the same cloud to have identical shapes. These kinds of observations emerge naturally from the study of isotopically substituted species, such as ^{13}CO or ^{18}CO, and their abundance ratios with the more abundant and widely observed CO (which we will use as conventional shorthand for $^{12}C^{16}O$). The line strengths of the rotational lines of these three molecules reflect the abundance ratios of 90:1 and 200:1 for $^{12}C/^{13}C$ and $^{16}O/^{18}O$ respectively. When saturation effects are taken into account, it is surprising that the line shapes of these three species (and even of $^{13}C^{18}O$) are normally geometrically similar. Further, the shapes themselves are generally quasi-gaussian, which is in apparent contradiction to the virtually isothermal temperature regimes, or the slow fall-off in temperature which can be shown to prevail within most clouds. There are no cases of "flat-topped" line profiles, even in the most saturated lines, and few cases of self-reversed shapes.

These characteristic line properties are strongly suggestive of lines where the photons are affected essentially by local rather than global cloud parameters. For this reason, Goldreich and Kwan (1974) used the somewhat artificial device of assuming that the clouds are in a state of continuous infall to apply the Sobolev (1960) condition to the transfer process. In their model, and the many subsequent models which follow essentially the same scheme, a cloud is deemed to be in a state of steady collapse, with a velocity gradient sufficiently steep that given a characteristic thermal velocity V_T and a velocity gradient V/R, a photon emitted within a CO line is either absorbed within a distance $d \simeq RV_T/V$ of its emission point, or it is transmitted out of the cloud without further interaction.

2. A simplified treatment with constant velocity gradient.

If we begin by assuming, as depicted schematically in Figure 1. that a GMC can be approximated by a spherically symmetric structure of uniform density falling towards its

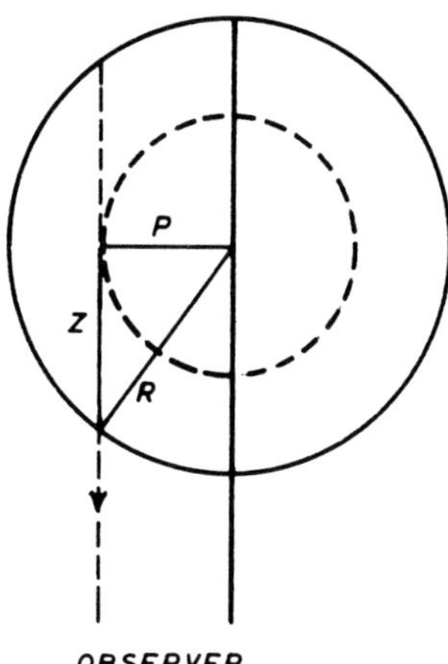

Fig. 1. Geometry of line formation in a molecular cloud.

centre of symmetry from an infinite radius, this produces a velocity law of form:

$$V(r) = (2GM/R^3)^{1/2}\ r = V/R\ r \qquad (1)$$

which produces a velocity V_Z in the line of sight of the observer given by:

$$V_2 = -V_z/R = V\ (r^2 - p^2/R^2)^{1/2} \qquad (2)$$

We express the energy of the Jth rotational level of the molecule, E_J, in the standard way:

$$E_J = hBJ\ (J+1) \qquad (3)$$

were B is the rotational constant, and we call the fraction of the molecules in the Jth level n_J, so that

$$\sum_{J=0}^{\infty} (2J+1)\ n_J = 1 \qquad (4)$$

In a cloud whose properties we wish to study the total number density of a given molecule P, summed over all its energy states, is N_P.

The Einstein coefficients for the rates of spontaneous and stimulated emission are $A_{J,J+1}$ and $B_{J,J+1}$, which are given respectively by standard forms for rotational transitions:

$$A_{J,J+1} = (16hB^3/c^2)\ (J+1)^3\ B_{J,J+1} \qquad (5a)$$

$$B_{J,J+1} = (32\pi^4\mu^2/3h^2c)\ (J+1)/(2J+3) \qquad (5b)$$

$$E_{J+1} - E_J = 2hB\ (J+1) \qquad (5c)$$

where μ is the dipole moment of the molecule, B is its rotational constant (not to be confused with $B_{J,J+1}$), and the expression $E_{J+1}-E_J$ is simply the difference in the energy terms.

We can also define an escape probability $\beta_{J,J+1}$ for a photon emitted during a transition from level J+1 to level J, and this is definable in terms of the optical depth $\tau_{J,J+1}$, using:

$$\beta_{J,J+1} = [1 - \exp(-\tau_{J,J+1})]/\tau_{J,J+1} \qquad (6)$$

where $\tau_{J,J+1}$ itself is defined by:

$$\tau_{J,J+1} = N_p\ (hc/4\pi)\ R/4\ B_{J+1,J}(2J+3)\ (n_J - n_{J+1}) \qquad (7)$$

Although equations (6) and (7) apply only to the case where $v(r) \propto r$, for other cases of spherical collapse somewhat more complex relations can easily be substituted.

To gain more physical grasp, we will rearrange (6) and (7) in terms of a collective variable DAU, defined by

$$\tau_{J,J+1} = \mathrm{DAU}\ (J+1)\ (n_J - n_{J+1}) \tag{8}$$

$$\mathrm{DAU} = (8\pi^3\mu^2 R N_p/(3hV)$$

In the simplest case we can assume that all collision rates between magnetic sublevels are equal. This is clearly not so, but the complexity of including individual values for each level is not warranted by present imperfect knowledge of the separate quantities. Thanks to the advent of very large, very rapid computers, it is now becoming possible to compute from quantum mechanical first principles the collision cross-sections for low energy interactions between H_2 and ion-molecules such as HCO^+; even neutrals such as CO and CS can be tackled. Here, we will use as an expression for the uniform rate:

$$C = N_{H_2} \langle\sigma v_T\rangle\ hB/kT \qquad \text{if} \qquad hB/kT > 1 \tag{9a}$$

$$\text{or} \qquad C = N_{H_2} \langle\sigma v_T\rangle \qquad \text{if} \qquad hB/kT > 1 \tag{9b}$$

In equation (9) $\langle\sigma v_T\rangle$ is termed the reduced collision rate, and T is the kinetic temperature of the molecular cloud. We can show that the expression kT/hB is equal to the total number of final states which can be reached by collisions at temperature T if kt >> hB. In that case:

$$\sum_{J=0}^{\infty} (2J+1)\ \exp[-hB/kT\ J(J+1)] =$$

$$= \int_0^{\infty} (2x+1)\ \exp[-hB/kT\ x(x+1)]\ dx = kT/hB \tag{10}$$

This is valid for most of the cases which are encountered in nature. It is also possible to approximate the collision rates by $\langle\sigma v_T\rangle = 7\times10^{-12}\ T\ \ (\mathrm{cm}^3\ \mathrm{s}^{-1})$.

We can express the condition of statistical equilibrium via

$$dn_J/dt = \beta_{J,J+1} A_{J,J+1} (2J+3)/(2J+1)$$

$$\{n_{J+1} - (n_J - (n_J - n_{J+1})/[\exp(2hB(J+1)/kT_{BB}) - 1]\}$$

$$- \beta_{J-1,J} A_{J,J-1} \{n_J - (n_{J-1} - n_J)/[\exp(2hB/kT_{BB} - 1]\}$$

$$+ C\left\{\sum_{L=J+1}^{\infty} (2L+1)(n_L - n_J \exp\{hB/kT[J(J+1) - L(L+1)]\} - \sum_{L=0}^{L=J-1} (2L+1)(n_J - n_L \exp\{hB/kT[L(L+1) - J(J+1)]\}\right\}$$

where the continuum temperature is taken to be T_{BB}, or the temperature of the cosmic background radiation, viz. 2.85K. This T_{BB} is in fact the only non-local quantity which aids in determining the rates of absorption or stimulated emission, since only those photons which actually escape allow a net transfer from level J to level J-1, and we have shown above that a photon is either absorbed within $d \sim V_T R/V$ of its emission point, or escapes from the cloud.

Given statistical equilibrium, $dn_J/dt = 0$, so that dividing (11) throughout by $3A_{1,0}$ we find:

$$0 = \beta_{J,J+1} (J+1)^4/(2J+1) \{n_{J+1} - (n_J - n_{J+1})/[\exp(2hB(J+1)/kT_{BB} - 1$$

$$- \beta_{J-1,J} J /(2J+1) \{n_J - (n_{J-\pounds} - n_J)/[\exp(2hBJ/kT_{BB})\ ^1]\}$$

$$+\varepsilon\left\{\sum_{L=J+1}^{\infty} (2L+1)\ (n_L - n_J \exp\{hB/kT\ J(J+1) - L(L+1)]\}) - \sum_{L=0}^{L=J-1} (2L+1)\ (n_J - n_L \exp\{hB/kT\ L(L+1) - J(J+1)]\})\right\} \quad (12)$$

where

$$\varepsilon = C/\ 3A_{1,0} \qquad (13)$$

For an optically thin line:

$$\beta_{J,J+1} \simeq 1/\tau_{J+\frac{1}{2}} = 1/\ DAU(J+1)(n_J - n_{J+1}) \qquad (14)$$

In this case DAU and ε enter (12) as combined variable εDAU, which is given by the expression:

$$\varepsilon DAU = 1/64\pi\ (C/B)^3(hB/kT)\ R/V\ \langle\sigma v_T\rangle\ N_H N_p \qquad (15)$$

which is independent of μ. Consequently the values of n_J and the brightness temperatures of the lines do not depend on μ.

It is of considerable interest to work out the energy loss rate per unit volume of cloud due to rotational emission lines, which comes from the number of transitions per unit time multiplied by the energy per transition of those photons which escape. This works out in numerical terms as:

$$dE/dt = 2hBN_p \sum_{J=1}^{\infty} (2J+1)J\ \beta_{J,J-1} A_{J,J-1}$$

$$\{n_J - (n_{J-1} - n_J)/[\exp(2hBJ/kT_{BB}) - 1]\} \qquad (16)$$

and using equations (5) and (8) to re-define the variables we have:

$$dE/dt = 128\pi hB\ (B/c)^3\ DAU\ V/R \qquad (17)$$

$$\sum_{J=1}^{\infty} J^5 \beta_{J,J-1}\ \{n_J - (n_{J-1} - n_J)/[\exp(2hBJ/kT_{BB}) - 1]\}$$

If we wish to know whether the gravitational energy made available to the cloud in its collapse is sufficient to maintain the output given by (17), we can compare the two directly. The former is found most easily from the internal kinetic energy per unit volume which is $3/2N_{H_2}kT$. Dividing (17) by this factor and multiplying by the collapse time R/V we find:

$$R/V\ 1/E\ dE/dt = 256\pi/3\ (hB/kT)(B/c)^3 DAU/N_{H_2}$$

$$\sum_{J=1}^{\infty} \beta_{J,J-1} J^5 \{n_J - (n_{J-1} - n_J)/[\exp(2hBJ/kT_{BB} - 1]\} \quad (18)$$

It gives useful insight into the physics to make the approximation $hB<<kT$ and $hB<<kT_{BB}$, when dE/dt simplifies to:

$$dE/dt = 5\ V/R\ (k/hB)\sigma/c[T^5 - T^5_{BB}] \quad (19)$$

where σ is the Stefan-Boltzmann constant. The T^5 dependence of dE/dt is accounted for when we realize that the line-widths of lines formed in this way are proportional to frequency.

3. Numerical results applied to real clouds.

In order to find the desired populations n_J, the J-ladder of equation (12) is solved together with (8). In order to truncate J, an upper level is chosen where $A_{Jmax+1,Jmax}>>C$, and by implication any collisional excitation from a level below J_{max} to a higher level is assumed to decay without time delay to J_{max}. Within the equations to solve, we can input B,T,$\mathcal{E}$ and DAU as independent input parameters.

Choosing value for R and V which are characteristic of large condensations within GMC's, i.e. a velocity gradient within the cloud of 6 km $s^{-1}pc^{-1}$ we obtain relatively straightforward expressions for $\mathcal{E}$ and DAU:

$$\mathcal{E} = 2.17\ T^{12}; \quad DAU = 1.3\ 10^7(N_{CO}/N_{H_2}) \quad (20)$$

for the CO molecule, where we have used the value of 0.1 for the dipole moment μ_{CO} of CO. Similar expressions are obtained for other molecules, whose dipole moments are characteristically at least an order of magnitude greater. This, in fact gives CO a rather large value of $\mathcal{E}$, which is to some extent compensated for in calculating the line intensities by rather small DAU. In fact the net result of the large value of $\mathcal{E}$ for CO is to make its rotational transitions easily excitable which, together with its great abundance, makes it the key molecule for diagnostics of GMC's. As H_2 is overwhelmingly the most abundant molecule within a GMC, values of N_P/N_{H_2} for a given molecule P can be treated as simple abundances with respect to hydrogen. In this instance, N_{CO}/N_{H_2} has values in the range 10^{-4} to 10^{-5}, and a value can be assigned without explicit measurement on a priori cosmic abundance grounds. However it is a parameter one can hope to derive from CO line intensity measurements.

From the solutions of the equations of transfer and statistical equilibrium using terms sufficiently high in the ladder of rotational transitions to be sure that 2hBJ>kT we can compute two key quantities: the excitation temperature, $T_{exc(J,J+1)}$, defined by

$$T_{exc\ (J,J+1)} = (2hBJ/k)\ \ln(n_{J-1}/n_J) \tag{21}$$

and the excess specific intensity $I_{ex(J,J-1)}$, defined by

$$I_{ex\ (J,J-1)} = \lambda^2_{J,J-1}/2k\ (I_{J,J-1} - I_{BB}) =$$

$$= c^2(I_{J,J-1} - I_{BB})/8kB^2J^2 \tag{22}$$

In equation (22), $I_{J,J+1}$ is the intensity measured by the observer, and I_{BB} is the intenisty of the cosmic background radiation at the relevant frequency.

In terms of the quantities defined, the excess specific intensity is expressed explicitly by:

$$I_{ex(J,J-1)} = 2hBJ/k\ [1 - \exp(-\tau_{J,J-1})]$$

$$\{[1/\exp(2hBJ/kT_{exc}) - 1] - [1/\exp(2hB/kT_{BB}) - 1]\} \tag{23}$$

The units of I_{exc} are those of temperature, and in the limit where $I_{exc} \gg 2hBJ/k$ is the brightness temperature of the source.

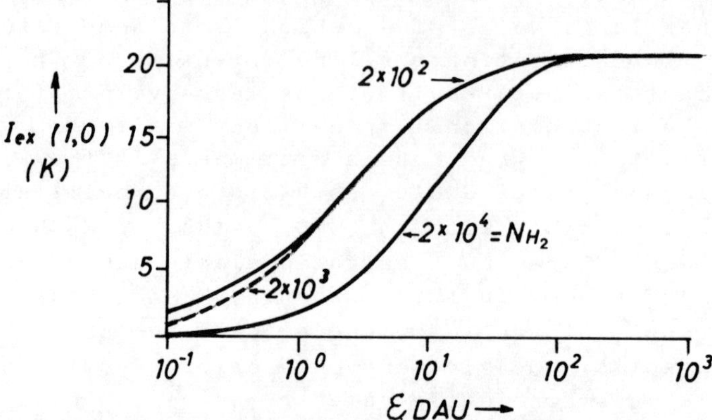

Fig. 2. Excess line temperature as a function of ε DAU for the J = 1—0 transition of CO

In figure 2 the results of a typical model calculation for CO are displayed. The value of ε has been retained as a constant while DAU has been effectively varied by changing N_{CO}/N_{H_2}. Plots are shown for two different values of the underlying cloud density N_{H_2}. The results show how the optical depth in a line varies with physical conditions, in a fairly realistic way. It can be used to a first order in any part of the line, from centre to wings, but this neglects line profile redistribution and is clearly too simplistic an attack on the problem. The results of a typical calculation of line intensities of this type are presented in Table 1. The occupation numbers, n_J, the optical depth, $\tau_{J+1,J}$ in each transition line, the excitation temperature T_{exc}, and the intesity I_{exc} are all shown, as is the quantity (R/V).dlnE/dt, which is the product of the cooling time and the collapse time of the cloud. The cooling rate due to each line is shown separately, and the sum for all the lines which radiate effectively is shown at the foot of column 6. This tabulation is for a single fixed value of the kinetic temperature of the cloud.

We should note here that what the observer detects is the quantity I_{exc}, and this falls steadily with increasing value of J, as the transitions become less and less thermalized. In fact in an ideal situation, a direct plot of I_{exc} against J would be a means of measuring N_{H2}, since the value of T_K can be found reasonably well either from the measured value of I_{exc} or from the computed value of T_{exc}. The observational problem is that, in practice, it is usually difficult to reach J levels much higher than 5 or 6, for the simpler molecules, such as CO, because of terrestrial atmospheric blocking. Acceptable fits to the data have been reached with a limited number of J levels, especially in those cases where thermalization fails for fairly small J. This implies that to use rotational lines to measure densities in molecular clouds, one must look for molecules whose intrinsic properties give rise to reasonable values of $\mathcal{E}$ and DAU. We must also note that in the case of CO presented in table 1. there is an apparent degree of inversion in the lowest J levels. This is not a realistic conclusion, because the result depends critically on the value of C chosen. In fact this demands greater precision than is currently available from quantum mechanics, (a situation which is in the course of being remedied), and in fact a matrix representation of C is needed, to account fully for collisional effects.

J	n_J	$\tau_{J+1,J}$	T_{exc}(K)	I_{exc}(K)	$\frac{R}{V}\frac{d\ \ln E}{dt}$
0	0.192	8.46	24.72	21.23	0.04
1	0.0873	24.29	24.77	19.46	0.30
2	0.0558	31.61	24.74	17.32	0.89
3	0.0285	26.25	24.42	14.99	1.83
4	0.0115	15.53	23.06	11.92	2.84
5	0.0035	6.57	19.39	7.29	3.00
6	0.00062	1.56	15.04	2.52	1.65
7	0.00047	0.14	16.77	0.43	0.42
8	3.5×10^{-6}	0.011	20.75	0.053	0.073
				Total	11.03

Figure 3 shows another direct use of this type of Sobolev modelling, which is to compare the cooling rate for the cloud, derived directly via the line fluxes in CO as shown in the single example at the end of Table 1, with the collapse time of the same cloud under gravitation. This can be plotted against DAU by varying the number corresponding to the abundance ratio $[CO]/[H_2]$, for a given value of temperature. In any real cloud the cooling is in fact dominated by CO because it is by far the most abundant species with strong lines.

4. Dust, gas, and energy balance.

For the molecular clouds observable within the galaxy, the temperatures vary between 5K and some 100K for the hottest, and this is reflected in the observed values of I_{exc} for CO. The corresponding values of I_{exc} for ^{13}CO are between 2 and 10 times less. We can use these factors to derive a rather general conclusion that values of DAU for CO must be greater than 100 (assuming that the abundance ratio $[^{12}C]/[^{13}C]$ is of the same order as the solar system value of 90). Since for values of DAU greater than 100 the model shows that CO is thermalized, we can infer that CO will in general be useful in the role of thermometer for measuring gas kinetic temperature.

It is easy to deduce from Fig. 3 that for values of N_{H_2} greater than 2×10^4, the cooling rate due to CO alone, to say nothing of the effects of other molecules, exceeds the gravitational energy available from collapse. We can generalize this conclusion to infer that any molecular cloud with a value of $I_{exc}(CO) > 15K$ must have an auxiliary source of energy. This can in fact be supplied by stars within the cloud, and it would be enough for the cloud to contain only a few O stars or protostars per few thousand solar masses to perform this job.

The identification of molecular clouds as principal star-formation regions suggest that the presence of young stars and protostars within them should indeed be expected (with suitable variations in population and hence in the intensity of individual sources shrouded within the cloud). These are almost by definition not directly observable due to the dust by which they are surrounded, except for the few which have burned their way to the edge of the cloud, and show an observable HII region. In the far infrared a dust cloud will be observable directly in terms of its emission, whose total luminosity can exceeded 10^6 solar luminosities, in the case of a giant molecular cloud.

In order to explain the observed line shapes of rotational molecular lines it is necessary to invoke clouds in which the temperature structure combines with the velocity structure to determine the profile, in such a way as to avoid either central self-absorption, which does not often appear, or too flat-topped a line with steep sides. This can be done by allowing the temperature to be controlled effectively by the dust, which itself acts as an effective trap for ultraviolet and visible photons emanating from the central star. The dust temperature itself is derived via a reasonably straightforward radiative transfer system, and depends monotonically on the distance from the heat source, which might be one or several stars. Then the gas temperature is derived from the dust value, T_D, taking the gas-dust interaction into account in terms of a relaxation time t_r, which is given by:

$$t_r = 4.5\ 10^{16}/N_{H_2} T^{1/2} S \qquad (24)$$

where s is the radius of a grain, which is of order 10^{-5}cm. To use equation (24) to calculate the gas temperature, one uses the relation

$$T_D = T_g\ [1 + t_r(d\ln E/dt - 2V/R)] \qquad (25)$$

where T_g is the gas temperature, the cooling rate is d lnE/dt, and V/R is the inverse collapse rate. Figure 4 shows the results of such cloud calculations, made assuming a constant gas/dust ratio, with variations in the hydrogen density, and also in the CO abundance taken into account.

5. A microturbulent model with PRD.

It is perhaps remarkable that such LVG (large velocity gradient) models, as this one by Goldreich and Kwan give excellent fits to line profiles, at least in general terms. The main features, viz. line-widths some ten times

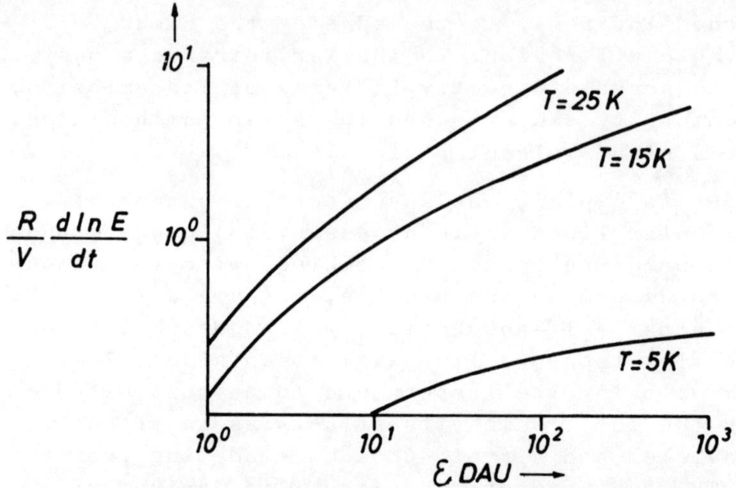

Fig. 3. Product of cooling rate and collapse time as a function of ε DAU for different valves of T

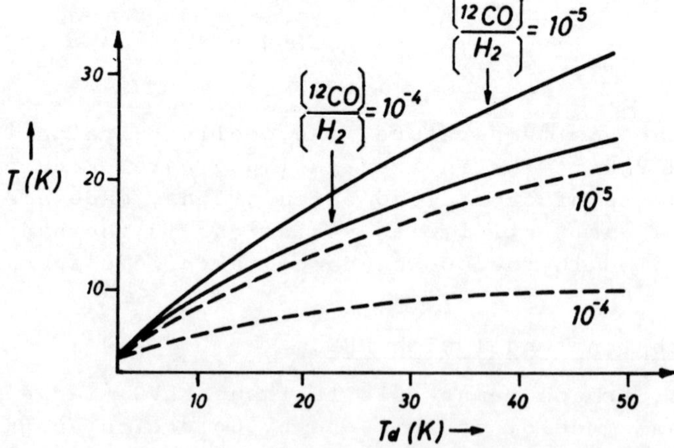

Fig. 4. Dust V. gas temperature for models where T is collision determined

the expected thermal value, and the lack of general self-absorption (although this is certainly seen in a number of clouds) can often be reproduced. In fact the similarity in line shapes between the optically thick CO and the optically much thinner ^{13}CO is one of the main reasons why models with systematic motion enjoy such success, as compared with turbulent models.

In fact one of the few serious attempts to introduce partial redistribution theory into molecular line formation problems was that of Deguchi and Kwan, in an unpublished preprint. They considered a uniform plane parallel cloud with uniform kinetic temperature, and pure microturbulent broadening, which has the same mathematical form as Doppler broadening. The reasons for introducing partial redistribution was in order to try to reduce the effects of self-absorption in the CO lines, and thereby bring their profiles more into line with those of their ^{13}CO counterparts. They used angle-averaged redistribution functions, and a five-level molecule, with populations determined for eight different ranges of speed varying from 0.1 to 2.5 times the thermal value. In each spatial step within the cloud, the radiation intensity in a transition was computed for eight frequency points and eight angle values. They achieved a convergence which was independent of initial guesses for spatial and velocity distributions within the level populations. Partial redistribution was assured by setting the collisional rates between molecules moving at different speeds to zero, so that the velocity distribution of the population depends solely on radiative ineractions.

The results of this effort were somewhat disappointing, as shown in Fig. 5. The difference between the partial and the complete redistribution cases are in no way big enough to allow the effects of partial redistribution to "save" the microturbulent model. In all cases some considerable degree of self-absorption is evident in the CO profiles, which is absent for ^{13}CO.

One might ask why it should be thought necessary to go to the trouble to compute PRD microturbulent models when LVG models seem good enough to account for observed profiles. The answer is that the collapse times implied by the LVG models, in the range of 10^6 years, are much shorter than the inferred lifetimes of the clouds, and are not compatible

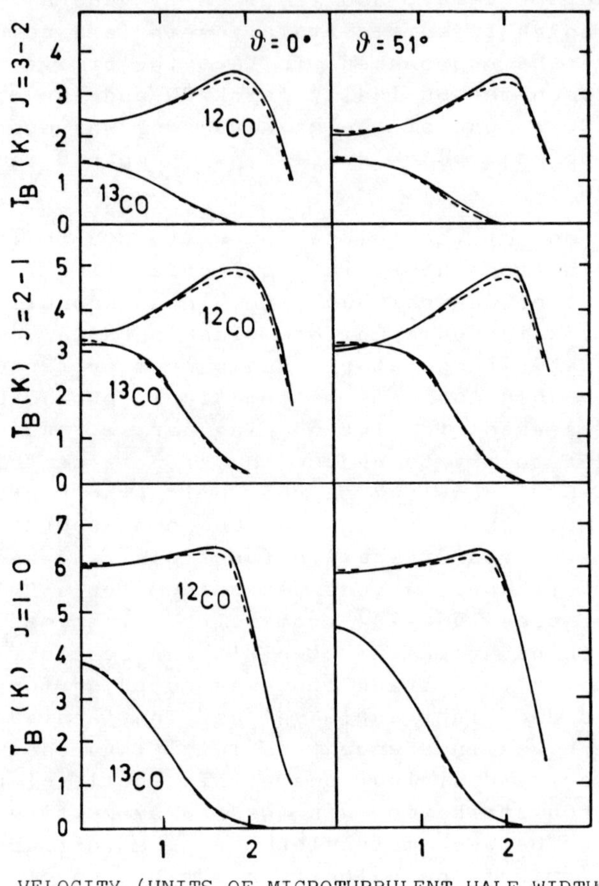

Fig. 5. Line profiles for plane parallel cloud viewed at two angles θ, for CO transitions formed in CRD and PRD cases.

with the presence of molecular clouds in such profusion throughout the spiral arms of the galaxy. There remains a more or less important lacuna in our understanding of the line formation process which no present set of theoretical constraints is easily able to resolue.

GENERAL REFERENCES.

1. Avery, L., 1983. A Monte Carlo program for multi-level cloud modelling. (Private communication).
2. Deguchi, S. and Fukui, Y., 1977, P.A.S. Japan, 29, 683.
3. Deguchi, S. and Kwan, J., 1980, (Unpubloshed preprint).
4. Goldreich, P. and Kwan, J., 1974, Ap. J. 189, 441, 1974
5. Leung, C.M. and Liszt, H.S., 1976, Ap. J. 208, 732.
6. Scoville, N.Z. ans Solomon, P.M., 1974, Ap. J. (Letters), 187, L67
7. Sobolev, A.M., 1982, Nauch. Inf. Vyp. (Russian), 50, 47.
8. Sobolev, V.V., 1960. Moving Envelopes of Stass (Cambridge, Harvard University Press).
9. Stenholm, L.G., 1980, Astron. Astrophys. Suppl. 42, 23.

APPENDIX

List of computer codes available at present.

A special session was devoted to a detailed discussion on computer codes presently used by people working in line radiative transfer. The underlying idea was to define principal guide lines with the aim of standardizing computational techniques in order to facilitate exchange of software.

At the same time a standard form was given to each participant, whose format was designed to abstract relevant feature of the computer codes on which they are presently working. The contents of these forms are presented here.

L.S. ANDERSON Ritter Observatory, Univ. of Toledo, Ohio 43606, USA.

PAM: Plane-parallel Atmosphere Modelling.

For line-blanketed model atmospheres in full radiative-collisional statistical equilibrium. The atomic physics is specified outside the code, which allows any model to be considered. The radiative and hydrostatic equilibrium constraints can be omitted to solve multi-level transfer problems in prescribed atmospheres. The code can handle several hundred bound-bound transitions (CRD) directly.

On the horizon: Generalization to include PRD, small ($\sim$ 2-3 $\Delta\nu_D$) motions, spherical geometry.

M. CARLSSON Institute of Theoretical Astrophysics, PO Box 1029, Blinderns, Oslo 3, Norway.

Radiative Transfer Code.

1-D, plane parallel geometry, CRD, multi-level atom, arbitrary velocity fields. Emphasis on ease of modifications to other geometries, PRD, etc.

J. E. Beckman and L. Crivellari (eds.), Progress in Stellar Spectral Line Formation Theory, 407–411.

Standard FORTRAN.

The code published as an "Uppsala Astronomical Observatory Report".

Comments:

Recommendation for programming practice: let the first non-blank comment line in every subroutine be a one-line description of the routine. A simple program can then extract an index of the routines in the format: routine name one-line description. (This program can be obtained from M. Carlsson; it is anyway easy to be written).

Another program is available, that extracts the calling structure of the program. It has proved to be a valuable tool for obtaining a nice overview of the program and for developing a "clearer" code.

S. DUMONT Institut d'Astrophysique, Paris. (Quoted by F. Praderie).

TULIPE.

Non-LTE line formation, CRD, plane parallel geometry, up to 8 or 10 levels, 1-D, static. Feautrier elimination scheme; use of Net Radiative Brackets.

Present users:

J. BORSENBERGER, F. PRADERIE (Obs. de Paris-Meudon). With alterations to compute radiative forces in view of diffusion problems in Ap-Bp stars.

R. FREIRE FERRERO (Obs. de Strasbourg). With alterations to include PRD.

P. MEIN, N. MEIN
Adapted to hydrogen spectrum.

Other users in the Meudon-Paris area.

P. GOUTTEBROZE L.P.S.P., PO Box 10, F-91370, Verrieres-Le-Buisson, France.

1) Code for computing semi-empirical solar atmosphere models (plane parallel)

- hydrostatic equilibrium
- ionization equilibrium (metals in LTE)
- hydrogen: 5 (or 3) levels plus continuum)
- transfer in lines (Feautrier method) Lyα, Lyβ and Hα, both CRD and PRD
- transfer in bound-free hydrogen continua (Feautrier method).

2) The same for stellar atmosphere models (A-type stars).

3) Codes for static, plane parallel atmospheres and ions (Mg II, Ca II, Si II, etc.) with N levels plus continuum. N depends on computer storage capacity; e.g. for Mg II 14 levels plus continuum, ~40 lines.

4) Codes for computing Mg II or Ca II lines in moving plane parallel atmospheres (e.g., chromospheric oscillations)
   - co-moving frame
   - PRD
   - N levels plus continuum ($N \leq 5$)

5) Code for computing some radiative transfer operators and their partial derivatives. (Applications: oscillations, weighting functions, MTF, effect of small changes in the atmosphere on line profiles. Only with CRD). See: Gouttebroze, 1983, JQSRT, 30, 193.

6) Miscellaneous for specific problems: coronal loops, Monte Carlo method in cylindrical or plane parallel atmospheres, etc.

All these codes run on CDC computers (7600 and Cyber 750); some of them on CRAY-1.

W.-R. HAMANN Institute fur Theoretische Physik und Sternwarte der Universitat, Kiel, FRG.

Code in CRAY-FORTRAN:

Line formation in a spherically-expanding atmosphere, co-moving frame method, two-level atom, generalized to blended doublets, CRD, no continuum.

See: Hamann, W.-R.: 1981, Astron. and Astrophys., 93, 353.

Code in FORTRAN IV (CDC):

Line formation in a plane parallel, differentially moving slab. Comoving-frame method, two-level atoms, pure Doppler broadening, CRD or PRD (angle-averaged).

See: Hamann, W.-R. and Kudritzki, R.P.: 1977, Astron. and Astrophys., 54, 525.

K. HEMPE Hamburger Sternwarte, Gojenbergsweg 112, D-2050 Hamburg 80, FRG.

Computer code for the solution of the radiative transfer in Zeta Aurigae systems. The code uses the two-level atom approximation and is a modification of the Castor escape probability method for the non- spherical geometry of Zeta Aurigae system. The code is written in FORTRAN 77.

D. HUSFELD Inst. fur Astron. und Astrophys. der Universitat Munchen, FRG.

Model atmosphere calculation for extreme He-rich composition.

Essentially a Auer and Mihalas type algorithm with complete linearization, newly written setting more helium levels in non-LTE: up to 5 He I levels ($n \geq 2$: hydrogenic), up to 10 He II levels. Bound-bound radiative transitions are not accounted for.

Internal and external documentation available. The code is now running on a CYBER 175 computer, and will be available by the end of October 1984.

B.W. LITES HAO/NCAR, P.O. Box 3000 , Boulder, CO 80307, USA

1) D. REES, Univ. of Sidney:
Code for computation of polarized radiative transfer (full Stoke representation) for a plane parallel atmosphere with variable magnetic field and thermodynamic

parameters.

2) B.W. LITES:
BLEND. Rybicki formulation for multi-level transfer, CRD in blended multiplets. Plane parallel geometry assumed.

3) B.W. LITES and A. SKUMANICH:
HCODE3A. Rybicki formulation of multi-level transfer with radiative transfer diagnostics. (See paper in these proceedings).

Coming in next few years:
Lites and Skumanich: A Scharmer-Carlsson code modified for blends, as described in these proceedings.

A. NORDLUND Astronomical Obs. Oster Volgade 3, DK-1350 Copenhagen K, Denmark.

1) Stellar Atmospheres Utility Subroutines.
This is essentially the code published by Gustafsson (Upsala Observatory Report), containing subroutines for LTE ionization equilibria, molecular equilibria, absorption coefficients, plus (vectorizable) code for Feautrier transfer and line profile diagnostic.
The structure is very flexible; most physical data are given as input data, very little is tied up in the code itself. In the current version, the input data cover cool stellar atmospheres. The code is well documented.

2) Two-level Atom Problems in 3-D.

The code used for the paper in these proceedings.

R. WEHRSE Institut fur Theoretische Astrophys. der Universitat, Im Neuenheimer Feld 294, D-6900 Heidelberg, FRG.

1) Grant-Hunt-Peraiah Method; arbitrary redistribution functions, spherical geometry, small velocities (observer's frame).

2) Discrete ordinate method (see paper in these proceedings); plane parallel geometry, arbitrary redistribution.

INDEX OF SUBJECTS.